Higher Education and Sustainability

Taylor and Francis Series in Higher Education and Sustainability

J. Paulo Davim, Professor

Department of Mechanical Engineering,
University of Aveiro, Portugal

This new series fosters information exchange and discussion on higher education for sustainability and related aspects, namely academic staff and student initiatives, campus design for sustainability, curriculum development for sustainability, global green standards: ISO 14000, green computing, green engineering education, index of sustainability, recycling and energy efficiency, strategic sustainable development, sustainability policies, sustainability reports, etc. The series will also provide information on principles, strategies, models, techniques, methodologies and applications of higher education for sustainability. It aims to communicate the latest developments and thinking as well as the latest research activity relating to higher education, namely engineering education.

Higher Education and Sustainability: Opportunities and Challenges for Achieving Sustainable Development Goals
Edited by Ulisses Manuel de Miranda Azeiteiro and J. Paulo Davim

For more information on this series, please visit: https://www.crcpress.com/Higher-Education-and-Sustainability/book-series/CRCHIGEDUSUS

For more information about this series, please visit: www.crcpress.com

Higher Education and Sustainability

Opportunities and Challenges for Achieving Sustainable Development Goals

Edited by

Ulisses Manuel de Miranda Azeiteiro and
J. Paulo Davim

CRC Press
Taylor & Francis Group
Boca Raton London New York

CRC Press is an imprint of the
Taylor & Francis Group, an **informa** business

CRC Press
Taylor & Francis Group
6000 Broken Sound Parkway NW, Suite 300
Boca Raton, FL 33487-2742

First issued in paperback 2021

ISBN-13: 978-0-367-77670-1 (pbk)
ISBN-13: 978-1-138-55653-9 (hbk)

Library of Congress Cataloging-in-Publication Data

Names: Azeiteiro, Ulisses, author. | Davim, J. Paulo, author.
Title: Higher education and sustainability : opportunities and challenges for achieving sustainable development goals / Ulisses Manuel de Miranda Azeiteiro and J. Paulo Davim.
Description: Boca Raton, FL : Taylor & Francis Group, 2020.
Identifiers: LCCN 2019016918 | ISBN 9781138556539 (hardback : acid-free paper)
Subjects: LCSH: Education, Higher—Environmental aspects—Cross cultural studies. | Environmental education—Cross cultural studies. | Sustainable development—Cross cultural studies.
Classification: LCC LB2324 .A94 2020 | DDC 378.01—dc23
LC record available at https://lccn.loc.gov/2019016918

Visit the Taylor & Francis Web site at
http://www.taylorandfrancis.com

and the CRC Press Web site at
http://www.crcpress.com

Contents

Preface

Education and research in the context of the Sustainable Development Goals (SDGs) are current challenges facing Higher Education Institutions (HEIs). This book's primary goal is the description of experiences from research and field projects and the role HEIs can play together with contributions presenting a variety of initiatives showing how regions around the world are implementing SDGs.

The objective was to publish material that promotes innovation and enhances understanding on this thematic and to disseminate knowledge and enhance international research and cooperation. The contributions cover the topics of the role of SDGs in advancing implementation of sustainable development, sustainability in higher education, the role of universities in sustainable development, new paths towards sustainable development and e-learning contributions.

Chapters are from Europe, Asia, Latin America and Africa, from emerging and developing nations facing great sustainability research and educational challenges presenting case studies, technological developments, outputs of research/studies and examples of successful projects.

Ulisses Manuel de Miranda Azeiteiro
J. Paulo Davim

Editors

Ulisses Manuel de Miranda Azeiteiro is a Senior Professor (Associate Professor with Aggregation/Habilitation and Tenure) and Coordinator of the Climate Change and Biodiversity Assets Unit from the Biology Department and Integrated Member/Senior Researcher of the Centre for Environmental and Marine Studies (CESAM) at University of Aveiro in Portugal. His main interests are the Impacts of Climate Change in the Marine Environment (Biology and Ecology of Global Change) and Adaptation to Climate Change in the Context of Sustainable Development (Social and Environmental Sustainability and Climate Change). He has written, co-written, edited or co-edited more than 200 publications, including books, book chapters, special issues of scientific journals and papers in refereed journals.

J. Paulo Davim received his Ph.D. in Mechanical Engineering in 1997, the aggregate title (Full Habilitation) from the University of Coimbra in 2005 and a D.Sc. from London Metropolitan University in 2013. Currently, he is the Professor at the Department of Mechanical Engineering of the University of Aveiro, Portugal. He has more than 30 years of teaching and research experience in Manufacturing, Materials and Mechanical Engineering with special emphasis in Machining & Tribology. He has also interest in Management & Industrial Engineering and Higher Education for Sustainability & Engineering Education. He is the Editor-in-Chief of several international journals, Guest Editor of journals, book Editor, book Series Editor and Scientific Advisory for many international journals and conferences. Presently, he is an Editorial Board member of 30 international journals and acts as a Reviewer for more than 80 prestigious *Web of Science* journals. In addition, he has also published as editor (and co-editor) for more than 120 books and as author (and co-author) for more than 10 books, 80 book chapters and 400 articles in journals and conferences.

List of Contributors

Rohit Agrawal
Department of Production Engineering
National Institute of Technology
Tiruchirappalli, Tamil Nadu, India

Luísa Aires
LE@D, DEED
Universidade Aberta
Porto, Portugal

Ulisses Manuel de Miranda Azeiteiro
Department of Biology & Centre for Environmental and Marine Studies
University of Aveiro
Aveiro, Portugal

Harsha Bake
Department of Energy and Environment
Symbiosis Institute of International Business, Symbiosis International University
Hinjawadi, Pune, India

Bojan Baletić
Faculty of Architecture
University of Zagreb
Zagreb, Croatia

J. Benayas
Department of Ecology
Universidad Autónoma de Madrid
Madrid, Spain

N. Blanco-Portela
Department of Environmental at the Faculty of Engineering
Universidad EAN Colombia
Bogotá, Cundinamarca, Colombia

Kathy Bui
International Development & Social Change
Clark University
Worcester, Massachusetts

Sandra Caeiro
Department of Sciences and Technology
Universidade Aberta
Porto, Portugal
and
Centre for Environmental and Sustainability Research
Universidade NOVA de Lisboa
Lisboa, Portugal

Elías Sanz Casado
INAECU (Research Institute for Higher Education and Science)
University Carlos III of Madrid
Getafe, Madrid, Spain
and
LEMI (Laboratory on Metric Information Studies), Department of Library and Information Science
University Carlos III of Madrid
Getafe, Madrid, Spain

Mariana Castellanos
École nationale d'administration publique
Montreal, Quebec, Canada

John Gelegenis
Department of Mechanical Engineering
University of West Attica
Aegaleo, Greece

A. Gomes Martins
Department of Electrical and Computer Engineering
INESC (Institute of Systems Engineering and Computers), University of Coimbra
Coimbra, Portugal

Catarina Isidoro
Department of Social, Political and Territorial Sciences
University of Aveiro
Aveiro, Portugal

Margarita Juárez-Nájera
Departamento de Energía
Universidad Autonoma Metropolitana
Azcapotzalco, México City, Mexico

Chhabi Kumar
Department of Sociology and Social Work
Rani Durgavati University
Jabalpur, Madhya Pradesh, India

S. Leal
School of Management and Technology & Life Quality Research Centre
Polytechnic Institute of Santarém
Santarém, Portugal

Rene Lisac
Faculty of Architecture
University of Zagreb
Zagreb, Croatia

Elba Mauleón
INAECU (Research Institute for Higher Education and Science)
University Carlos III of Madrid
Getafe, Madrid, Spain
and
LEMI (Laboratory on Metric Information Studies), Department of Library and Information Science
University Carlos III of Madrid
Getafe, Madrid, Spain

Albert Mawonde
College of Agriculture and Environmental Sciences
University of South Africa
Muckleneuk, Pretoria, South Africa

Evanthie Michalena
Sustainability Research Centre
University of the Sunshine Coast
Maroochydore, Queensland, Australia

José Carlos Mota
Department of Social, Political and Territorial Sciences
University of Aveiro
Aveiro, Portugal

Ana Pinto de Moura
GreenUPorto, DceT
Universidade Aberta
Porto, Portugal

Frederico Moura e Sá
Department of Social, Political and Territorial Sciences
University of Aveiro
Aveiro, Portugal

Will O'Brien
Associate Professor of Practice
Graduate School of Management
Clark University
Worcester, Massachusetts

Carla Padrel de Oliveira
Department of Sciences and Technology
Universidade Aberta
Porto, Portugal
and
Centro de Química Estrutural
Instituto Superior Técnico, CQE/IST/Universidade de Lisboa
Lisboa, Portugal

Umesh Chandra Pandey
Regional Evaluation Centre
Indira Gandhi National Open University (IGNOU)
Bhopal, Madhya Pradesh, India

Morana Pap
Faculty of Architecture
University of Zagreb
Zagreb, Croatia

Bernardo Campos Pereira
Department of Social, Political and Territorial Sciences
University of Aveiro
Aveiro, Portugal

Núria Bautista Puig
INAECU (Research Institute for Higher Education and Science)
University Carlos III of Madrid
Getafe, Madrid, Spain
and
LEMI (Laboratory on Metric Information Studies), Department of Library and Information Science
University Carlos III of Madrid
Getafe, Madrid, Spain

Prakash Rao
Department of Energy and Environment
Symbiosis Institute of International Business, Symbiosis International University
Hinjawadi, Pune, India

F. Seabra
ISCAL (Instituto Superior de Contabilidade e Administração de Lisboa)
Lisboa, Portugal
and
Lisbon Accounting and Business School
Instituto Politécnico de Lisboa
Lisboa, Portugal
and
Lisbon Polytechnic Institute & IJP Instituto Jurídico Portucalense
Lisboa, Portugal

Muchaiteyi Togo
College of Agriculture and Environmental Sciences
University of South Africa
Muckleneuk, Pretoria, South Africa

Jorge Trindade
Department of Sciences and Technology
Universidade Aberta
Porto, Portugal
and
Centre of Geographical Studies
Instituto de Geografia e Ordenamento do Território
Universidade de Lisboa
Lisboa, Portugal

S. Vinodh
Department of Production Engineering
National Institute of Technology
Tiruchirappalli, Tamil Nadu, India

chapter one

The role of SDGs in advancing implementation of sustainable development

The case of University of South Africa, South Africa

Albert Mawonde and Muchaiteyi Togo
University of South Africa

Contents

1.1 Introduction

Sustainable development is the moral imperative of satisfying needs, ensuring equity and respecting environmental limits (representing constraints on human activities) including efforts to maximise economic value (Holden et al., 2014). Robinson (2004) emphasised that sustainable development is a collective institutional response, efficiency gains and a social responsibility. The authors maintained that sustainability is a consistent set of concepts more usefully thought of as an approach or process of community-based thinking. Considering the university as a community, the term sustainability can be seen as anthropocentric where the community gives it value through deciding the route to follow and policies to adopt and implement so that an institution can be regarded as sustainable. However, Sanchez et al. (2013) noted that, despite lack of agreement on an unequivocal interpretation of the concept of sustainable development, there is a general agreement that it involves simultaneous satisfaction of economic, environmental and social goals. The commonly used definition is the one proposed by the Brundtland Commission (1987) which states that sustainable development is a concept which allows economic development while considering environmental limits and equity. The chapter leans more towards the Brundtland Commission (1987) definition.

In 2008, in the South African context, the government developed a National Framework on Sustainable Development by listing specific strategies and critical areas aimed at guiding the country's development on a more sustainable path (DEAT, 2008). To accomplish these interventions, all stakeholders needed to fulfil their roles. One sector that can play an essential role in the education sector is the university; by providing training and role modelling. Universities can teach and demonstrate the theory and practice of sustainability by taking action to understand and reduce the negative environmental impacts of their activities (Lotz-Sisitka, 2014).

Following the Rio+20 Summit in Brazil (2012), 17 Sustainable Development Goals (SDGs) were adopted to address challenges resulting from environmental catastrophes (natural and anthropogenic). SDGs were adopted as a follow up to the just-ended Millennium Development Goals (MDGs) as a way of escalating sustainable development globally (Hak et al., 2016). University campus sustainability challenges identified by GUPES and other sustainability stakeholders; especially the United Nations; resulted in universities embracing SDGs. Out of the 17 SDGs, some of the goals naturally fit the purpose and missions of specific institutions, and in the case of universities, this happens to be SDG 4: to ensure inclusive and equitable education and to promote lifelong learning opportunities for all.

SDGs can be seen as an opportunity that can be and is being embraced by the higher education sector to further its quest to become more sustainable. According to Lozano (2006), implementing campus sustainability is not an easy task. Leal Filho (2011) supported this when he said that campus sustainability implementation is confusing, time-consuming and creates many uncertainties among various stakeholders. The promulgation of SDGs demystified the confusion surrounding campus sustainability implementation. The pre-existing lack of direction regarding the role of universities in sustainable development can be answered by embracing SDGs which are linked to the mission statement of the institution.

The role of the university in addressing global challenges of sustainability has been cross-examined by several authors (Cortese, 2003; Moore, 2005; Edelstein, 2009; Lozano et al., 2015). Tilbury (2011) argued that universities can change the world through training and expanding young minds, researching answers to challenges and informing public policy by demonstrating their understanding and commitment through careful campus management. Stephens et al. (2008) claimed that the role of a university is pivotal in society since universities are manifestations of longevity and social stability with a non-profit focus. Universities can foster long-term thinking, which is critical for sustainability planning and policy review; which in turn is critical for sustainability.

Policy aims proposed by the United Nations General Assembly in education and lifelong learning gave impetus to reflections on the role of universities and their potential contribution in moving towards a low carbon and resource efficient economy. According to the 2012 UNCSD outcome document (resolution 234), universities were strongly encouraged to consider adopting good practices in sustainability management on their campuses and surrounding communities, with the active participation of students, teachers and local partners.

Many declarations challenged universities to embrace campus sustainability. Some of these were the Talloires Declaration (1990), the Halifax Declaration (1991), the Kyoto Declaration (2000) and the Turin Declaration (2009). The United Nations Decade of Education for Sustainable Development (2005–2014) was proposed to promote sustainability mainstreaming, and it culminated in the UNESCO World Conference on Education for Sustainable Development (2014). The Global Action Programme (2015) also promoted the same agenda. These declarations and fora have one significant similarity, a call for higher education to utilise the weapon of education, research and training in tackling environmental "wicked problems" (Dentoni and Bitzer, 2015: p. 2).

Velazquez et al. (2006) defined a sustainable university as a higher education institution as a whole or as a part, which addresses, involves and minimises adverse environmental, economic and health effects generated in the use of its resources. Universities fulfil its function of teaching, research, outreach, partnership and stewardship in ways which help society make a transition to sustainable lifestyles. Cole (2003) also defined a sustainable campus community as one that acts upon its local and global responsibility to protect and enhance the health and well-being of humans and ecosystems. It actively engages the knowledge of the university community to address the ecological and social challenges which we face currently and in the future.

SDG 4 which focusses on 'promoting inclusive and equitable quality education and lifelong learning opportunities for all' can be realised through students' involvement in campus greening initiatives. Regarding sustainable development, educational quality is promoted through provisioning of relevant curricula which address current and topical environmental and sustainable development issues. Through learning about sustainable development, students are equipped with lifelong learning skills applicable not only for examination purposes but also for their entire life, especially at home, in their communities and at workplaces.

Several other campus greening initiatives were implemented worldwide by various institutions with the goal of promoting equitable quality education and lifelong opportunities for all. Examples are listed and briefly discussed below:

- Oberlin College promoted lifelong learning and quality education using green campus initiatives (Orr, 2011). It had green energy supply initiatives, green procurement, carbon sequestration and sustainability alliances. In these initiatives, equity, an integral part of sustainable development, was realised through the college partnerships with students, community members and the City of Oberlin in tackling green energy and carbon offsetting Projects. Irrespective of gender, creed, occupation and status in society, the project involved all stakeholders in finding lasting solutions to environmental challenges. Lifelong learning was enhanced since these skills were imparted to all the people involved, and they used them as part of their daily lives; not only for the project.
- The partnership of Lund University, Malmo University and Swedish university of Agricultural Sciences is another example which showed how universities were embracing SDGs (Trencher et al., 2016). The partnership saw an opportunity to involve various stakeholders in re-orienting the university operations division. Quality education and lifelong learning were promoted through the exchange of ideas and sharing of different experiences by each institution. Staff members, students and sustainability champions all benefited from putting their minds together to find lasting solutions to environmental challenges.
- Rhodes University, through community participation, involved staff members, students, councillors and the local people in an HIV-AIDS Resistance Campaign (Togo, 2009). The community benefited immensely through relevant information, which enhanced their awareness of HIV-AIDS infections and how to practice home-based care of infected patients. The process of gaining knowledge on how to treat and take care of infected patients and to prevent stigma was part of the learning process and plays a pivotal role in reducing HIV-AIDS infections and early demise of infected patients. The campaign was open to the community members of Makana District in Grahamstown and reached people with diverse backgrounds (equity).
- According to Govender (2005), at the University of KwaZulu Natal (UKZN), the Faculty of Engineering implemented an Energy Management Programme to improve energy efficiency. The energy

saving programme served as an excellent example of how sustainable campus greening initiatives can be implemented (Smit, 2009). According to Knox (2013), the estimated amount of money UKZN saved through energy retrofitting reached almost 4 million rands per year which is approximately equivalent to US$300,000.

Analysing the aforementioned literature on campus sustainability as well as how various institutions implemented campus greening initiatives aligned with SDGs, one can conclude that most of the institutions which are at the forefront of campus sustainability initiatives are traditional universities where students reside on campus. Almost all reviewed universities have initiatives which address aspects of goal number 4 (inclusivity, lifelong learning and quality education). The question that arises in this chapter pertains to how Unisa, an Open Distance Education university with students who do not frequent the campus is promoting inclusivity, lifelong learning and quality education through campus greening initiatives.

To provide a brief background, Unisa was founded in 1873 as the University of the Cape of Good Hope situated in Gauteng province, South Africa. It was formed through the merger of the old Unisa, Technikon Southern Africa and the Vista University. Unisa boasts a student enrolment of about 400,000 which represents over a third of South Africa's tertiary students which stands at 1,200,000 (Unisa, 2015). In 2008, Unisa became a member of the United Nations Global Compact (UNGC) which is the most extensive corporate citizen network that draws participants from business, national and civil fraternities (Unisa, 2015).

By becoming a member and a signatory of the UNGC, Unisa committed to contribute towards sustainable development since principles 7–9 of the UNGC promotes sustainable utilisation of resources and sound sustainable management of organisations. The UNGC is a principle-based framework for business, stating ten principles in the areas of human rights, labour, the environment and anti-corruption (Unisa, 2015).

Unisa has made significant success in making sure that its academic program qualification mix promotes sustainable development values underscored by the UNGC. The university takes pride in a myriad of programmes that drive sustainable development in the education sector in Southern Africa such as The Chancellor's Sustainability Programme spearheading most of Unisa's sustainability initiatives namely: The Green Economy Campus Initiatives (GECI) and the Green Economy Sustainability Engagement Model (GESEM). According to Unisa (2015), the GESEM model integrates policy development focussing on various areas of environmental sustainability, for example, carbon footprint, energy and carbon policy, pollution and wastewater policy, ISO 14001 certification and waste control policy.

It is, however, vital to note that sustainability is not a one-size-fits-all framework (Barkemeyer et al., 2011). The principal focus of the chapter is to establish how Unisa, an ODeL institution, is involved in the implementation of SDG 4. A review of other literature sources has been explored, and it outlined how various universities are embracing sustainability and SDG 4 in their contexts. In the next section, a brief overview of the research methods is provided. Results and discussion follow the methods section. The final section is a conclusion to the chapter.

1.2 Methodology

The research was informed by a mixed methods design and was based on a case study of Unisa. According to Yin (2003), case studies allow the use of multiple data sources which enhance validation of results, credibility and dependability. Data was collected through questionnaires, interviews, document analyses and campus observations (Creswell et al., 2003) between March 2017 and September 2017.

Telephonic interviews and face-to-face interviews were conducted. Semi-structured interviews (Alverson, 2003) were employed to interrogate the Unisa Florida Campus Director, the Waste Controller, Sustainability Champions, the Campus Operations Manager and the Student Representative Council President on how the institution is embracing SDGs. BSc Honours Environmental Management students responded to an online questionnaire. Interview questions asked varied across respondents. The Sustainability Office members were asked questions related to Unisa Sustainability Policies and their implementation, sustainability planning and projects. Interview questions for Unisa Operations Manager and Campus Director were on sustainability projects present at the campus, for example, recycling, which were aligned to the SDGs and how the projects were involving students. The Student Representative Council interview guide interrogated how students were involved in sustainability initiatives and SDGs at Unisa. The respondents were selected purposively. The Campus Director, Sustainability Office Personnel and Operations Manager were interviewed because they have the technical knowledge of campus sustainability initiatives and SDGs as campus sustainability is part of their work. The questionnaire for BSc Honours Environmental Management and BSc Honours Environmental Monitoring and Modelling students was meant to solicit information on how their programme curriculum is contributing to SDG 4. During campus observations (discussed later), ten students were informally engaged to establish their involvement in greening practices on campus (Table 1.1).

A variety of documents were analysed including the Unisa Environmental Sustainability Policy of 2012, Unisa Waste Management Policy

Table 1.1 Methods which involved human participants and their roles

Method	Participant(s)	Socio-demographics of the participant	Participant role at the university
Interview	Campus Director	Male aged 40 years + PhD holder	Oversees university functioning and planning of physical infrastructure, technological resources and facilities management of the university.
Interview	Sustainability Office Manager	Female aged 40 years + PhD holder	Oversees university sustainability policy formulation, planning and implementation of campus greening initiatives in partnership with all universities departments and external stakeholders.
Interview	Operations Manager	Male aged 35 years + MSc holders	The university core which plans facilities management, procurement, estates and gardens section, physical planning and university green development.
Informal interviews	Unisa students engaged informally during campus observations	Six females and four male students doing BSc Honours Environmental Monitoring and Modelling	Students studying various programmes at Unisa Florida Campus
Questionnaire	CAES BSc Honours Environmental Management and BSc Honours Environmental Monitoring and Modelling students (n = 114)	79 female and 35 male students doing BSc Honours Environmental Management	Students studying in the College of Agriculture and Environmental Sciences at Florida campus.

of 2017, Unisa Energy and Carbon Policy of 2016, *Younisa* 2018 Magazine, 2015, 2017 and 2018 Unisa Annual reports, Unisa website and the UnisaWise magazine of summer 2011. These documents were selected because they have information, which addresses the research questions.

Campus observations provided a means for methodological triangulation to enhance the quality of the collected data (Patton, 2002: p. 247). They, however, resulted in incorporation of new data which had not been revealed through the other methods. Trustworthiness was enhanced through member checking (Johnson and Waterfield, 2004). Content and thematic analysis were used to analyse documents, interview transcripts and campus observation notes (Braun and Clarke, 2006: p. 79).

1.3 Results

This section presents data and analyses and discusses the findings of the study. The findings were discussed within the context of literature on past studies for convergence and corroboration. Lifelong learning themes and quality education themes were discussed simultaneously in brief paragraphs.

1.3.1 Unisa campus energy greening initiatives

Unisa hosted a Zero Carbon Emissions Electric Cars roadshow in 2011. This is an annual meant event to increase sustainability awareness at the campus and the surrounding communities. The roadshow was hosted in partnership with the Department of Environmental Affairs and Nissan Clover Leaf which supplied electric vehicles. The event involved all stakeholders including Unisa teaching staff, students, government officials and members of the community and all these stakeholders benefited from the knowledge gathered. Based on BSc Honours Environmental Management students' data, 75 students out of 114 (66%) respondents are aware of the Unisa energy initiatives. According to the Unisa Sustainability Manager, further education and awareness campaigns were carried out to empower the local community and students on the benefits of adopting green energy.

According to Trencher et al. (2016), knowledge co-creation through university partnerships is a way of generating new knowledge and escalating sustainability solutions and awareness. In this context, new knowledge generated is equated to quality education. The lessons drawn by the stakeholders is applied at various levels ranging from household level to multinational level. Electric vehicles help to tackle climate change impacts through reducing carbon dioxide emissions produced by vehicles. Lessons drawn from this initiative supports SDG 13 (take urgent action to combat climate change and its impact). Electric vehicles are considered "green vehicles" with less environmental impacts compared to the traditional vehicles.

According to the Unisa Environmental Sustainability Policy of 2012, the university has an Energy and Carbon Policy Masterplan. Under section 4.1 a of the master plan, Unisa advocates for the significant reduction of electricity consumption using alternatives energy sources. Out of 114 students interviewed, 20 students (18%) knew about the master plan. Below is one of the responses from the Sustainability Manager:

> Unisa has a solar installation program in place and a solar-powered car charging station for its staff members and students. However, the solar project is very successful in cutting down energy costs, but the solar car charging initiative is not fully used to its capacity due to the high prices of solar-powered cars……,
>
> *Sustainability Office Manager, Unisa (2018).*

Use of solar energy as an alternative source of energy is in line with Unisa Energy Policy and South Africa's National Climate Change Response White Paper (2011) which pledged the reduction of carbon emissions by 34% by 2020 and 42% by 2025. Through the solar initiative, lifelong learning is enhanced through lessons learnt from adopting cleaner energy sources (SDG 13). Quality education (SDG 4) is enhanced when stakeholders like students and the greater community surrounding the university get an opportunity to understand the rationale behind cleaner sources of energy. In the long run, communities will adopt these strategies in their own lives. The solar car charging project serves to demonstrate the need for innovation and awareness (SDG 9, industry, innovation and infrastructure) of the necessity of renewable energy. Initiatives like the solar energy installation contribute towards lifelong learning in sustainable development from benefits accrued and lessons drawn.

Evidence gathered from the UnisaWise (Unisa 2011) through content analysis shows that Unisa adopted a vehicle fleet management system in 2011 as a way of carbon offsetting and saving fuel costs. A Global Positioning System tracks staff vehicles to cut down on unnecessary trips and fuel expenses. The initiative resulted in 17% savings on fuel expenses in the 2012 financial year, and it also led to a reduction in carbon emissions. It was estimated that approximately 3,400 tonnes of carbon dioxide equivalence (CO_2-eq) were offset by the project (Nhamo and Ntombela, 2014). Unisa outsourced buses for staff transportation between Johannesburg and Pretoria which is cheaper and is a way to offset carbon emissions. From 10% users of public staff transport, the percentage increased to approximately 30% due to the efficiency and strategic pick and drop off points which were introduced. Following the introduction of the "Gautrain" in Johannesburg, Unisa linked its shuttle services to the Gautrain pickup points which increased the number of staff who uses public transport. Gautrain is equivalent to electric trains found in most developed countries, and it has a high standard service and good security (Unisa, 2015). This initiative complimented South Africa's adoption of the Bus Rapid Transport system which is considered a greener mode of transport. Nhamo and Mjimba (2014) noted that the Gauteng province where Unisa is located is a heavily polluted province with the highest greenhouse gas emissions in South Africa due to a high concentration of vehicles and industries. The use of public transport and cutting off of unnecessary trips by Unisa's transport department served as a lifelong learning lesson to Unisa staff, students and community members. Vehicle tracking is a powerful way institutions are adopting to minimise unnecessary fuel costs and carbon pollution.

A new theme emerged from the "Younisa Issue 1 of 2018" document through content analysis. It was found that Unisa implemented its energy policy outside its campus in Melane, a rural community in the Eastern Cape Province in July 2016 as part of its community engagement. A biogas digester project was implemented through a partnership of the Unisa College of Agriculture and Environmental Sciences; College of Economic and Management Sciences and College of Science, Engineering and Technology as a transdisciplinary research programme The programme witnessed the installation of a 10m^3 biogas digester which has potential to generate 4.3 kg of liquified petrolium gas (LPG) per day for the vulnerable rural populace. Waste from cattle, goats and pigs is being used as feedstock while the by-product (digestate slurry) is used as a source of fertiliser for growing crops. Like the Oberlin College partnership which exhibited specific knowledge generation (Trencher et al., 2016), the project aligns with the SDG 4 theme of lifelong learning. The community taps into the project using animal waste to solve their challenges of electricity shortages in rural areas and high fertiliser costs. Besides the socio-economic benefits of the project, communities gained crucial and relevant knowledge and practical skills applicable in life beyond the project.

In addition to the Unisa biogas project, the Unisa Sustainability Office stated that Unisa is in partnership with the South African Medical Research Council (SAMRC) to research and provide clean sources of energy. The energy is in the form of solar energy, biogas and cool coating to reduce health risks and deaths in informal settlements around Gauteng in which people rely on kerosene and paraffin. These energy sources are a fire hazard and can cause respiratory diseases. Figure 1.1 shows the Unisa demonstration site for clean energy solutions.

Figure 1.1 Unisa demonstration site for three clean-energy solutions for households: solar, biogas and cool coatings. (Source: Kimemia, 2019.)

1.3.2 Unisa campus recycling and conservation initiatives

Content analysis of the Unisa Annual Report (Unisa, 2015) revealed that the Ethiopian Graduate Office (EGO), under Unisa College of Graduate Studies, held a tree planting event on 29 August 2014 on the premises of the Ethio-Japan Primary School, which is located at the front of the EGO campus at Akaki in Ethiopia. Together with the Department of Geography, the EGO responded to SGD 4 by developing a community engagement project entitled 'Ethiopian Children Embracing Green Living' (Unisa, 2015). Children were taught sustainability initiatives such as recycling, tree planting, water harvesting, organic farming, healthy living and climate literacy. Such activities promote lifelong learning for sustainability especially in desert-prone areas of Ethiopia where desertification is threatening to decimate the populace. The Ethiopian school together with the local community had an opportunity to learn about strategies for averting the impacts of desertification and climate change. The project is most likely going to grow in scale as the community can implement the same initiatives at different levels, for example, their households and farms. The project involved a diversity of people including males and females, different age groups as well as people from different religions and creed, hence, it addressed equity of access to information. There could be a multiplier effect as well whereby those involved in the project can teach others, hence, more community members will also learn to solve environmental challenges.

Thematic analyses of campus observation notes and data gathered from other sources revealed that Unisa has operational practices to plant and promote indigenous plants as part of its landscaping. Indigenous plants, due to their better adaptation to local environmental conditions, tend to utilise less water. They do not disrupt normal ecosystem dynamics compared to alien plants which tend to have characteristics of a different climate. They also provide a natural habitat for various animal species. These plants use less water compared to alien plants and they promote ecosystem balance. Through this initiative, many sustainability lessons were learnt, and these could be adopted by university students, staff and surrounding communities for implementation in other contexts. Many BSc Honours Environmental Management and BSc Honours Environmental Monitoring and Modelling students (60% of respondents) were engaged in off-campus sustainability initiatives (recycling, water harvesting, eco-schools' gardens) at workplaces, in the community or at home. Due to the nature of distance education, Unisa students are not that involved in practices that respond to SDGs on campus, although 50% indicated a willingness to participate. University students are doing research on the importance of indigenous trees and the role they play in the ecosystem and the environment. Unisa Annual Report (Unisa, 2017) revealed an increase in student funding in research from 42,200 to 43,500 students valued at R540 million rand which is equivalent to approximately US$38 million. The number of Unisa students' research projects on sustainability/sustainable development was 101 in total (Unisa, 2017). Unisa Press published 45 journal articles on campus, community and sustainability-related research. Such research reaches out to the entire university and beyond when it results in publications – carrying with it important lifelong lessons.

UnisaWise (Unisa, 2011) revealed that the Sunnyside North aloe garden located at Unisa Sunnyside Campus in Pretoria is the most prolific, unique aloe garden with 15 species of aloes which flower at different times in the year. The Grounds and Gardens Section of the university is looking for innovative ways to weave teaching and learning into the garden and use it for research purposes. The garden is meant to increase the knowledge and understanding of aloes as they have potential in medicinal use, and this is linked to knowledge creation and lifelong learning. Through research there is potential for more discoveries where various uses of aloe plants are concerned. This means the learning process will continue, and the vital knowledge regarding how the plants are used to better lives will continue to be generated. From the aloe project, Unisa staff members, students and the surrounding community are now aware that aloe cures numerous diseases including dermatitis, and it lowers blood pressure among others. According to the Department of Agriculture, Forestry and Fisheries (2015), South Africa increased the growth of aloe for export purposes. Below is a quote from the Department of Agriculture, Forestry and Fisheries about the importance of aloe for South Africa:

> Aloe industry arose from the need to explore new opportunities for the agricultural sector that could lead growth and exports. For the future, industries are required to manufacture more complex tradable goods, provide multiple high-value product opportunities, provide decent jobs and match the natural resource environment. Aloe was found to suit many of the challenges for future growth.

Job creation, poverty alleviation and industrial innovation are part and parcel of the agenda 2030 SDGs. Aloe research at Unisa is contributing to the much needed data and innovation by South African aloe farmers to grow their aloe industries and export markets. The Unisa aloe gardens are small, but the initiative set a baseline for learning opportunities at the university.

The Unisa Estates division partnered with the College of Agriculture and Environmental Sciences (CAES) to undertake the rehabilitation of the Cycad Garden. All plants were labelled in 2014 to add knowledge to the enjoyment of the visitors. This was done to make the garden a place of botanical interest, with the vegetation serving as a valuable educational and research tool. The Unisa Cycad collection at Muckleneuk campus is internationally renowned for the number and variety of South African Encephalartos species which opens the door for quality learning and lifelong learning. The Cycad Garden offers the visitor a unique opportunity to learn about the 28 cycad species of South Africa, growing among other indigenous trees, shrubs and bulbous plants. Figure 1.2 is an image of one of the gardens with labelled tree species at Unisa.

Figure 1.2 Labelled Unisa Garden, Muckleneuk campus. (Source: Sbahle Landscapes, undated.)

Content analysis of the Unisa Waste Management Policy of 2017 revealed that Unisa has a sound and functional recycling policy. Unisa approved the Bokhasi wet waste recycling programme as part of its Environmental Sustainability Plan (2012) and Solid Waste Management Plan (2017). The project is housed at Florida Campus in Johannesburg and Muckleneuk Campus in Pretoria and is managed by Earth Probiotic Recycling Solutions. The Bokhasi wet waste recycling process stabilises waste by eliminating pathogens while prohibiting the production of greenhouse gases like methane which is 25 times more toxic than carbon dioxide. The waste is in turn used as organic matter for the gardens and estates section. It is estimated that the project diverted 8,500 kg and 3,480 kg food waste from Muckleneuk and Forida campus respectively, from landfill disposal per month (Unisa, 2015). One interviewee had this to say about the project:

> Bokhasi is still operational at Florida, and Muckleneuk campuses and the project has been very successful...
>
> *Sustainability Office Manager, Unisa (2018)*

Brinkhurst et al. (2011) posit that universities have already engaged themselves in restructuring their curriculum, the research agenda and community services to focus more on sustainable development as well as incorporating sustainability into campus development and daily operations. In line with Brinkhurst's observations, Unisa practises dry waste recycling and the separation of waste at source. This is in tandem with its Waste Management Policy hinged in the National Environmental Management Waste Act, 59 of 2008. Unisa utilises the services of Oricol environmental services and Nampak for removal and recycling of wastepaper since 2011. These initiatives offer an opportunity for lifelong learning among Unisa staff and students as well as the community through research and publications. Figure 1.3 shows labelled igloo bins at Unisa Florida campus for separating waste at source.

1.3.3 Student involvement in campus greening initiatives that address SDG 4

Unisa also responds to SDGs through research and the curriculum. Unisa being a distance education learning institution which uses computers and the internet, creates opportunities for students' involvement in sustainability learning and research. While in traditional universities only those students who can study fulltime have an opportunity to access certain degrees or programs, with ODeL anyone, anywhere who

Figure 1.3 Labelled igloo bins for waste separation at sources Unisa Florida Campus.

qualifies can study. Furthermore, the programmes are relatively more affordable over and above most regular programmes offered by traditional universities.

Based on a student survey (n = 114), 77 students (68%) were aware of sustainability initiatives at Unisa. Out of the 77, 44 students (58%) noted involvement in sustainability initiatives through projects at work and at home. Recycling of waste, composting, water harvesting, planting of trees for medicinal use and energy saving projects are some of the initiatives these students were practising off-campus. Through work integrated learning, two students from those interviewed informally, are involved in a solar project at Greenpeace, in Gauteng, a non-governmental organisation which advocates for cleaner, cheaper energy supply to save resources and to avert impacts of climate change.

The Unisa Student Representative Council was involved in HIV-AIDS projects through raising HIV-AIDS awareness on the need for abstinence, the use of protection during sexual intercourse or HIV testing before indulging with a partner and encouraged students and the community to avoid promiscuity (Unisa, 2018). The students together with Unisa staff observed the World AIDS Day on 1st December 2017 by offering free HIV-AIDS testing for students, staff members and the community. They celebrated the day under the banner of ending infections, stigma and isolation of HIV-AIDS patients. Through this project, many people were taught how to take care of patients affected with HIV-AIDS (home-based care initiatives). All this is in observation of the principle of quality education and lifelong learning as encapsulated in SDG 4; through knowledge sharing and community consultations with professional HIV-AIDS counsellors. This also aligns with SDG 3 (ensure healthy lives and promote well-being for all at all ages).

There was no direct link or synergy between Unisa Operations Management projects and student involvement in campus greening initiatives. From the survey of 114 students, 77 (68%) students were involved in sustainability initiatives through research at the workplace or through individual research projects aligned with the Unisa programme qualification mix. There are opportunities for students to be involved in sustainability initiatives and implementation of SDGs within the operations and management section of the university. Many operational projects can be done with students to equip them with valuable skills to apply at future workplaces, homes, communities and in their lives. However, distance is barrier as students do not attend classes on campus. It is important to note that SDG 4 influences all other 16 SDGs because education makes all other SDGs more feasible to implement (El-Jardali et al., 2018).

1.3.4 *Challenges*

Evidence gathered from interview data suggested that Unisa is struggling to involve students in practices which respond to SDGs, the reason being that it is a distance education institution. The institution as a whole, is not providing opportunities for students to get more exposure to SDGs initiatives on campus. From the sample of students surveyed (n = 114), 58% of them portrayed interest to participate in SDGs implementation at the campus, but they cited lack of such opportunities due to the distance mode of learning.

Through the questionnaire, students cited financial challenges to travel to the campus to attend sustainability awareness workshops, seminars and symposia. Sixty-six percent of the students further cited the unavailability of internet services and the high cost of data as prohibiting them from participating in campus sustainability initiatives. Lack of funding of most sustainability initiatives and green projects at Unisa were cited by 40% of the students as a significant setback to their involvement in SDG implementation.

Eighty percent of the students pointed out that there is lack of a sustainability framework to expose students to greening practices especially in such faculties where sustainability and SDGs are not a prerequisite of their learning programmes. In addition, 20% of the students viewed Unisa as an institution that does not recognise students as essential stakeholders;

hence, they felt left out in the SDG's matrix and campus sustainability initiatives.

There is controversy and confusion about what constitutes sustainability among students, and 58% of those surveyed are failing to appreciate SDGs. In addition, 40% of the students argued that they could not participate in concepts they do not understand. Lozano et al. (2015) argued that most universities are engaged in sustainable development, but there is still much confusion about the concept and how universities must embrace it. The main barriers identified by Leal Filho (2011) including limited resources, lack of student participation and misconceptions of the sustainable development concept seem to apply in the context of Unisa.

Unisa is putting much effort towards addressing operations to meet most environmental management standards. However, it does not have many opportunities for directly involving students in SDG initiatives unlike in tertiary institutions in developed countries. In October 2016, for example, at Monash University, the Monash Sustainable Development Institute undertook an initiative which brought together student leaders and key staff members in deliberating student actions towards SDGs (SDSN Australia/Pacific, 2017).

1.4 Conclusion

At Unisa, there is evidence of the adoption of SDG 4 especially in campus operations and estates division. The adoption of UNGC principles and their operationalisation at Unisa paved way for a greener campus. However, student involvement in sustainability initiatives is still a matter of concern. Unisa is struggling to maximise on available opportunities to involve students in campus environmental management. The distance between Unisa and its students is the main barrier hindering involvement of students in practices which respond to SDGs. Most students graduate without having visited the Unisa campuses even once.

Unisa is applauded for its programme qualification mix which is reaching out to a broad range of students irrespective of age, providing them with a chance to learn and appreciate sustainable development and SDGs. Distance education caters for thousands of students who access study material via the internet and courier services, thereby facilitating and breaking the boundaries of traditional institutions which enrol fewer students who can stay and attend classes on campus. The affordability of Unisa tuition fees compared to fulltime universities is another factor attracting more students, thus providing more students an opportunity for lifelong learning.

References

Alvesson, M. 2003. Beyond neopositivists, romantics and localists: A reflective approach to interviews in organizational research. *Academy of Management Review* 28(1), 13–33, Emerald Group Publishing.

Barkemeyer, R., Holt, D., Preuss, L., and Tsang, S. 2011. What happened to the 'development' in sustainable development? Business guidelines two decades after Brundtland. *Sustainable Development*. Published online in Wiley Online Library. DOI: 10.1002/SD.521.

Braun, V. and Clarke, V. 2006. Using thematic analysis in psychology. *Qualitative Research in Psychology* 3, 77–101.

Brinkhurst, M., Rose, P., Maurice, G., and Ackerman, J.D. 2011. Achieving campus sustainability: Top-down, bottom-up or neither? *International Journal of Sustainability in Higher Education* 12(4), 338–354.

Bruntland, G. (Ed.). 1987. *Our Common Future*. Oxford: Oxford University Press: The World Commission on Environment and Development.

Cole, L. 2003. *Assessing Sustainability on Canadian University Campuses: Development of a Campus Sustainability Assessment Framework*. Victoria: Royal Roads University.

Cortese, A.D. 2003. The critical role of higher education in creating a sustainable future. *Planning for Higher Education* 31(3), 15–22.

Creswell, J.W., Plano Clark, V.L., Gutmann, M.L., and Hanson, W.E. 2003. Advances in mixed methods research designs. In *Handbook of Mixed Methods in Social and Behavioural Research*, by A. Tashakkori & C. Teddlie (Eds.), pp. 209–240. Thousand Oaks, CA: Sage Publications.

DEAT. 2008. *A National Framework for Sustainable Development in South Africa*. Pretoria: Department of Environmental Affairs and Tourism.

Dentoni, D. and Bitzer, V. 2015. The role(s) of universities dealing with global wicked problems through multi-stakeholder initiatives. *Journal of Cleaner Production* 106, 68–78.

Department of Agriculture, Forestry and Fisheries. 2015. *A Profile on the Aloe Industry for Export: A Focus on South Africa*. Pretoria: Directorate International Trade.

Edelstein, M.R. 2009. Sustainable campuses and institutions: The need to lead and the transformation to a new social paradigm. *Research Gate*. www.virtuniv.cz/images/6/6c/Edelstein Slunecnice2009.pdf (Accessed 10 December 2017), 37–42.

El-Jardali, F., Ataya, N., and Fadlallah, R. 2018. Changing roles of universities in the era of SDGs: Rising up to the global challenge through institutionalising partnerships with governments and communities. *Health Research Policy and Systems* 16(1), 38.

Govender, P. 2005. Energy audit of the Howard College Campus of the University of KwaZulu Natal. *Masters dissertation*. University of KwaZulu Natal.

Hak, T., Janouskova, S., and Moldan, B. 2016. Sustainable development goals: A need for relevance indicators. *Ecological Indicators* 60, 565–573. DOI: 10.1016jecolind.2017/10/10.

Holden, E., Linnerud, K., and Banister, D. 2014. Sustainable development: Our common future revisited. *Global Environmental Change* 26, 130–139.

Johnson, R. and Waterfield, J. 2004. Making words count: The value of qualitative Research. *Physiotherapy Research International* 9(3), 121–131.

Kimemia, D. 2019. Safe energy demonstration site at Unisa Institute for Social and Health Sciences. Lenasia: Johannesburg.

Knox, A. 2013. UKZN Energy management earns R4 milin savings. KZN. Energyhttp://www.kznenergy.org.za/ukzn-energy-management-earns-r4mil-in-savings (Accessed 12 December 2017).

Leal Filho, W. 2011. About the role of universities and their contribution to sustainable development. *Higher Education Policy* 24(4), 427–438.

Lotz-Sisitka, H. 2014. UNESCO world conference on education for sustainable development, *Conference Report by the General Rapporteur.* Okayama (Japan).

Lozano, R. 2006. Incorporation and institutionalisation of SD into universities: Breaking through barriers of change. *Journal of Cleaner Production* 18, 637–644.

Lozano, R., Ceulemans, K., and Seatter, C.S. 2015. Teaching organisational change management for sustainability: Designing and delivering a course at the University of Leeds to better prepare future sustainability change agents. *Journal of Cleaner Production* 106, 205–215.

Moore, J. 2005. Seven recommendations for creating sustainability education at the university level: A guide for change agents. *International Journal of Sustainability in Higher Education* 6(4), 326–339.

Nhamo, G. and Mjimba, V. 2014. Biting the hand that feeds you: Green growth and electricity revenues in South African metropolitans.

Nhamo, G. and Ntombela, N. 2014. Higher education institutions and carbon management: Cases from the UK and South Africa. *Problems and Perspectives in Management* 12, 208–217.

Orr. 2011. The Oberlin Project: What do we stand for? *Oberlin Magazine,* Fall 2011. Oberlin College, Oberlin. 19–28.

Patton, M.Q. 2002. *Qualitative Evaluation and Research Methods* (3rd ed). Thousand Oaks, CA: Sage Publications.

Robinson, J. 2004. Squaring the circle? Some thoughts on the idea of sustainable development. *Ecological Economics* 48(4), 369–384.

Sanchez, F.G., Bernaldo, M.O., Castilleio, A., and Manzanero, A.M. 2014. Education for sustainable development in higher education: State of art, barriers and challenges. *Higher Learning Research Communications* 4(3), 3–11.

Sbahle Landscapes. Undated. Unisa Mariam Makeba. https://www.sbahlescapes.co.za/2015/08/02/unisa-mariam-makeba/ (Accessed 12 December 2017).

SDSN Australia/Pacific. 2017. *Getting Started with the SDGs in Universities: A Guide for Universities, Higher Education Institutions, and the Academic Sector. Australia, New Zealand and Pacific Edition.* Melbourne: SDSN Australia/Pacific.

Smit, G. 2009. Sustainable energy solution for the residences of Stellenbosch University. *Masters dissertation.* University of Stellenbosch.

Stephens, J.C., Hernandez, M.E., Roman, M., Graham, A., and Scholz, R.W. 2008. Higher education as change agent for sustainability in different cultures and contexts. *International Journal of Sustainability in Higher Education* 9, 317–338.

Tilbury, D. 2011. Are we learning to change? Mapping global progress in education for sustainable development in the lead to Rio Plus 20. *Global Environmental Research* 14(2), 101–107.

Togo, M. 2009. A systems approach to mainstreaming environment and sustainability in universities: The case of Rhodes University. *Unpublished Doctoral Thesis,* South Africa: Rhodes University.

Trencher, D., Rosenberg, D.D., McCormick, K., Terada, T., Peterson, J., Yarime, M., and Kiss, B. 2016. *The Role of Students in the Co-creation of Transformational Knowledge and Sustainability Experiments: Experiences from Sweden, Japan and the USA*, pp. 191–215. Cham: Springer International Publishing.
UNCSD. 2012. *The Future We Want*. Rio de Janeiro: UNCSD.
Unisa. 2015. *Unisa Annual Report*. Pretoria: Unisa.
Unisa. 2017. *Unisa Annual Report*. Pretoria: Unisa.
Unisa. 2018. *Unisa Annual Report*. Pretoria: Unisa.
Unisa. 2011. *Unisawise: Into the sustainability era*. Pretoria: Unisa.
Velazquez, L., Munguia, N., Platt, A., and Taddei, J. 2006. Sustainable university: What can be the matter? *Journal of Cleaner Production* 14(9–11), 810–819.
Yin, R.K. 2003. *Applications of Case Study Research* (2nd ed). Thousand Oaks, CA: Sage Publications.
Younisa. 2018. *Magazine for Unisa students, alumni and friends*. Pretoria: Unisa.

chapter two

Evolution of the actions of Latin American universities to move towards sustainability and the SDGs

J. Benayas
Universidad Autónoma de Madrid

N. Blanco-Portela
Universidad EAN

Contents

2.1 Introduction

The complex challenges facing the world today – increasing poverty, unemployment, hunger, social and gender inequalities, increasing world population, the degradation of ecosystems, depletion of resources, the urgent need to develop strategies in the different sectors that allow them to mitigate and adapt to climate change, among many others – require unified global efforts towards resolution.

In addition to the abovementioned challenges, the Global Action Plan (GAP) of Education for Sustainable Development (ESD) (UNESCO 2014) defines five priority areas of action, one of which directly addresses the transformation of educational environments by "incorporating sustainability into campus operations, governance, policies and administration" (UNESCO 2014, p. 18). The fulfilment of this action has been previously addressed (Sterling and Thomas 2006; Sterling, Warwick, and Wyness 2015; Lozano 2006; Barth 2013). However, it has also been highlighted that the number of universities that have incorporated sustainability into their institutions is inadequate (Tilbury 2011; Shiel 2013; Sterling, Warwick, and Wyness 2015).

Higher education helps shape the future by training future decision makers, professionals and citizens (Sterling, Maxey, and Luna 2013). Eighty percent of the decisions made worldwide regarding industry, the economy, the market and politics are made by university graduates (Tilbury 2011). However, if universities are to continue shouldering this responsibility to train future generations in the face of profound societal transformations, universities must also transform (UNESCO 2015).

In this chapter, we present information on the evolution of Latin American universities in their efforts to integrate sustainability into their policies and dynamics by transforming their curricula, teaching practices and research and discuss the different strategies they are currently implementing to fulfil their commitments to the global agenda.

2.2 Sustainability in higher education institutions in Latin America and the Caribbean, an approach based on a systematic review of the relevant literature

To identify which actions higher education institutions (HEIs) have carried out in the past few years in the effort to integrate sustainability, a systematic review of the relevant literature (SRRL) pertaining to the incorporation of sustainability into the region's universities was conducted. SRRLs are based on the basic principles of being a systematic, transparent, replicable and synthesized (Tranfield et al. 2003; Briner and Denyer 2012). The SRRL followed the five consecutive phases proposed by Denyer and Tranfield (2009) and applied by Garza-Reyes (2015): (1) formulation of search terms, (2) location of studies, (3) selection of studies and evaluations, (4) analysis and synthesis, and (5) presentation and use of the results.

The SRRL was conducted in 2018 using the Thomson Reuters Web of Science and Scopus. These databases were selected due to their large compendium of documents available from the most widely recognized sustainability journals. The terms used for the search were Sustainability and Universities, or Higher Education or Latin America Universities. After this selection, filters for the type of document were applied so that only research articles were selected. Time-of-publication filters were not applied, and articles were accepted with no restriction for the year of publication. Subsequently, the filters for country/territory were applied to select research papers from Latin America and those referring to a specific Latin American country. The results of the search were contrasted to eliminate any duplicate documents reported in both databases. Finally, a total of 73 research papers regarding HEIs in Latin America and the Caribbean (LAC) was obtained.

Subsequently, the main contents or topics of the selected research articles were assessed via analytical readings of the abstracts of the 73 filtered articles. The main criterion used to select the articles was that the studies should have been carried out within LAC HEIs. Therefore, research on campus sustainability, education for sustainability (EfS) at any undergraduate or postgraduate level and studies of sustainable behaviours and changes in attitude or leadership towards sustainability were selected. Finally, 57 articles were retained, and each was assigned a code to facilitate referencing, Table 2.1.

Table 2.1 Analysis of studies in LAC HEIs related to the integration of sustainability

Code	References	Tematic
P1	Brito, Rodriguez, and Aparicio (2018)	EfS
P2	Berchin et al. (2018)	Campus
P3	Friman et al. (2018)	EfS
P4	Leal Filho et al. (2018)	Campus
P5	Velasco et al. (2018)	Campus
P6	de Andrade et al. (2018)	EfS
P7	Aguilar-Virgen et al. (2017)	Campus
P8	Martins et al. (2017)	Sustainable behaviour
P9	Nagy et al. (2017)	Research
P10	Casarejos, Frota, and Gustavson (2017)	Campus
P11	Callejas et al. (2017)	EfS
P12	Vasconcelos de Oliveira et al. (2017)	Campus
P13	Berchin et al. (2017)	EfS
P14	Rodríguez-Solera and Silva-Laya (2017)	EfS
P15	Rodrigues and Payne (2017)	EfS

(Continued)

Table 2.1 (Continued) Analysis of studies in LAC HEIs related to the integration of sustainability

Code	References	Tematic
P16	Cronemberger de Araujo and Magrini (2016)	Campus
P17	Charli-Joseph et al. (2016)	EfS
P18	Mattos, Gomes, and Ribeiro (2016)	Campus
P19	Dutra et al. (2016)	EfS
P20	Ashton, Hurtado-Martin, and Anid (2017)	EfS
P21	Bormann et al. (2016)	EfS
P22	Machado et al. (2016)	Campus
P23	de Lima et al. (2016)	Campus
P24	Cabello et al. (2015)	EfS
P25	Ramos et al. (2015)	Campus
P26	Velazquez et al. (2015)	EfS
P27	Vásquez, Iriarte, and Almeida (2015)	Campus
P28	Martínez-Fernández and González (2015)	Campus
P29	Czykiel, Figueiró, and Nascimento (2015)	EfS
P30	Machado et al. (2015)	EfS
P31	Rauen, Rojas, and da Silva (2015)	Campus
P32	Nobre (2015)	EfS
P33	Mattos et al. (2015)	Campus
P34	de Mello and Pamplona (2015)	EfS
P35	Spenassato et al. (2015)	Sustainable behaviour
P36	Avilés, Pérez, and Rosano (2014)	EfS
P37	Lozano and Lozano (2014)	EfS
P38	Bursztyn and Drummond (2014)	EfS
P39	da Silveira et al. (2013)	Campus
P40	Marinho, Gonçalves, and Kiperstok (2014)	Campus
P41	Chiappetta et al. (2013)	EfS
P42	Velazquez, Munguia, and Ojeda (2013)	Campus
P43	Espinosa et al. (2013)	Campus
P44	Rieckmann (2012)	EfS
P45	Palma, Oliveira, and Viacava (2011)	EfS
P46	Yánez and Zavarce (2011)	EfS
P47	Juárez-Nájera, Rivera-Martínez, and Hafkamp (2010)	Sustainable behaviour

(*Continued*)

Table 2.1 (Continued) Analysis of studies in LAC HEIs related to the integration of sustainability

Code	References	Tematic
P48	Chiappetta (2010)	EfS
P49	Correia et al. (2010)	EfS
P50	Ramos and Avila (2010)	EfS
P51	Junyent and Geli de Ciurana (2008)	EfS
P52	Orozco and Cole (2008)	EfS
P53	Mitsch et al. (2008)	EfS
P54	Juárez-Nájera, Dieleman, and Turpin-Marion (2006)	EfS
P55	Skewes, Cioce Sampaio, and Conway (2006)	EfS
P56	Geli de Ciurana and Leal Filho (2006)	EfS
P57	Fehr (2003)	EfS

An analysis of the chronological evolution of these papers (Figure 2.1) showed that between 2003 and 2018 there was a progressive increase in the number of publications; however, the number increased significantly in 2014, and 2015 was the year with the greatest number of investigations (12), followed by 2017 (10). These data show that the interest in these issues within the region is quite recent.

An analysis of the themes of these studies (Figure 2.2) showed that the largest number (34) were related to EfS. Among these, the vast majority (16) referred specifically to studies on the integration of sustainability into academic programmes (P6, P13, P15, P17, P19, P20, P21, P29, P32, P37,

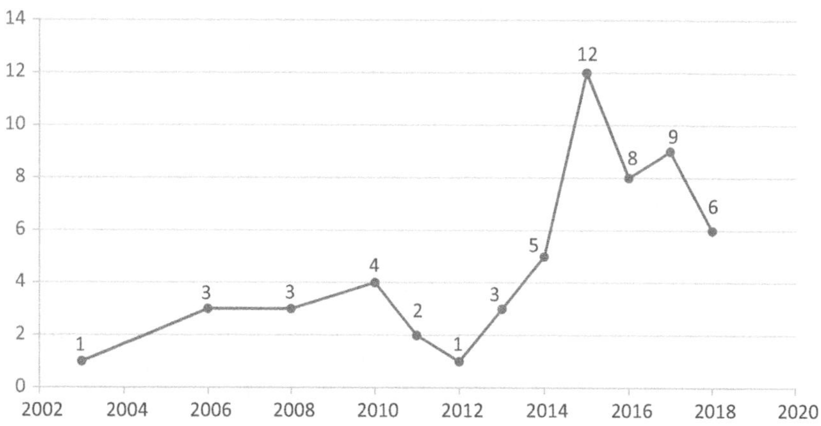

Figure 2.1 Distribution of the 57 articles by year.

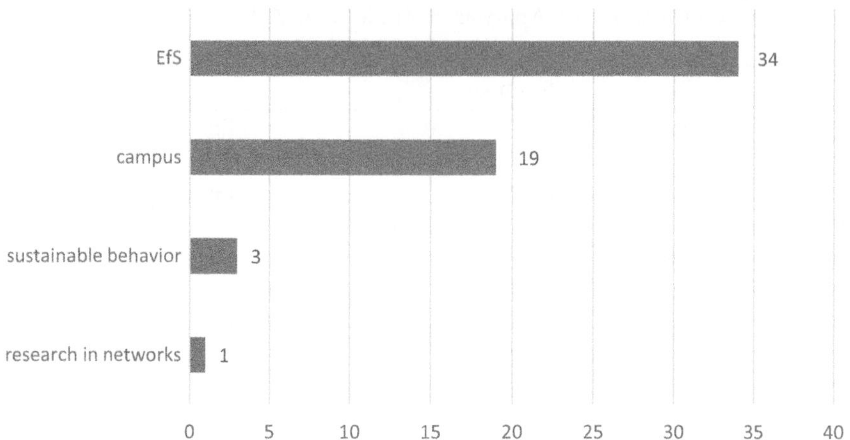

Figure 2.2 Distribution of the 57 research papers by topic.

P41, P48, P50, P52 and P57). Relating the above results with Sustainable Development Goal 4 (SDG 4) quality education and your target 4.7, which relates the ESD, we can infer that interest has grown in the area for EfS and, in some, measured the decade of the ESD (2005–2014) have contributed to some extent in mobilizing the interest of researchers and university professors in the region.

Other works were more related to local environmental problems in relation to study plans (P53 and P55) or focused on the relationship between the university and industry (P24). Other EfS articles evaluated the progress of integrating this concept into the curriculum (P3, P38, P45, P46, P51, P54 and P56). Others went further, evaluating training models based on the sustainable performance of graduates in their professional fields (P14 and P30). Other studies claimed to address ESD (P34 and P49), the study of ESD-related competencies (P36 and P44) and the professional development of university professors in terms of teaching styles (P11). This research trend seems to be very similar to the trends identified in other regions of the world, where the incorporation of sustainability issues into education for different careers is of major importance (Mulà et al. 2017). In this area, there seems to be a good relationship with research teams from other countries (Junyent and Geli 2008; Benayas, Alba and Justel 2014; Blanco-Portela 2017).

In the same figure, 19 articles reporting more classical studies on the integration of sustainability into different university campus management areas are identified. The topics of these articles included general evaluations of progress (P2, P5, P10, P16, P22, P23, P25, P33 and P39), evaluations of sustainability policies on campuses (P4 and P28), the mention of sustainability in the university's mission (P18) and the evaluation of some specific environmental dimension (P7, P12, P27, P31, P40, P42 and P43). Finally, some articles were related to the study of the sustainable behaviours of university graduates (P8, P35 and 47), and one was related to the efforts of cooperative networks of universities to address climate change research (P9).

When the research was analysed by country (Figure 2.3), Brazil emerged as the most active country in the region with the highest number of publications (35). This figure clearly reflects the commitment and progress of Brazilian universities towards the institutionalization of sustainability on their campuses. The result for Brazil differs significantly from the results of the other countries in the region. Mexico was at quite a distant second place, with 12 articles. These two Latin American countries have the largest university systems in the region and the largest number of universities; furthermore, some of their universities, such as the University of Sao Paulo or the National Autonomous University of Mexico, hold prominent positions in international rankings. Argentina, Chile and Costa Rica were third place as countries with slightly more research activity in this field.

It is important to point out that the total number of studies in this graph (70) exceeds the number of included articles (57) because several of the articles presented the results of studies carried out in more than one LAC country.

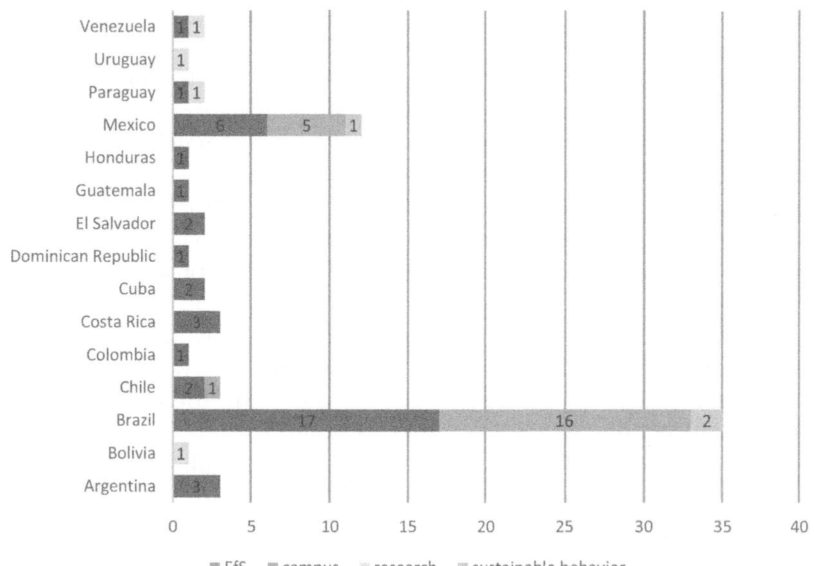

Figure 2.3 Distribution of studies by country, grouped by themes.

This analysis allowed us to obtain an overall view of the research activity examining the different approaches to incorporating sustainability in the region's universities. These results serve as a framework for contextualizing the advances obtained by more specific projects. When relating these results with the requirements of SDG 4 and your target 4.7, we can find a first approach to the contribution of universities in the region to fulfil with target 4.7, which is not enough compared to the number of universities and countries in the region. This result may serve to encourage teams of professors and researchers to take a more active role in training teachers and students in ESD and thus align the efforts of the global agenda.

2.3 Latin American universities face the challenge of incorporating the environment and sustainability into their practices

The historical background of the integration of sustainability into Latin American and Caribbean HEIs is based on the broad tradition of environmental education in the region. An article by Sáenz and Benayas (2015) on higher education, the environment and sustainability in LAC presents a review of the evolution of this issue in the region. The authors define three stages of what they call the development of environmental higher education in LAC, although later they refer to the same process as the environmentalization of HEIs or the incorporation of the environmental dimension into higher education.

The first stage they describe is characterized by the "emergence of technical and professional training for the use of natural resources and the conservation of nature". They warn that "Unfortunately, this is still a phase of the process of incorporation of the environmental dimension in higher education about which very little is known" (Sáenz and Benayas 2015, p. 194); however, they refer to a work carried out in Colombia in 1950 that cites the creation of 26 environmental academic programmes in 14 different HEIs.

At this stage, the region's participation in the international meetings that took place in the 1970s and 1980s, including the International Seminar on Environmental Education in Belgrade 1975, the Intergovernmental Conference on Environmental Education and Training in Tbilisi 1977 and the International Congress on Environmental Education and Training 1987, stands out. A highlight is the meeting held in 1976 on "Environmental Education in Latin America and the Caribbean" that contributed two important working documents that reflected "the importance of higher education regarding the environment in the region" (161).

The second stage is called "Higher-level environmental education and international organizations in LAC". This stage began in 1976, and it refers to the important role played by two international organizations in promoting the advances of HEIs in the field of environmental education: the United Nations Programme for the Environment (UNEP) and the International Centre for Training and Education in Environmental Sciences (CIFCA), the latter of which was created in agreement between the UNEP and the Government of Spain. A highlight is a study that was carried out in this centre in 1977 called "Panorama of environmental higher studies in Latin America", which CIFCA considered "the first diagnosis of the progress in incorporating the environmental dimension into HEIs in the region". This stage was also characterized by the creation of networks and the preparation of the first diagnostic report in the region, titled "Incorporation of the Environmental Dimension into Higher Education in Latin America and the Caribbean" (162). Both of these forums and the resulting diagnostic strategies will be discussed later.

Saenz and Benayas call the third stage "Higher education for sustainable development in Latin America and the Caribbean". They define it as the current stage, which is centred on concepts such as "education for sustainable development", "education for sustainability" and "education for sustainable societies" (163). They recognize that in this stage, processes for incorporating environmental issues into the region's HEIs have accelerated compared with the two previous stages, and they report an increased number of academic programmes for environmental training.

Together with the historical route of this process, it is important to highlight that in the great majority of the countries in the region, there is some resistance to using the term "sustainable development". As Ángel Maya argues, the term "carries behind it the ideological accumulation of developmentalism and continuous progress" (Maya 1999, p. 108), later adding that "it carries with it the connotation of 'continuous growth'" (Maya 2003, p. 14). This idea of development as associated with growth and wealth implicitly evokes the reality of unequal distribution and the widening of the gap between the poor and the rich, that is, those who accumulate wealth at the expense of marginalization and unequal distribution to others, opening the way to poverty. As Maya states:

> The poverty of the South is not explained by the cultural laziness engendered in the geographical conditions of the tropics. There are five hundred years of domination and exploitation of their wealth - gold, sugar, minerals, oil - without forgetting the accumulation of human labour in the slave systems.
>
> *Maya (2003, pp. 19–20)*

As a result of the previous reality and reclaiming of the concept of sustainable development, discontent regarding the concept as a way of solving the region's environmental, social and economic crisis persists. However, discussion of this concept is not finished; it remains open, but for now, the efforts have moved towards the field of action. However, it should be noted that in the region, the terms "environmental education" and, more recently, "sustainability" are most commonly used; these terms identify actions that involve working towards a sustainable, economically viable world that is socially fair and uses natural resources responsibly.

Finally, returning to the present, sustainability has been integrated into LAC HEIs to varying degrees, from systemic proposals that act as an umbrella over all the institution's actions to efforts aimed at a single area, such as environmental management or integration into a curriculum within an academic programme. In a systemic context, sustainability is institutionalized, permeating not only the substantive functions of teaching, research and extension but also their operations. Examples of this integrative perspective are sustainable campuses, green campuses or eco-campuses. In comparison, proposals focusing on one area, such as solid waste management or 14001 certification, constitute a gateway for sustainability in universities; however, such efforts are not enough, and we expect these universities to move towards institutionalization, integrating sustainability into all the functions of the campus and transforming the educational community's behaviour towards the environment.

2.4 *Alliance of university networks for sustainability and the environment (Alianza de Redes Universitarias por la Sostenibilidad y el Ambiente, ARIUSA)*

In 2007, the Alliance of University Networks for Sustainability and the Environment (ARIUSA) was created to promote and support cooperation among the different university networks that had been working on issues of environment and sustainability in the region's HEIs. In 2014, UNESCO, in a report closing the decade of ESD, recognized ARIUSA as one of the best university environmental networks worldwide (Buckler and Creech 2014).

According to ARIUSA's official website (2018), a total of 22 environmental university networks made up of more than 350 universities in 15 Ibero-American countries are identified in the Alliance: Colombia, Guatemala, Mexico, Cuba, Spain, Argentina, Brazil, Costa Rica, Peru, Dominican Republic, Venezuela, Chile, Ecuador, Portugal and Nicaragua. Recently, the university network of Honduras (2018) was incorporated.

ARIUSA's main objectives also stand out on the alliance's official website. They are as follows:

a. Broadly disseminate the agendas of each of the member networks, allowing access and their possible coordination in terms of dates and topics of common interest.
b. Formulate and carry out collaborative and interdisciplinary research projects on environmental problems or environmental management and education in Ibero-America.
c. Support the creation of new joint academic programmes and the strengthening of existing postgraduate programmes on the environment and sustainability in Latin American universities.
d. Promote the strengthening of the teaching and research capacity at the universities of LAC to sustainability and the environment.
e. Become an instrument of action and collective institutional representation of environmental university networks that integrate it to enhance influence on the institutional and governmental agencies that define and approve academic and environmental policy programmes in the region.
f. Promote the implementation of practices and systems of institutional management among the universities in the associated networks and the ordering of university campuses as important aspects of their institutional sustainability.

In 2012, within the framework of the United Nations Summit on Sustainable Development, RIO+20, representatives from 25 university networks worldwide met in Brazil to analyse advances in the universities' commitments over the 20 years since the RIO 92 Summit. The representatives of the 25 networks jointly developed and approved. "The People's Sustainability Rio+20 Treaty on Higher Education: Engaging communities of learning in change for sustainability" (Rio+20 2012). This treaty defined different strategies that involved developing specific short-, medium- and long-term commitments to sustainability in HEIs around the world. This meeting was an important impetus for some of the projects that initiated at ARIUSA.

This document presented within its strategic lines the need to evaluate the sustainability policies of HEIs in different regions of the world. This is how the proposal to develop a specific project for investigating the Latin American and Caribbean region came about: The definition of indicators for the evaluation of sustainability policies in Latin American Universities, a project called RISU.

The project was created by the Advanced Research Institute on Science and University Evaluation (Instituto de Investigación Avanzada sobre Evaluación de la Ciencia y la Universidad, INAECU) in close collaboration with the Network of University Sustainability Indicators (RISU) of ARIUSA and with the financial support of the Centre for Latin American Studies of the Autonomous University of Madrid (UAM) and Santander Bank. Since its inception, the programme has come under the auspices of the Latin American Chapter of the World Alliance of Universities on Environment and Sustainability (GUPES-LA) of the United Nations Environment Programme (UNEP) (Benayas, Alba and Justel 2014).

The project has the following main objectives (Benayas, Alba and Justel 2014):

- Strengthening the joint work and network actions being developed by the universities in the region who are concerned about sustainability and social responsibility.
- Defining an analytical framework for evaluating sustainability and social responsibility policies in Latin American universities.
- Training university leaders to apply indicator systems for evaluating commitments to the sustainability of their universities.
- Reflecting on the deficiencies or strengths of the application of indicator systems designed as internal self-assessment processes in Latin American universities.
- Promoting sustainability and social responsibility in Latin American universities through a proposal to develop a regional strategy for improvement.

2.5 RISU project (Phase 1): diagnosing the starting point

In 2014, a questionnaire was created based on the experience of applying indicators within the context of Spanish universities (CADEP-CRUE). A total of 175 indicators for Spanish universities were analysed in intense debate and consensus with the leaders of Latin American networks participating in the Viña del Mar Forum (2014). Previous iterations of defining indicators used in different countries of the region, such as Mexico, Costa Rica and Colombia, were included. Finally, the RISU questionnaire was adapted to the specific language of the Latin American context. It included 114 indicators and 11 areas of analysis:

- Sustainability policy (15 indicators)
- Awareness and participation (12 indicators)
- Socio-environmental responsibility (10 indicators)
- Teaching (13 indicators)
- Research and transfer (13 indicators)
- Urban planning and biodiversity (7 indicators)
- Energy (10 indicators)

- Water (10 indicators)
- Mobility (8 indicators)
- Waste (11 indicators)
- Responsible contracting (5 indicators).

The RISU questionnaire was applied to 65 universities in ten Latin American countries: Argentina, Brazil, Chile, Colombia, Costa Rica, Guatemala, Mexico, Peru, Dominican Republic and Venezuela. The results of its application revealed the compliance of the HEIs in the region with the 114 indicators and highlight the important deficiencies Latin American universities face when incorporating the environmental dimension into their campuses. Only three of the 11 indicator areas obtained values above the minimum recommended level. The dimensions of socio-environmental responsibility, sustainability policies and waste management obtained the most favourable results, although they were still far from satisfactory. The rest of the eight dimensions obtained average scores below the approved values. The incorporation of responsible hiring criteria and the application of measures to control traffic and mobility within the university obtained the lowest scores. Thus, this report marked a low starting point that the universities had to overcome.

For this reason, the questionnaire was administered not only to identify the universities' compliance with the 114 indicators but to prompt reflection within the university community regarding the indicators on which they scored unfavourably. For this reason, at the end of the questions for each area, the questionnaire asked whether the university "would be willing to develop concrete actions to implement and develop indicators within 3–5 years to address items that were answered negatively". Thus, the questionnaire went from being a still photo of a specific action to being a starting point for identifying deficiencies or weak points that allowed the universities to establish an improvement plan for addressing unfavourable indicators while strengthening the indicators of those that were previously working on the HEIs. The questionnaire went from being a diagnostic tool to being a tool for facilitating changes.

In turn, the questionnaire, as a meta-evaluation, incorporated four questions that highlighted the role of each university in this phase of the RISU project. The results of these questions reinforced the processes that begun within the universities:

- A total of 74.5% of the universities indicated that participation in the RISU project-initiated processes on their campuses to improve their involvement in sustainability and the environment in the short and medium term.
- A very high percentage of the universities, 84.3%, indicated that involvement in the project allowed the participants at their university

to acquire new knowledge or strategies to deepen sustainability measures. In this sense, the universities noted that the training process was of great interest for the professionals, each of whom was involved in sharing experiences to generate changes and new proposals.
- A slightly lower percentage, 76.5%, indicated that they had gained the involvement of university authorities responsible for sustainability issues.
- All the universities that participated showed a clear interest in continuing to work on new initiatives to promote the continuity of the objectives defined in the project.

These data suggest that the project met its initial objectives. Data were obtained that permitted an assessment of the region; furthermore, it was possible to generate change within Latin American universities through the involvement of university authorities, who became involved in streamlining processes that would incorporate environmental goals and objectives into their universities' strategic plans (Benayas, Alba and Justel 2014).

These findings revealed that the questionnaire was an effective instrument for triggering processes of change within the universities. Thus, in 2015, a second phase of the RISU project was proposed to learn more about the extent of the transformations and improvements that were induced within the participating universities during the first phase of the project. The second phase aimed to analyse changes in the areas of management, teaching and research and in the application of sustainable policies within the institution and in relation to society. However, we also sought to identify the main factors that promoted these changes and the main barriers that slowed or hindered them (in Blanco-Portela et al. 2018).

2.6 RISU project (Phase 2): identifying the processes of change

While the first phase of the RISU project asked what universities were doing to integrate sustainability, the second phase sought to analyse how they were integrating sustainability and what processes of change had been unleashed (Hoover and Harder 2015). To the same extent, the study aimed to analyse the barriers and success factors reported by the surveyed universities, the results are found in the study by Blanco-Portela et al. (2018).

This second phase of the RISU project was initiated by contacting the 65 sustainability leaders of the universities in the ten countries that had participated in the first phase of RISU. Forty-five of these leaders agreed to participate in this second phase of the investigation. Therefore, both projects involved the same countries but a different number of universities.

Information was collected through a virtual interview lasting more than 1h with each of the 45 sustainability leaders from the participating Latin American universities. Prior to the interview, a questionnaire was sent that addressed the 11 areas of analysis of RISU 1 and other questions regarding the transformations that the institutions had achieved after the application of the RISU 1 indicators. The advance mailing of the questionnaire allowed that the leaders to arrive at the interview with all the necessary information regarding the level of progress in each area on their campuses. The interviews were recorded, and the responses were qualitatively and quantitatively analysed.

The main results showed that in 87% (39) of the participating universities, changes were indeed triggered. The university professionals who reported changes said that participation in RISU was "valuable and positive" and that it guided their way forward, prompting them to strengthen existing programmes or identify areas of deficit where new efforts must be focused. The RISU project was valued as a space in which to gain understanding of the scope of all the areas within the university where sustainability efforts should be present as well as a reflection point on the way to scaling and obtaining commitments from the different actors in the academic community.

Other universities considered participation in RISU part of the region's benchmarking, as a space that would allow them to interact with other universities and learn about other valuable experiences and initiatives for the integration of sustainability. Additionally, the universities recognized that RISU allowed them to order and systematize actions that they had been carrying out in a timely manner and with low degrees of institutionalization.

In general, the universities considered RISU a process for diagnosis or performance evaluation that triggered improvement actions. These results reinforced the value of RISU as an instrument of transformation and change in the integration of sustainability within universities.

In conclusion, this first round of results confirmed that the RISU questionnaire was an instrument of institutional evaluation that promoted improvement. One benefit of the RISU evaluation process is the awareness that the universities are making efforts, but it goes further: In the interviews, the sustainability leaders expressed as they became familiar with their individual reports they developed improvement plans that included short- and long-term projects for each of the areas with low results or on which they had not been working.

The sustainability policy is the "umbrella" institutional instrument for all regulations that guide good sustainability practices in HEI operations. The results showed improvements in the universities where the instrument has not only been approved but is in operation. Other HEIs are in the process of receiving approval. However, some have not contemplated seeking approval. A good practice mentioned in the interviews was gaining agreement from the different interest groups within the institution to participate in the construction of the policy. However, a necessary point to address in future research is how universities will ensure that the document is a mobilizer of change and is not regarded as simply something that fulfils a requirement.

In the analysis of the interviews with the coordinators of these processes in each university, it was possible to identify the main barriers to these internal changes as well as the main success factors (in Blanco-Portela et al. 2018).

Among the most important barriers were disciplinary boundaries that limited research and holistic learning. The competitive environment of the university favours departmental boundaries and hinders the overall vision and cooperative work. Inertia, another barrier to integration, takes numerous forms: disinterest among academics, who are busy with their work, student disinterest, administrators and conventional teachers who are resistant to change, contentious worldviews that reinforce work barriers, the lack of an institutional action plan for integrating sustainability, low unification of institutional criteria for sustainability, unclear institutional priorities, a lack of policies for integrating sustainability into teaching and research, no time allotted for human resource capital to help with sustainability efforts, a lack of incentives and independent departments and faculties (Blanco-Portela et al. 2017, 2018).

The success factors reported were related to cooperation among students, teachers, researchers and interested personnel in the region to solve real problems; abandoning the paradigm of the teacher as a "provider" of information and students as "consumers" of this information; global challenges that encourage universities to set goals and make plans; pressure from internal and external actors; external financial programmes that support sustainability initiatives; leadership by sustainability champions; teaching staff and students to act collectively as agents of change for sustainability initiatives; the promotion of shared governance, transparency and open communication; the integration of key sustainability issues into the institution's strategic plan; the presence of language related to care for the environment in the mission, goals and objectives of the institution; support from top management and defined policies on how to integrate sustainability (Blanco-Portela et al. 2017, 2018).

One area of RISU 2 in which the coordinators identified unfavourable and poor results was the incorporation of sustainability issues into the field of university teaching. The incorporation of these topics is limited to very motivated teachers taking very specific actions. The previous results, compared to the results presented in Section 2.2, in which the largest number of publications (34), belonged to the issue of EfS, a non-enough and low result considering the number of universities and countries of the region. However, the result obtained in RISU 2, related to the low results in the incorporation of the sustainability to the teaching processes, confirms to some extent that the issue in the region has not been addressed with the enough urgency that the global agendas demand, especially in the SDG 4 and your target 4.7.

This analysis inspired the start of the third phase of the RISU 3 project. This new phase aimed at finding alternatives that would address these problems within universities by developing the Academy Programme in the Latin American context, which will be explained below.

2.7 *RISU project (Phase 3): generating change processes*

The new phase of the RISU project is the Latin American version of the "Academy UE4SD" project, which was developed (2014–2016) with the involvement of 55 universities from 33 countries of the European Union. During the development of the project, each country was assessed, and good practices were identified. This process gave rise to different publications and materials, which are available on the project website: https://platform.ue4sd.eu/.

The last stage of this project consisted of the launch of a pilot programme, from July 2015 to April 2016, to boost institutional initiatives for university teaching staff to incorporate sustainability into their teaching practices. This project was carried out at Spanish universities: The Autonomous University of Barcelona, University of Girona, University of the Basque Country and the University of Granada; the project was coordinated by the UAM. Once this project was completed, its excellent results spurred great interest in extending the initiative to other regions or continents. Specifically, its application to universities in Latin American in 2017 was proposed.

In 2017, a new pilot programme called RISU 3: Academy Latin American (Academy LA) was launched in Colombia. Four institutional teams from the University Agustiniana, University El Bosque, University EAN and University Ciencias Aplicadas y Ambientales universities participated, with the support of advisors from the four Spanish universities: The Autonomous University of Barcelona, University of the Basque Country/Euskal Herriko Unibertsitatea, University of Girona and University of Granada. The pilot study was coordinated in Colombia by the Universidad Agustiniana; in Spain, general coordination was provided by the UAM.

In the pilot, the promoting universities (advisors) committed themselves to helping and guiding the participating Colombian universities. Each advising university guided a Colombian university in a series of monthly monitoring meetings via Skype between each pair. In turn, the Colombian universities committed to carrying out the selected initiative with the institutional involvement of a member of the university's government team. In this way, it was intended that the actions of the project would continue.

It is important to highlight that the Academy LA is a programme in which different institutional perspectives converge in a collaborative manner with a common goal: the professional development of university professors in EfS. The programme is geared towards change in universities through the support of teachers so that they can develop their own EfS competences. At the same time, it seeks to develop the skills of leadership and training agents of change among the participants, a necessary condition to facilitate the processes that lead to sustainability. For this, the 17 SDGs are taken as objectives and goals defined by the United Nations for the 2030 strategy.

The objectives of the Academy LA programme for both the pilot programme in Colombia and the general application in Latin America are oriented towards supporting the strategies for professional development in EfS at the participating universities under the guidance of the advisors. At the same time, the academy intends to promote the pedagogical transformation and integration of EfS into the curricula of the universities and promote institutional strategies that generate impact and change within the participating universities and allow those changes to continue and progress in subsequent years. Finally, the Academy LA seeks to facilitate the exchange of information, strategies, lessons learnt and good practices for EfS among the network of participating Ibero-American universities, thus providing a setting for collaborative work.

As a synthesis of this process, a final meeting was held in Bogotá (Figure 2.4) with the participation of the institutional teams and the support teams from the Spanish universities. The main objective of this meeting was to evaluate the reported experiences and to work collaboratively to identify the lessons learnt that could guide new actions or practices at other universities in the region. The project aimed to generate a snowball effect that would allow the initiative to be progressively extended to other Colombian universities and to other countries in the region.

The invited universities from Mexico, Peru, Cuba and Colombia participated in this meeting (Academy LA). A second objective of the meeting was to bring together the new universities that were invited to participate in the next phase of RISU 3. The presence of representatives of the new institutions at the meeting in Bogota allowed them to interact, learn about the pilot experiences and finally, together, identify the obstacles

Figure 2.4 Meeting of the Project Academy in Bogotá.

and success factors that could affect the application of the project at their institutions.

Within this exercise, participatory analysis and reflection on the process, a weaknesses, threats, strengths and opportunities matrix was constructed. The main results are presented below:

Weaknesses

1. Disjointed efforts between academic programmes and other units
2. Lack of commitment and motivation among some academic directors and professors
3. Difficulties of identifying an institutional implication for the project
4. Lack of support for directives
5. Lack of a sustainability policy among the general guidelines of the institution
6. Diversity of concept of ESD
7. Lack of means and budgets linked to the loan system
8. Ineffective dissemination of activities and projects
9. Little time to develop the project.

Threats

1. Lack of coherence between what is taught at universities and the needs of business and government

2. Unsustainable model of Colombian society
3. Lack of Ministry of Education policies that guide the training of teachers in sustainability.

Strengths

1. Motivated human team
2. Enriching work with experts
3. The incorporation of key multidisciplinary actors
4. Initiatives that facilitate teacher training
5. Institutional support, policies and guidelines at some universities
6. Time allotted for teachers to develop the project
7. Compulsory subjects at some transversal universities on subjects of sustainability.

Opportunities

1. Universities working in networks
2. Working with the Spanish universities that are already applying the ES
3. The 2030 Agenda and the 17 SDGs
4. Influencing the public education policies
5. Projection of the university towards its surroundings
6. Global transformation of the university
7. Creating interdisciplinary and transdisciplinary research lines.

The RISU 3 Colombia Project (Academy LA) has been involved in joint learning about overcoming the difficulties that universities face when reorienting the university curriculum towards sustainability and involving university teaching staff. The advising universities lend support in terms of experience with the process followed at their own universities. The new participating universities are engaged in reflection and networking to implement actions related to the university's change towards sustainability and involve the teaching staff to lead that change. In addition, the participation of the institution is essential for guaranteeing that the change has a long-term impact.

In 2018, the Academy LA project incorporated new universities in Colombia and initiated processes in other Latin American countries. Specifically, in Peru, a new experience has begun at four universities: Cayetano Heredia, San Marcos, Nacional de Barrancas and San Martín de Porres. The project starts with virtual training for the professors and leaders participating in the institutional initiatives on the requirements of the ESD so that they can have more tools with which to carry out their individual proposals.

For the advisors and experts guiding the academy at Latin American universities, they find in this project an opportunity to exchange educational experiences that will help them guide the strategies proposed by their university institutions within the framework of EfS. At the same time, collaboration between the different universities opens new scenarios of joint lines of work in the pedagogical, didactic and evaluative research fields.

Additionally, for the universities participating in the Academy LA, the initiative provides an opportunity to respond to the challenges of the 2030 Global Agenda, which requires universities to take a more active and significant role in EfS. Universities in the academy find a path that facilitates assuming the training of their students, preparing them to face and manage with greater efficiency social, economic and environmental challenges that are characterized by uncertainty and complexity. Although the 17 SDGs present a challenge, they provide an opportunity at the same time. The collaborative work that takes place among participants from different universities and countries transforms the enriching experience into the exchange of good practices and opportunities for joint research. The contribution of the academy to the processes of institutional change adds to other efforts universities are making to integrate sustainability into other areas, such as environmental management, research and social responsibility.

Without a doubt, universities must become the beacon that guides the societal transformations necessary to reach the 17 SDGs. The initiatives described in this study are firm steps by Latin American universities moving along the path of sustainability. This road is still long and tortuous, but with collaboration and mutual support, it will be easier to overcome all the obstacles that may appear, making the great challenge of sustainability a much more attainable goal.

2.8 Conclusion

The universities are an excellent strategic ally in the promotion and help to fulfil the 17 SDGs. The actions that Latin American universities have carried out over the years have contributed to some extent to the global agenda. In the region where universities predominate, their actions focus on environmental management, creation of institutional sustainability policies and social responsibility with a focus on sustainability. However, the training of university professors to integrate sustainability into their curricula is low. The barriers to involve more teachers of all the disciplines in EfS coincide with those who are found in other regions. The drivers described by the leaders of sustainability have favoured the integration of sustainability and have allowed to involve more university professors. These can be considered by the decision makers of the universities of the region, to drive change. University networks, joining forces and sharing experiences and good practices are a good strategy that continues to consolidate among Latin American universities. The indicators of SDG that are being built from RISU will be able to provide an overview of the degree of contribution of universities to the regional and global agenda.

References

Aguilar-Virgen, Q., P. Taboada-González, E. Baltierra-Trejo, and L. Marquez-Benavides. 2017. Cutting GHG emissions at student housing in central Mexico through solid waste management. *Sustainability*, 9, 1415.

Ashton, W.S., M. Hurtado-Martin, N.M. Anid, N.R. Khalili, M.A. Panero, and S. Mcpherson. 2017. Pathways to cleaner production in the Americas I: Bridging industry-academia gaps in the transition to sustainability. *Journal of Cleaner Production*, 142, 432–444.

Avilés, K.L., B. Pérez, and G. Rosano. 2014. Relaciones entre variables sociales de aprendizaje en el desarrollo sustentable: el caso del Instituto Tecnológico de Tláhuac, D.F., México. *Revista Internacional de Contaminación Ambiental*, 30(4), 407–416.

Barth, M. 2013. Many roads lead to sustainability: A process-oriented analysis of change in higher education. *International Journal of Sustainability in Higher Education*, 14(2), 160–175.

Benayas, J.,D. Alba, and A. Justel. 2014. RISU project. Development of indicators to assess the implementation of sustainability policies in Latin American universities. Madrid: Universidad Autónoma de Madrid and Alliance of Ibero-American Networks of Universities for Sustainability and the Environment, ARIUSA. 52pp.

Berchin, I.I., V. dos Santos, G. Almeida, L. Corseuil, and J.B. Salgueirinho. 2017. Strategies to promote sustainability in higher education institutions: A case study of a federal institute of higher education in Brazil. *International Journal of Sustainability in Higher Education*, 18(7), 1018–1038.

Berchin, I.,M. Sima, M.A. de Lima, S. Biesel, L.P. Santos, R.V. Ferreira, J.B. Salgueirinho, O.A. Guerra, and F. Ceci. 2018. The importance of international conferences on sustainable development as higher education institutions' strategies to promote sustainability a case study in Brazil. *Journal of Cleaner Production*, 171, 756–772.

Blanco-Portela, N. 2017. Análisis de impacto del proyecto RISU: un estudio desde las transformaciones y mejoras en las estructuras y dinámicas de las universidades latinoamericanas frente a la sostenibilidad. *Ph.D. Thesis*, Universidad Nacional de Educación a Distancia, España. Available online: http://e-spacio.uned.es/fez/view/tesisuned:Educacion-Nblanco

Blanco-Portela, N., J. Benayas, L.R. Pertierra, and R. Lozano. 2017. Towards the integration of sustainability in Higher Education Institutions: A review of drivers for and barriers to change. *Journal of Cleaner Production*, 166, 563–578

Blanco-Portela, N., L. R-Pertierra, J. Benayas, and R. Lozano. 2018. Sustainability leaders' perceptions on the drivers for and the barriers to the integration of sustainability in Latin American higher education institutions. *Sustainability*, 10, 2954.

Bormann, H., J. Steinbrecher, I. Althoff, H. Roth, J. Baez, C. Frank, …, and I. Sanchez. 2016. Recommendations for capacity development in water resources engineering and environmental management in Latin America. *Water Resources Management*, 30(10), 3409–3426.

Briner, R.B. and D. Denyer. 2012. Systematic review and evidence synthesis as a practice and scholarship tool. In D.M. Rousseau (Eds.). *Handbook of Evidence-Based Management: Companies, Classrooms and Research* (pp. 112–129). New York: New York University Press.

Brito, R.M., C. Rodríguez, and J.L. Aparicio. 2018. Sustainability in teaching: An evaluation of university teachers and students. *Sustainability*, 10, 439.

Buckler, C. and H. Creech. 2014. Shaping the future we want: UN decade of education for sustainable development. *Final Report*. UNESCO.

Bursztyn, M. and J. Drummond. 2014. Sustainability science and the university: Pitfalls and bridges to interdisciplinarity. *Environmental Education Research*, 20(3), 313–332.

Cabello, J.J., A. Sagastume, D. García, J.B. Cogollos, L. Hens, and C. Vandecasteele. 2015. Bridging universities and industry through cleaner production activities. Experiences from the cleaner production center at the university of Cienfuegos, Cuba. *Journal of Cleaner Production*, 108, 1–10.

Callejas, M.M., N. Blanco-Portela, Y. Ladino-Ospina, R.N. Tuay, and K. Ochoa. 2017. Professional development of university educators in ESD: A study from pedagogical styles. *International Journal of Sustainability in Higher Education*, 18(5), 648–665.

Casarejos, F., M.N. Frota, and L.M. Gustavson. 2017. Higher education institutions: A strategy towards sustainability. *International Journal of Sustainability in Higher Education*, 18(7), 995–1017.

Charli-Joseph, L., A.E. Escalante, H. Eakin, M.J. Solares, M. Mazari-Hiriart, M. Nation, P. Gómez-Priego, C.A. Domínguez, and L.A. Bojórquez-Tapia. 2016. Collaborative framework for designing a sustainability science programme: Lessons learned at the National Autonomous University of Mexico. *International Journal of Sustainability in Higher Education*, 17(3), 378–403.

Chiappetta, C.J. 2010. Greening of business schools: A systemic view. *International Journal of Sustainability in Higher Education*, 11(1), 49–60.

Chiappetta, C.J., J. Sarkis, A.B.S. Jabbour, and A.B. Lopes de Sousa. 2013. Understanding the process of greening of Brazilian business schools. *Journal of Cleaner Production*, 61(15), 25–35.

Correia, P.R.M., B.X. Valle, M. Dazzani, and M.E. Infante-Malachias. 2010. The importance of scientific literacy in fostering education for sustainability: Theoretical considerations and preliminary findings from a Brazilian experience. *Journal of Cleaner Production*, 18(7), 678–685.

Cronemberger de Araújo, H. and A. Magrini. 2016. Higher education institution sustainability assessment tools: Considerations on their use in Brazil. *International Journal of Sustainability in Higher Education*, 17(3), 322–341.

Czykiel, R., P.S. Figueiró, and L.F. Nascimento. 2015. Incorporating education for sustainability into management education: How can we do this? *International Journal of Innovation and Sustainable Development*, 9(3–4), 343–364.

da Silveira, G., C.H. Jabbour, S.V.W. Borges, and A. Alves. 2013. Greening the campus of a Brazilian university: Cultural challenges. *International Journal of Sustainability in Higher Education*, 15(1), 34–47.

de Andrade Guerra, J.B.S.O., J. Garcia, M. de Andrade Lima, S.B. Barbosa, M.L. Heerdt, and I.I. Berchin. 2018. A proposal of a balanced scorecard for an environmental education program at universities. *Journal of Cleaner Production*, 172, 1674–1690.

de Lima, R.G., H. Nunes, E. Dahmer, J. Garcia, A. Suni, J. Baltazar, and F.C.R. Delle. 2016. A sustainability evaluation framework for science and technology institutes: An international comparative analysis. *Journal of Cleaner Production*, 125, 145–158.

de Mello, D. and D.A. Pamplona. 2015. "Knowledge" of morin on legal education: Pathways to education for sustainable development | ["Saberes" de morin na educação jurídica: Caminhos à educação para o desenvolvimento sustentável]. *Revista de Ciencias Humanas y Sociales*, (3), 446–469.

Denyer, D. and D. Tranfield. 2009. Producing a systematic review. In D.A. Buchanan and A. Bryman (Eds.). *The SAGE Handbook of Organisational Research Methods* (pp. 671–689). Thousand Oaks, CA: Sage Publications.

Dutra, M.M., A.C. Bampi, J.O. Diel, and M.R. Kohler. 2016. Higher education activities in the Lower Araguaia/MT: The course of agronomy and sustainable family farming | [Ações de educação superior no Baixo Araguaia/MT: O curso de agronomia e a sustentabilidade na agricultura familiar]. *Espacios*, 37(8), 2.

Espinosa, R.M., S. Turpin, R.C. Vázquez, A. Vázquez, A. De La Luz, A. De La Torre, and B.A. García. 2013. Environmental management in an institution of higher education-related practices of separation and recovery of waste. *Revista Internacional de Contaminación Ambiental*, 29(suplemento 3), 49–57.

Fehr, M. 2003. Overcoming established thinking models. The role of engineering education in environmental sustainability. *Industry and Higher Education*, 17(4), 283–289.

Friman, M., D. Schreiber, R. Syrjänen, E. Kokkonen, A. Mutanen, and J. Salminen. 2018. Steering sustainable development in higher education – Outcomes from Brazil and Finland. *Journal of Cleaner Production*, 186, 364–372.

Garza-Reyes, J.A. 2015. Lean and green–A systematic review of the state of the art literature. *Journal of Cleaner Production*, 102, 18–29.

Geli de Ciurana, A.M. and W. Leal Filho. 2006. Education for sustainability in university studies: Experiences from a project involving European and Latin American universities. *International Journal of Sustainability in Higher Education*, 7(1), 81–93.

Hoover, E. and M.K. Harder. 2015. What lies beneath the surface? The hidden complexities of organisational change for sustainability in higher education. *Journal of Cleaner Production*, 106, 175e188. Doi:10.1016/j.jclepro.2014.01.081.

Juárez-Nájera, M., H. Dieleman, and S. Turpin-Marion. 2006. Sustainability in Mexican Higher Education: Towards a new academic and professional culture. *Journal of Cleaner Production*, 14, 1028–1038.

Juárez-Nájera, M., J.G. Rivera-Martínez, and W.A. Hafkamp. 2010. An explorative socio-psychological model for determining sustainable behavior: Pilot study in German and Mexican Universities. *Journal of Cleaner Production*, 18(7), 686–694.

Junyent, M. and A.M. Geli de Ciurana. 2008. Education for sustainability in university studies: A model for reorienting the curriculum. *British Educational Research Journal*, 34(6), 763–782.

Leal Filho, W., L.L. Brandli, D. Becker, C. Skanavis, A. Kounani, C. Sardi, ... and R.W. Marans. 2018. Sustainable development policies as indicators and pre-conditions for sustainability efforts at universities: Fact or fiction? *International Journal of Sustainability in Higher Education*, 19(1), 85–113.

Lozano, R. 2006. Incorporation and institutionalization of SD into universities: Breaking through barriers to change. *Journal of Cleaner Production*, 14, 787–796.

Lozano, F.J. and R. Lozano. 2014. Developing the curriculum for a new bachelor's degree in engineering for sustainable development. *Journal of Cleaner Production*, 64, 136–146.

Machado, D.Q., F.R.N. Matos, A.M.C. Sena, and A.S.R. Ipiranga. 2016. Framework for sustainability analysis for higher education institutions: Application in a case study | [Quadro de Análise da Sustentabilidade para Instituições de Ensino Superior: Aplicação em um Estudo de Caso]. *Arquivos Analíticos de Políticas Educativas*, 24(115), 2–29.

Machado, D., L. Rejane, L.F. Días, L. Veiga, and T. Kader. 2015. Sustainability: A study of the level of ecological behavior of postgraduate students in Brazil. *Environmental Quality Management*, 25(2), 71–89.

Marinho, M., M.S. Gonçalves, and A. Kiperstok. 2014. Water conservation as a tool to support sustainable practices in a Brazilian public university. *Journal of Cleaner Production*, 62, 98–106.

Martínez-Fernández, C.N. and E. González. 2015. Las políticas para la sustentabilidad de las instituciones de educación superior en México: Entre el debate y la acción. *Revista de la Educación Superior*, 44(174), 61–74.

Martins, V., R. Tezza, G. Días, and L.M.S. Campos. 2017. Future professionals: A study of sustainable behavior. *Sustainability*, 9, 413.

Mattos, R., A. Gomes, and G.H. Ribeiro. 2016. Sustainability insights from the mission statements of leading Brazilian Universities. *International Journal of Educational Management*, 30(3), 403–415.

Mattos, R., R. De Castro, K. Rabelo, A.E. Leite, and C.J. Chiappetta. 2015. The journey to sustainable universities: Insights from a Brazilian experience. *International Journal of Business Excellence*, 8(2), 146–159.

Maya, A. 1989. Programas Ambientales Universitarios. Diagnóstico. Instituto Colombiano para el Fomento de la Educación Superior (ICFES) e Instituto Nacional de Recursos Naturales y Ambiente (INDERENA). Bogotá.

Maya, A. 2003. Desarrollo sostenible o cambio cultural. Universidad Autónoma de Occidente.

Mitsch, W.J., A. Tejada, A. Nalhlik, B. Kohlmann, B. Bernal, and C.E. Hernandéz. 2008. Tropical wetlands for climate change research, water quality management and conservation education on a university campus in Costa Rica. *Ecological Engineering*, 34, 276–288.

Mulà, I., D. Tilbury, A. Ryan, M. Mader, J. Dlouha, C. Mader, and D. Alba. 2017. Catalysing change in higher education for sustainable development: A review of professional development initiatives for university educators. *International Journal of Sustainability in Higher Education*, 18(5), 798–820.

Nagy, G., C. Cabrera, G. Coronel, M. Aparicio-Effen, I. Arana, R. Lairet, and A. Villamizar. 2017. Addressing climate adaptation in education, research and practice: The CLiVIA-network. *International Journal of Climate Change Strategies and Management*, 9(4), 469–487.

Nobre, F.S. 2015. Sustainability-centric learning: A case study in management. *The International Journal of Sustainability Education*, 11(3), 1–10.

Orozco, F. and D.C. Cole. 2008. Development of transdisciplinarity among students placed with a sustainability for health research project. *EcoHealth*, 5(4), 491–503.

Palma, L.C., L. M. de Oliveira, and K.R. Viacava. 2011. Sustainability in Brazilian federal universities, *International Journal of Sustainability in Higher Education,* 12(3), 250–258.

Ramos, K. and E. Avila. 2010. Higher education in management: Reinventing the paradigm to gain the capacity to handle today's complexity. *On the Horizon,* 18(1), 45–52.

Ramos, T.B., M. Montano, J.J. Melo, M.P. Souza, C.C. Lemos, A.R. Domingues, and A. Polido. 2015. Strategic environmental assessment in higher education: Portuguese and Brazilian cases. *Journal of Cleaner Production,* 106, 222–228.

Rauen, T.R.S., A.G. Rojas, and V. da Silva. 2015. Environmental management: An overview in higher education institutions. *Procedia Manufacturing,* 3, 3682–3688.

Rieckmann, M. 2012. Future-oriented higher education: Which key competencies should be fostered through university teaching and learning? *Futures,* 44(2), 127–135.

Rodrigues, C. and P.G. Payne. 2017. Environmentalization of the physical education curriculum in Brazilian universities: Culturally comparative lessons from critical outdoor education in Australia. *Journal of Adventure Education & Outdoor Learning,* 17(1), 18–37.

Rodríguez-Solera, C.R. and M. Silva-Laya. 2017. Higher education for sustainable development at EARTH University. *International Journal of Sustainability in Higher Education,* 18(3), 278–293.

Sáenz, O. and J. Benayas. 2015. Ambiente y sustentabilidad en las instituciones de educación superior en América Latina y el Caribe. *AMBIENS Revista Iberoamericana Universitaria en Ambiente, Sociedad y Sustentabilidad,* 1(2), 193–224.

Shiel, C. 2013. Leadership. In S. Sterling, M. Larch, and H. Luna (Eds.). *The Sustainable University: Progress and Prospects* (pp. 110–131). London (UK): Routledge.

Skewes, J.C., C.A. Cioce Sampaio, and F.J. Conway. 2006. Honors in Chile: New engagements in the higher education system. *Honors in Practice,* 2, 15–26.

Spenassato, D., A.C. Trierweiller, A.C. Bornia, B. Marcondes, R.H. Erdmann, and L.M.S. Campos. 2015. Development of a sustainable behavior measurement scale of undergraduate students. *Espacio,* 36(9), 1.

Sterling, S. and I. Thomas. 2006. Education for sustainability: The role of capabilities in guiding university curricula. *International Journal of Innovation and Sustainable Development,* 1(4), 349–370.

Sterling, S., L. Maxey, and H. Luna. (Eds.). 2013. *The Sustainable University: Progress and Prospects.* London and New York: Earthscan/Routledge.

Sterling, S., P. Warwick, and L. Wyness. 2015. Understanding approaches to ESD research on teaching and learning in higher education. In M. Rieckmann, et al. (Eds.). *Handbook of Higher Education for Sustainable Development.* Abingdon: Routledge.

Tilbury, D. 2011. Higher education for sustainability: A global overview of commitment and progress. In GUNI (Eds.). *Higher Education's Commitment to Sustainability: From Understanding to Action* (pp. 18–28). Barcelona: GUNI Higher Education in the World 4.

Tranfield, D., D. Denyer, and P. Smart. 2003. Towards a methodology for developing evidence-informed management knowledge by means of systematic review. *British Journal of Management*, 14(3), 207–222.
UNESCO. 2015. Rethinking education: Towards a global common good?
United Nations Educational, Scientific and Cultural Organisation (UNESCO). 2014. UNESCO roadmap for implementing the global action programme on education for sustainable development.
Vasconcelos de Oliveira, O.A., F.H. Monturil, C. de Sousa, J.R. Araújo, and W.V. Clementino. 2017. Use of energy bills for energy management in Multicamp universities [Utilização das faturas de energia para a gestão energética em universidades Multicampi]. *Espacios*, 38(12), 20–34.
Vásquez, L., A. Iriarte, M. Almeida, and P. Villalobos. 2015. Evaluation of greenhouse gas emissions and proposals for their reduction at a university campus in Chile. *Journal of Cleaner Production*, 108, 924–930.
Velazquez, L.E., N.E. Munguia, A.G. Herrera, and E. Picazzo. 2015. Designing a distance learning sustainability bachelor's degree. *Environment, Development and Sustainability*, 17(2), 365–377.
Velasco, A., M. Valencia, S. Morrow, and V. Ochoa-Herrera. 2018. Understanding the limits of assessing sustainability at Universidad San Francisco de Quito USFQ, Ecuador, while reporting for a North American system. *International Journal of Sustainability in Higher Education*, 19(4), 721–738.
Velazquez, L., N. Munguia, and M. Ojeda. 2013. Optimizing water use in the university of Sonora, Mexico. *Journal of Cleaner Production*, 46, 83–88.
Yánez, R. and M.C. Zavarce. 2011. Sustainable development and scientific research in autonomous Venezuelan universities: Challenges and contradictions. *Revista Venezolana de Gerencia*, 16(53), 89–100.

chapter three

Conflict, ambiguity and change 10 years working towards sustainability

The case of a Mexican University

Margarita Juárez-Nájera
Universidad Autonoma Metropolitana

Mariana Castellanos
École nationale d'administration publique

Contents

3.1 Introduction

Sustainable development promotes life, limits to consumption, interdependence between the natural and the human world, and equity for all (van Weenen 2000). Since the 1990s, there has been a steady increase in the number of empirical studies on sustainability policies and practices in higher education institutions (HEIs; Beveridge et al. 2015). These studies explore how sustainability issues have been addressed and how processes to achieve sustainability have been implemented in different universities around the world. While Trowler (2002) compares processes of institutional change in educational policy in general in three countries (England, Norway, and Sweden) from 1980 to 1990, using the dynamic regime methodological approach, other authors present holistic interdisciplinary approaches to reviewing sustainability issues in higher education. Such evaluations often use quantitative methods based on statistical techniques (Barnard and Van der Merwe 2016, Disterheft et al. 2015).

The studies of policies to implement sustainability in higher education may be grouped into five general themes: (1) evaluations of sustainability-related courses in curriculums and campus audits or sustainability evaluations of campus operations using indicators and language from a business context, (2) sustainability guidelines followed by institutional policies, such as Brundtland's intergenerational orientation or the three pillars of sustainability (economic, environmental, and social), (3) general information on sustainability policies (e.g. years in which policies were implemented; terminology used in policy names), (4) the extent to which municipal or regional policy regarding application of sustainability measures in HEI concords with international declarations, and (5) efforts to establish an administrative position or office in charge of environmental or sustainability issues (Barnard and Van der Merwe 2016, Beveridge et al. 2015, Disterheft et al. 2015, Vaughter et al. 2016).

The objective of the present study was to provide a detailed description and analysis of the process of implementing the Institutional Plan Toward Sustainability (Plan Institucional Hacia la Sustentabilidad or

PIHASU) of the Autonomous Metropolitan University (UAM) in Mexico City. We examine sustainability-oriented actions implemented by the General Provost's office (GP) from 2006 to 2017 in order to understand the following: the process of implementing PIHASU at the UAM, what actions key actors have been involved in during implementation, and what mechanisms these actors have used to further or oppose this plan.

This study also differs from previous studies in that it uses qualitative, descriptive, and inductive methods based on the ontological and epistemological approach of social constructivism. The methodology used is the case study focusing on archival documents and interviews. Most studies addressing sustainability policies in higher education—such as those mentioned above—use deductive quantitative approaches. Furthermore, they fail to specify the ontological and epistemological foundations of the methods they use, which is essential for transparency of the study's findings.

In order to respond to the abovementioned questions, this chapter first presents how the UN's Sustainable Development Goals affect the UAM community as Latin American citizens, as well as how they affect a Mexican University that implements a sustainability plan. Secondly, we apply the advocacy coalition framework (ACF)—developed by Jenkins-Smith and Sabatier (1994)—to the process of implementing PIHASU at the UAM. Thirdly, we describe the case study methodology by which we examined implementation of PIHASU. Finally, we analyse the process of implementing PIHASU.

3.2 Context

3.2.1 International sustainability goals

In 2015, the United Nations General Assembly adopted the resolution *Transforming our world: the 2030 Agenda for Sustainable Development* as "a plan of action for people, the planet and prosperity [which] also seeks to strengthen universal peace in larger freedom" (UN 2015, p. 1). This resolution describes how governments can collaborate to eradicate poverty, heal and secure the planet, and redirect the world towards a sustainable, resilient path (UN 2015, p. 1). The resolution presents 17 Sustainable Development Goals and 169 specific targets, addressing the three dimensions of sustainability: economic, social, and environmental. Fulfilling these goals and targets will allow for implementing the resolution through a global partnership for sustainable development based on strengthening global solidarity.

Through this study, we aimed to identify goals which could be pursued in higher education to promote this resolution in the context of Mexico. We believe that two of the UN's Sustainable Development Goals in particular are critical to resolving some difficulties of Latin American and Caribbean citizens. Goal 4 is "to ensure inclusive and equitable quality education and promote lifelong learning opportunities for all" (UN 2015, p. 14). This means that all humans should work towards ensuring that all learners acquire the knowledge and skills needed to promote sustainability, and that nations should increase the number of scholarships they provide for higher education, including for women and members of disadvantaged social groups, which in turn will increase enrolment (UN 2015, p. 17). Goal 17 is "to strengthen the means of implementing and revitalizing the global partnership for sustainable development" (UN 2015, p. 14). To accomplish this, the UN resolution proposes that all HEI improve their policy coordination and coherence, that nations improve their partnership relations, and that both improve data monitoring and accountability with respect to sustainable actions they undertake (UN 2015, p. 27).

3.2.2 *How to incorporate international sustainability goals: the case of the UAM in Mexico City*

To understand how the UN's Sustainable Development Goals may be furthered, we examined the case of the UAM in Mexico City, which joined a national initiative to promote sustainable development among HEI and contribute to a culture that emphasizes the values of academic freedom, social responsibility, and innovation (Juárez-Nájera, Dieleman, and Turpin-Marion 2006, p. 1029). The present study seeks to help determine how the UN's international goals #4 and #17 may be implemented in Latin American HEI.

The UAM is one of the Mexico City metropolitan area's largest public universities, offering 77 undergraduate degrees, 62 masters programs, and 37 PhD programs in wide variety of academic fields. The UAM's five campuses serve 53,900 students and employ 3,163 professor-researchers. The UAM was founded in 1974 because the city's two other major universities—the Universidad Nacional Autónoma de México, and the Instituto Politécnico Nacional—were unable to meet Mexico City's demand for higher education (Juárez-Nájera, Dieleman, and Turpin-Marion 2006, p. 1031).

In 2007, in collaboration with Mexico's federal environmental agency (SEMARNAT according to its Spanish initials) and Mexico's National Association of Universities and Higher Education Institutions (ANUIES according to its Spanish initials), the UAM carried out a diagnostic study of their operations with respect to education, research, campus management, and community outreach as part of ANUIES's goal for 2020 to determine the current state of sustainable development in Mexico's HEI (Juárez-Nájera, Dieleman, and Turpin-Marion 2006, p. 1029). This study revealed that most universities did not include environmental education and sustainable development in their course curriculum.

Therefore, ANUIES launched a campaign to develop an institutional environmental plan (IEP) for all Mexican HEI to promote "establishment of networks among institutions to stimulate collaboration and increase the scope and improve the quality of programs and services offered by public universities" (Juárez-Nájera, Dieleman, and Turpin-Marion 2006, p. 1029). In line with this campaign, the UAM aims to incorporate environmental topics in all academic fields.

3.2.3 *Application of PIHASU as a sustainability plan, not just an environmental plan*

The IEP was important to incorporating sustainable development into the UAM's environmental strategy. In June 2005, the provosts of four UAM campuses selected a group of researchers to design this plan; these researchers decided to promote greening of the university and increase the university's commitment to sustainability. The IEP established mechanisms for guiding the UAM towards building a culture of sustainability within the university as well as in the broader society (Chávez Cortés and Chávez Cortés 2006, Juárez-Nájera, Dieleman, and Turpin-Marion 2006).

In October 2006, at a national conference of those designated to develop environmental plans for their universities, the UAM presented its IEP which they termed the Institutional Plan Toward Sustainability (or PIHASU according to its Spanish initials). This plan is based on three key principles—systems thinking, participatory planning, and sustainable development, and conceives the world as a complex interconnected system. Therefore, multidisciplinary collaboration is essential to implementing this plan (Juárez-Nájera, Dieleman, and Turpin-Marion 2006, p. 1031, UAM 2006). PIHASU calls for a bottom-up decision-making process involving actors with various levels of responsibility. The UAM agreed that fomenting sustainable development should be a participatory process involving joint decision-making (Juárez-Nájera, Dieleman, and Turpin-Marion 2006, p. 1031) and called upon the university community to think about their vision of the UAM's future, as well as their understanding of their responsibility to nature and to society (Juárez-Nájera, Dieleman, and Turpin-Marion 2006, p. 1032). However, professor-researchers acknowledged that the current focus of university sustainability policies overly emphasizes the environment, and a more comprehensive vision of sustainability was necessary.

The researchers who designed PIHASU presented it as an initiative by which the GP had the responsibility of encouraging actions to implement the plan (UAM 2006). This is significant, given that the GP's functions include general planning of the university's functioning and development (DOF 1973, UAM 1981), and therefore, the provosts of all five campuses would promote implementation of PIHASU.

3.3 *Conceptual framework for reviewing the PIHASU implementation process*

In order to understand how PIHASU has been implemented at the UAM, we used the ACF, which, on a global level, is one of the most widely used approaches to studying policy implementation processes (Weible 2007).

3.3.1 *Characteristics of the ACF theoretical framework*

ACF was originally developed to analyse policy implementation in the United States (Sabatier 1998), but it has also been used in Western Europe, Asia, Australia, and Canada to analyse a variety of types of policy, including that regarding nuclear safety, air pollution, and climate change (Weible and Jenkins-Smith 2016, p. 16), as well as human rights, gender policies, and weapons control (Sabatier 1998, p. 122). We believe that ACF can be helpful in analysing implementation of PIHASU at the UAM.

ACF was developed by Sabatier in the early 1980s and modified by Jenkins-Smith in 1994. Development of this analytical approach responded to three main concerns (Sabatier 1998, p. 98): the need for an alternative to the heuristic approaches that dominated policy framework at the time, a desire to combine the best features of top-down and bottom-up policy implementation approaches (Sabatier 1986), and a commitment to stress the role of technical information in theories of policy processes.

3.3.2 *The goal of ACF*

The goal of ACF is to provide an understanding of the principle procedures involved in the overall policy process, which includes defining the problem as well as policy development, implementation, and evaluation. This process is typically carried out over the course of a decade or more (Sabatier 1998, p. 98). PIHASU fits this timeframe, as it began to be developed in 2005 and was implemented in 2007, and now—over 10 years later—it is possible to identify those factors which have affected PIHASU, those university processes that it has addressed, and modifications to PIHASU that would benefit the UAM.

ACF is capable of analysing complex situations. Its model of the individual, derived from psychology, makes it attractive to scholars seeking an alternative to institutional rational choice models, which currently dominate research on policy and programs (Sabatier 1998, p. 122). In addition, ACF reconciles "bottom-up" and "top-down" approaches by taking into account that policy goals may be ambiguous and that they may be incompatible with policy objectives (Matland 1995, p. 146).

ACF also provides a more realistic method for analysing policy change as compared to other methods (Weible and Jenkins-Smith

2016, p. 17), given that it conceives an organization as being made up of human beings who use various methods to analyse their own policy changes as well as those of other organizations. That is, members of an organization use a variety of approaches to analyse their own organizational processes, including to identify the stages and cycles of change in their organization as well as the causes of those changes, and even to analyse differences among their own processes of analysis. They also analyse new processes which are underway by methods which allow for characterizing collective contexts, including dynamic agent-based modelling.

3.3.3 *ACF's structure*

ACF consists of three main components (Weible and Jenkins-Smith 2016, Weible and Sabatier 2007): a system of external events, relatively stable parameters, and a policy subsystem, which vary according to the degree of public consensus needed for policy change, as well as the constraints and resources of the actors.

ACF's structure allows for studying implementation of—and change in—a program or policy based on five assumptions (Weible and Sabatier 2007, pp. 127–130): (1) Individuals are rationally motivated but are bounded by their imperfect cognitive ability to learn. (2) The tendency of most political participants is to join together in advocacy coalitions which share similar core policy beliefs and coordinate to a significant extent in order to engage in action. (3) Policy brokers seek reasonable compromise among hostile coalitions, although few policy brokers are likely to remain neutral. (4) Coalitions employ a variety of resources that enable them to develop strategies to influence policy—including public opinion, financial resources, skilful leadership, and legal information to make decisions. (5) Coalitions make use of venues where they may influence policies; these include legislatures, courts, regulatory agencies, and the media. These five assumptions describe a policy subsystem—that is, a program, plan, or statute. Figure 3.1—based on a diagram by Weible and Jenkins-Smith (2016, p. 18)—provides a summary of the ACF conceptual framework as understood in the present study.

In this study, the policy subsystem is the UAM's Institutional Plan Toward Sustainability (PIHASU), which involves territorial boundaries, a topic, participants (with varying levels of participation), multiple interest groups, the media, and research institutions (Weible and Jenkins-Smith 2016, Weible and Sabatier 2007).

This study focuses on two types of policy changes that have taken place since PIHASU was developed and implemented: major policy changes, in which alterations occur throughout the subsystem and minor policy changes, consisting of modifications in a subcomponent of

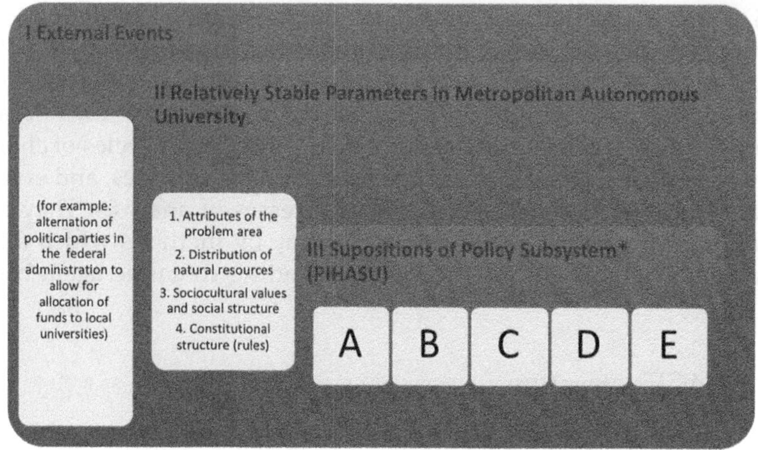

*The letters A – E correspond to the assumptions of the policy subsystem.

A. Individuals are rationally motivated, with imperfect learning ability.
B. Political participants unite in advocacy coalitions.
C. Policy brokers seek compromise but are rarely neutral.
D. Coalitions use resources to influence policy.
E. Coalitions use venues to influence policy.

Figure 3.1 Diagram analysing the UAM's Institutional Plan Toward Sustainability. (Adapted from Weible and Jenkins-Smith 2016.)

the system (Weible and Jenkins-Smith 2016, Weible and Sabatier 2007). Implementation of PIHASU is also analysed through three mechanisms that allow for change: external shocks, a stalemate in negotiations, and accumulation of scientific or technical evidence (Weible and Jenkins-Smith 2016, Weible and Sabatier 2007).

3.4 *Methodology used to analyse implementation of PIHASU at the UAM*

The objective of this study was to provide a detailed description and analysis of the process of implementing PIHASU at the UAM. As its ontological basis, this inductive, descriptive, qualitative study (Bourgeois 2016, p. 68, Fortin and Gagnon 2016, p. 65) uses social constructivism, which is an epistemology that allows for researchers and participants of the study to work jointly, based on their individual experiences as well as multiple processes that are constructed through interactions (Creswell 2013).

3.4.1 Method

The method of the case study was used for data collection and analysis. The unit of analysis was the process of implementing PIHASU, focusing specifically on actions that took place in the GP, which was responsible for managing planning processes on the five campuses, providing funds, and coordinating management processes among campuses. We aimed to understand the interventions of all actors involved in the process of implementing PIHASU. Also, we examined the tools that actors used and the means to which they resorted to oppose or favour creation and implementation of the plan.

According to Flyvbjerg, the case study allows for understanding the relationship between causes and results in a process (2011, p. 306), exploring in detail mechanisms of interrelations, developing and testing explanations, understanding the level of explanatory power of concepts related to the context studied, and formulating new questions raised by the case. "The advantage of the case study is that it provides a close-up view of real-life situations and tests perspectives directly in relation to phenomena as they unfold" (Flyvbjerg 2011, p. 309).

Some authors point out that the case study allows for detecting specific social patterns and making generalizations (Rule and John 2015, p. 3). Rule and John (2015, p. 4) call this process "type theory from the case"—"where one might begin with the case to generate a new theory from it (...) from concrete data through careful bottom-up analysis." Such an approach is used in the present study as it allows for describing and abstracting that which occurs in a given social context (Fouche 2002, p. 274, cited in Rule and John 2015, p. 6).

According to Creswell, "the researcher explores a bounded system (a case) [...] over time [... and] focuses on key themes, not in order to generalize beyond the case, but to understand its complexity" (Stake 1995, p. 123 cited in Creswell 2013, p. 75). The present case has been selected given that it illustrates several problems that arise upon implementing a new program in a Latin American HEI.

3.4.2 Data collection

Various types of documents were used to describe and analyse the process of implementing PIHASU. Several documents provided information on university rules and procedures or about PIHASU: The Organic Law (DOF 1973) and Organic Regulations of the University (UAM 1981), the GP's annual and 4-year reports from 2007 to 2017 (UAM 2017), the UAM's final version of PIHASU (UAM 2006), records from nine meetings of the PIHASU Coordinating Group, and *Environmental Planning in Mexico* (Chávez Cortés et al. 2014), a book coordinated by the researchers that developed PIHASU. We also carried out interviews of 18 university members: four provosts, four executive directors, eight deans, and two researchers involved in developing the plan.

3.4.3 *Method of analysis*

In order to examine our results, we carried out a thematic analysis, as according to Aguinaldo (2012, p. 768), an initial step in analysing results is to organize data into themes. The case study method of analysis, which has been used for a great variety of cases in different contexts (Aguinaldo 2012), was adapted to the present case. Thematic analysis consists of three stages that allow for standardizing and comparing results: (1) identifying data content, (2) reducing redundancy, and (3) grouping data into categories that describe a particular phenomenon (Aguinaldo 2012).

Data was coded and analysed by both researchers using Nvivo software. Three thematic coding matrices were generated at the end of the study. As a result of the first stage of analysis, a matrix composed of eight topics was formulated: governance, budget, campus management, teaching, research, outreach, challenges in incorporating sustainability, and values. These topics were included in the Coordinating Group's original proposal to monitor and evaluate PIHASU. The majority of these topics are mentioned in the University's Organic Law (DOF 1973), which establishes regulations for the UAM's administrative functioning.

As these eight topics did not allow for fully evaluating the PIHASU implementation process, a second analysis was carried out in which data was recoded under six new categories: vision, perception, actions, actors, support, and difficulties. These new categories originated from a second stage of analysis, as well as from a presentation by the PIHASU Coordinating Group titled *Characterization of Current Visions towards Sustainability from Students to Top Staff Members at a Mexican University* (Juárez-Nájera et al. 2010). While this second analysis provided a greater level of understanding, several topics did not fit any of these categories. Therefore, data was again rearranged chronologically into four stages. This new classification was combined with the first eight topics and the second set of six topics, thereby allowing for determining those actors involved and describing the plan's implementation process over the past 10 years.

3.5 *Analysis of results of the PIHASU implementation process*

As a result of data analysis, we divided PIHASU's lifetime into four stages—background, conception, implementation, and future—each of which includes some of the first set of eight topics and the second set of six topics mentioned above.

3.5.1 Background stage of PIHASU

This first stage, occurring in 2005 and 2006, consisted of preparing the ground for PIHASU. We identified five components at that time: national context, predominant actors, incorporation of the concept of sustainability, actions, and values.

With regard to the national context, in 2002, several Mexican federal agencies created an action plan for sustainable development of Mexican HEI. This initiative sought to promote changes in institutional dynamics of Mexico's universities in order to respond to pressing socio-environmental problems (Mercado 2014). In the following years, many HEI agreed to participate in this initiative, and in 2005, the UAM committed to a process of institutional transformation upon incorporating the concept of sustainability into its vision and practices (Chávez Cortés et al. 2014).

Key actors during this "background" stage were the federal agencies coordinating the national initiative and Dr. Lema Labadie—the UAM's provost from 2005 to 2009. According to reports by Dr. Lema, in 2005 and 2006, the university was focused on defining its geographical scope as well as its social function and its identity within its campuses, in the broader community, and in Mexico's higher education system (UAM 2017).

During this period, the PIHASU Coordinating Group interviewed 16 senior officials of the UAM regarding their concept of sustainability and how sustainability might be incorporated into their daily tasks. Respondents were familiar with the concept of sustainability, and their definitions coincided with that presented in the Brundtland Report (UN 1987). However, they pointed out that lack of a university policy clearly oriented towards sustainability, as well as lack of funding, made it difficult to implement sustainability on campus.

Regarding the actions and values mentioned by interviewees during this stage, sustainability appeared to be a new ideal or a utopia that offered opportunity and hope to review, redefine, and even reconstruct the concepts, beliefs, values, and paradigms involved in sustainability (Mercado 2014). Furthermore, Mercado (2014, p. XVIII) has pointed out that the university must not only participate in construction and diffusion of a new culture but, above all, exercise moral authority, which will only be possible if the university becomes an example of what it teaches.

3.5.2 Conception stage of PIHASU

The second stage of PIHASU was its conception, in 2007 and 2008, when PIHASU was formally written up and made public. We defined seven components of the context at this stage: redefinition of sustainability, perceptions, actors, actions, values, support, and difficulties.

During this stage, senior officials identified legislative mechanisms that might be used to promote sustainability in the university's daily tasks. Two central questions initially guided development of an institutional sustainability plan: "What is the current level of sustainability of universities?" and "How might a sustainability plan contribute to the university's pre-existing social commitment?" (Chávez Cortés et al. 2014).

Despite growing global consensus regarding the multiple facets (social, environmental, and economic) of the concept of sustainability (Curran 2009), such a vision had not yet completely permeated the perceptions of university decision makers. Furthermore, promotion of sustainability within the UAM depended on the particular interest and political will of senior officials.

In an arduous dialectical multidisciplinary process in 2005 and 2006, a total of five professor-researchers from the four UAM campus worked jointly to create the plan. With the participation of the directors of each campus as well as other personnel, these professor-researchers developed a document that went beyond greening the university to include cultural, political, economic, and social aspects of sustainability and named responsibility, commitment, and ethics as guiding principles of sustainability. They stressed the importance of participation by the entire university community (students, professors, directors, and other administrative staff) to build a sustainable university and proposed a vision of sustainability that incorporates the concepts of complexity and interdisciplinarity and transdisciplinarity in order to understand the current context and search for solutions to challenges posed by sustainability (Meppem and Gill 1998).

Interviews with two members of the PIHASU Coordinating Group highlight two significant ways in which they supported creation of the plan. First, they provided their time, effort, and interest. As one stated, "we all met to be able to discuss, comment… We read. Someone provided articles or readings. We read them and discussed the topic" (Interview 17 with a researcher). Second, the GP, as well as the provost of each campus, contributed to administrative aspects of the plan—for example, by appointing a PIHASU representative and financially by budgeting for related projects.

Several difficulties arose during the conception of PIHASU. For example, the provosts of some campuses wanted to unilaterally direct the plan's content and development. As a result, some researchers decided not to participate; one explained: "they wanted [me to draw up] a document that (…) was an academic document—an academic exercise (…) to conceptualize sustainability and not to make it a political issue that we would have to work on" (Interview 17). Furthermore, when the plan was published, it was labelled as "an initiative of the general provost" (UAM 2006, p. 1) which did not provide it with legal standing. Rather, the plan "was subject to whether the next provost considered it to be important enough to continue providing support" (Interview 18 with a researcher).

3.5.3 *Implementation of PIHASU*

The third stage—implementation— from 2009 to 2012, took place just after Dr. Lema finished his term, while Dr. Fernández Fassnacht was GP (2010–2012). We identified six topics related to this stage: campus management, outreach, actors, actions, support, and difficulties.

As the plan was an initiative of the GP, each campus prepared its own PIHASU and was responsible for implementing it. Additionally, the GP implemented a PIHASU for itself, in which the PIHASU Coordinating Group continued to be involved: "We keep (...) advancing on implementing sustainability, thinking about projects (...) that were just an echo of what we had proposed in the plan" (Interview 17). Furthermore, administrative positions responsible for sustainability were established, and financing was provided—principally indirectly for sustainability-related research projects. Various sustainability-related outreach activities were carried out, although none specifically addressed the importance of PIHASU to the university. For example, in 2009, a sustainability forum was held at the UAM. In 2010, professors and researchers met to examine national challenges such as health, childhood development, education, sustainability, and university development. In 2012, the International Rio+20 Summit, which highlighted trends in sustainable development, was held in Mexico.

Also, during this stage, a high level of administrative and financial support was granted for sustainability-related projects. As one interviewee stated, "I believe that I had all the support that one could wish for. I prepared the program and we had a meeting with the general provost. (...) He gave us his time, which is very difficult because he has many activities. He gave us his time, he received us in his office, and we talked, not only about my project, but about all the others, about what would be done" (Interview 17).

During this stage, difficulties arose due to lack of time, effort, commitment, and interest: "I think the hardest thing was coordination. (...) The people were willing to discuss what should be done, but when it came time to taking responsibility for activities, everyone wanted others to be responsible. They wanted to give their opinion, but not commit to action" (Interview 18).

In 2010, the PIHASU Coordinating Group was disbanded, and no multi-campus or interdisciplinary group has since been established to continue the discussion process.

3.5.4 PIHASU's future

A fourth stage—addressing what at the time was the future of the plan, from 2013 to 2017, while Dr. Vega y León—one of a second generation of leaders following creation of the plan—was provost, consisted of planning the future, as the first fruits of PIHASU became evident. During this stage, the GP's annual reports considerably increased their level of sustainability-related information, suggesting that the university community considered sustainability to be important. Thanks to creative collaboration by all campuses, sustainability took hold at the UAM. Five characteristics were defined regarding this stage which demonstrates a new context: actions, perception, limitations, values, and the future.

Reports by GP Dr. Vega y Leon highlight the importance of sustainability and environmental protection. For example, 2013 marked the 40th anniversary of bachelor's programs in environmental engineering, hydrologic engineering, and energy engineering, and department anniversary celebrations emphasized the existence of almost a thousand classes and other projects related to sustainability and environmental protection. In 2014, the GP reported sustainability to be a value of the university and presented the UAM as a pioneer on the subject, as the university participated in several advisory councils of Mexico City's natural areas. Furthermore, in 2015, the UAM obtained fourth place for Mexican universities in the University of Indonesia's Greenmetric World University Ranking for environmentally committed universities.

Interviews with two PIHASU Coordinating Group members showed that sustainability is now on the minds of the university community. For example, one stated, "I feel that PIHASU has left its mark in some way. If we had not done the work of PIHASU, the topic of sustainability would not ever have been placed on the agenda, or maybe it would have taken longer, because there were teachers who covered the subject, but there was no prior initiative or interest in fomenting it within the institution, on campus. As a result of our work, it was placed on the agenda, and so far, it has not been removed" (Interview 18).

However, we have identified some limitations, such as the fact that PIHASU has never been formally institutionalized. With respect to this, we witnessed a lack of interest or sense of importance among university administrators. This presents a great challenge for the plan, as "everyone addresses it if they want to, in the way they want" (Interview 18 with a provost/executive director/dean/researcher), which leads to irregularities and lack of continuity. Another limitation is the fact that most actions carried out with respect to the plan to date have been related to the environmental aspect of sustainability, "but [they should] also look at the plan with [a broader perspective] of sustainability and not only the environmental part—that is, culture, politics… and all of that should permeate research, teaching, cultural diffusion (...) within the university" (Interview 17).

Despite this, a significant positive change in understanding of the concept occurred during this stage, as sustainability was presented in GP annual reports under a new heading called "values." As one interviewee stated, "it seems to me that someone must again take up the issue [and] develop an initiative again in the new [era, taking into account] the interests of the provost, making him interested" (Interview 18).

This outline of the 10-year implementation process of PIHASU shows the various actions and reactions to sustainability by the actors, most of whom were in favour of this plan, although not all understood the breath of its content at all times.

3.6 Discussion and final thoughts on the PIHASU implementation process

3.6.1 Five conditions for overcoming deficiencies in implementing PIHASU

Sabatier and Mazmanian (1979, p. 485) point to the following five conditions for overcoming deficiencies in any policy implementation process involving incomplete information, conflicts of goals, or individuals or groups who veto the plan.

1. The plan should be based on a sound theory related to changes in target group behaviour in order to achieve the desired objectives.
2. The basic policy decision should contain unambiguous policy directives for the implementation process so as to maximize the likelihood that target groups will perform as desired.
3. The leaders of the agencies or organizations implementing the policy should possess substantial managerial and political skills.
4. The plan should actively be supported by organized constituency groups as well as by a few key legislators (or the director of the organization implementing the policy) throughout the implementation process, with the director of the organization being neutral or supportive of the policy.
5. The relative priority of policy objectives should not be significantly undermined over time due to development of conflicting public policies or changes in socioeconomic conditions that could weaken political support.

With respect to the first of these conditions, PIHASU is based on the principles of systems theory, participatory planning, and sustainability applied to teaching, research, cultural diffusion, and campus management. However, following the reasoning of Sabatier and Mazmanian (1979), the PIHASU Coordinating Group would have benefited from incorporating elaboration of the plan into a learning process through experimental projects, extensive research and development, and project evaluations which involves as many professors and researchers as possible. However, the PIHASU implementation process only included UAM members designated by campus provosts.

With respect to the second condition, the UAM community is likely to achieve PIHASU's objectives given the existence of legal indications to establish consensus by the UAM's Organic Law and Organic Regulation. However, not all university actors participate in—or feel responsible for policy actions, as indicated by interviewee 18.

With respect to the third condition, UAM leaders (provosts, executive directors, and deans) have limited managerial skills with respect to sustainability. Although they adequately manage finances and promote a high level of morale among staff, they did not show an impartial attitude towards sustainability. They also did not adequately handle the dissent of a PIHASU Coordinating Group member, according to interviewee 17.

With respect to the fourth condition, PIHASU was supported by UAM provosts from the moment of its creation to its implementation, as observed in the four stages of implementing the plan. However, the plan has not yet been institutionalized or formally provided with permanent financial and human resources; nor has sufficient outreach been carried out so that the entire university community might understand the importance of the plan to the future of the university. Whether or not PIHASU is supported depends on the GP in turn, as mentioned by interviewee 18. There is a need for reflection and consensus by the entire university community regarding the meaning of sustainability so that all members will comprehend the relationship of the UN's Sustainable Development Goals to teaching, research, cultural diffusion, and campus management.

With respect to the final condition, the GP has appointed personnel to manage the UAM's "greening" process, as indicated by the fourth stage of PIHASU's implementation. However, if the university hopes to effectively implement the plan, the university will have to make a greater effort to effectively train these personnel. According to Sabatier and Mazmanian (1979), there is a need for careful monitoring of implementation of policies as well as of human and financial resources to carry out this task. Hiring personnel capable of monitoring the implementation process is also important. While PIHASU's objectives are not threatened by crises or changes in personnel in the national government, reaching these objectives could be slowed down by funding cuts or appointment of new provosts. Nonetheless, the past three GPs have supported PIHASU's objectives, as indicated by their annual reports.

3.6.2 *The ACF diagram and consequences of the assumptions involved in PIHASU*

As Figure 3.1 shows, the ACF indicates the ability of policies to alter the social behaviour of organizations. Assumption A indicates that the beliefs of individual system are rationally motivated. An individual's beliefs are based on the basic values. These beliefs are also based on that person's perception in the severity and the causes of a problem of implementation and the policy changes. The beliefs of UAM directives are one of the most important issues because the directives prepare, propose, and implement a change in policy directly related to a policy subsystem. PIHASU Coordinating Group members and the UAM's senior staff were familiar with the concept of sustainability, as shown in PIHASU's background, but were unable to explain it for the whole university community.

The beliefs of UAM participants in PIHASU policy subsystem influenced assumption B—that political participants unite in advocacy coalitions. Coalitions are established because participants seek other actors with similar political beliefs. Perhaps PIHASU was not institutionalized because the multidimensional vision of sustainability had not permeated decision makers' perceptions. Promotion of sustainability within the university depended on the political will of the provosts and their interest in this issue, as shown in analysis of PIHASU's conception.

Assumption C regards policy disagreements among coalitions, such that conflicts may escalate if no effective intermediary or facilitator is designated. PIHASU was presented by the GP, who did not institutionalize the plan; therefore, it may be said that he did not play the role of political broker with respect to sustainability within the UAM.

Assumption D—that coalitions use resources to influence policy—such as UAM decision makers having leadership skills or the legal authority to make decisions—indicates the strategies that individuals may develop to influence policy. As PIHASU was not institutionalized, human and financial resources to implement it were insufficient.

Assumption E states that human and financial resources for implementing PIHASU could be carried out through potential arenas within which interest groups have the opportunity to influence participants' beliefs and policies. From PIHASU's creation to the present, no GP has used the UAM's top administrators as the arena for institutionalizing the plan because university legislation does not clearly mention the concept of sustainability.

3.6.3 *The relationship of the UN Sustainable Development Goals to the UAM*

For the UAM to further the UN's Sustainable Development Goals, the entire university community must behave sustainably. We observed that the UAM's weakness with respect to these goals is that thus far it has only identified with, and acted upon, the environmental aspect of sustainability. As PIHASU's creators did not broadly define the concept of sustainability to include all three aspects included in the Brundtland Report, there has been no connection among social, economic, environmental, and cultural activities carried out in UAM.

Furthermore, the UAM's statutes, developed in the 1970s, inhibit changes and incorporation of new activities related to the concept of sustainability within the university's three principal functions, which are teaching, research, and outreach. Sustainability can be seen as a university value, as well as a transversal theme that links teaching, research, and outreach. Incorporation of the concept of sustainability in the university allows for linking its various activities in a holistic manner, as mentioned in the 2016 report of the GP and in PIHASU itself. Finally, interpretations by the UAM's directors, their advisors, and other collaborators in developing the university's approach to sustainability have not been completely evaluated or understood by the university community.

With respect to what should be done to achieve the UN's Sustainable Development Goals, we believe that university members must have a clear understanding of the concept of sustainable behaviour. While the concept of sustainability is an utopia, sustainable behaviour is a set of effective, deliberate, anticipated actions aimed at accepting responsibility for prevention, conservation, and preservation of physical and cultural resources, which include the integrity of animal and plant species, individual and social well-being, and safety of present and future human generations (Juárez-Nájera 2015, p. 9). This concept may contribute to transforming values and morals, the relationship between humans and the rest of nature, and economic processes, thereby putting an end to spiraling growth of unlimited aspirations, consumption, and material wealth, and rather focusing on adequately meeting the needs of all (Juárez-Nájera 2015, p. 118).

3.7 *Conclusion*

This study has presented conflict, ambiguity, and change that has taken place in the past 10 years at Mexico City's UAM as it moves towards sustainability by implementing PIHASU.

We first presented the context in which these changes have occurred, as well as a manner in which two of the UN's international sustainability goals could be incorporated in the case of the UAM. We reviewed the ACF, with which we evaluated the process of implementing PIHASU, and provided a diagram of the process used to analyze the plan. We then examined the case study methodology, the data collection process, and the thematic analysis used to explore implementation of PIHASU at the UAM. By examining the four stages of PIHASU—background, conception, implementation, and future—we explored the process of implementing this policy. Finally, we described five requisites for overcoming deficiencies in the process of implementing PIHASU and suggested how the UN's Sustainable Development Goals might be pursued at the UAM.

This study allows for understanding how PIHASU has been implemented at the UAM. It demonstrates that most actors' actions have furthered implementation of PIHASU over the past 10 years, as most were in favour of the plan, although some lacked understanding of its content.

Further research would be necessary to explore the concept of sustainable behaviour in order to identify mechanisms which might allow for incorporating PIHASU in the university's daily academic and administrative work.

References

Aguinaldo, Jeffrey P. 2012. Qualitative analysis in gay men's health research: Comparing thematic, critical discourse, and conversation analysis. *Journal of Homosexuality* 59(6):765–787. doi: 10.1080/00918369.2012.694753.

Barnard, Zenia and Derek Van der Merwe. 2016. Innovative management for organizational sustainability in higher education. *International Journal of Sustainability in Higher Education* 17(2):208–227. doi: 10.1108/IJSHE-08-2014-0120.

Beveridge, Dan, Marcia McKenzie, Philip Vaughter, and Tarah Wright. 2015. Sustainability in Canadian post-secondary institutions: The interrelationships among sustainability initiatives and geographic and institutional characteristics. *International Journal of Sustainability in Higher Education* 16(5):611–638. doi: 10.1108/IJSHE-03-2014-0048.

Bourgeois, Isabelle. 2016. La formulation de la problématique. In *Recherche sociale: de la problématique à la collecte des données*, edited by Benoît Gauthier, Isabelle Bourgeois, André Forget, and Jean Turgeon, xi, 670 p. Québec: Presses de l'Université du Québec.

Cortés Chávez, Martha Magdalena and Juan Manuel Chávez Cortés. 2006. Introduciendo la perspectiva de la sustentabilidad en las actividades sustantivas de la UAM. *Contactos. Revista de Educación en Ciencias e Ingeniería* 61:16–22.

Cortés Chávez, Martha Magdalena, Juan Manuel Chávez Cortés, Octavio Francisco González Castillo, Enrique Mendieta Márquez, and Margarita Juárez Nájera. 2014. La experiencia de elaborar un plan institucional hacia la sustentabilidad: retos y lecciones aprendidas (Universidad Autónoma Metropolitana). In *La planeación ambiental en México. Experiencias desde las instituciones de educación superior*, edited by María Teresa, Bravo Mercado, and Octavio Francisco González Castillo, pp. 65–83. México: UAM-UNAM.

Creswell, John W. 2013. *Qualitative Inquiry & Research Design: Choosing Among Five Approaches*. Third edition. Qualitative Inquiry and Research Design. Los Angeles, CA: Sage Publications.

Curran, Mary Ann. 2009. Wrapping our brains around sustainability. *Sustainability* 1(1):5–13. doi: 10.3390/su1010005.

Disterheft, Antje, Ulisses M. Azeiteiro, Walter Leal Filho, and Sandra Caeiro. 2015. Participatory processes in sustainable universities — what to assess? *International Journal of Sustainability in Higher Education* 16(5):748–771. doi: 10.1108/IJSHE-05-2014-0079.

DOF. 1973. Ley Orgánica de la Universidad Autónoma Metropolitana. Honorable Cámara de Diputados, Last modified: 17-12-1973. www.diputados.gob.mx/LeyesBiblio/ref/louam.htm.

Flyvbjerg, Bent. 2011. Case study. In *Handbook of Qualitative Research*, edited by Norman K. Denzin and Yvonna S. Lincoln, pp. 301–316. Thousand Oaks, CA: Sage Publications.

Fortin, Marie-Fabienne and Johanne Gagnon. 2016. *Fondements et étapes du processus de recherche: méthodes quantitatives et qualitatives*. 3e édition. Montreal: Chenelière éducation.

Jenkins-Smith, Hank C. and Paul A. Sabatier. 1994. Evaluating the advocacy coalition framework. *Journal of Public Policy* 14(2):175–203. doi: 10.1017/S0143814X00007431.

Juárez-Nájera, Margarita. 2015. *Exploring Sustainable Behavior Structure in Higher Education, Management and Industrial Engineering*. Switzerland: Springer International Publishing.

Juárez-Nájera, Margarita, Hans Dieleman, and Sylvie Turpin-Marion. 2006. Sustainability in Mexican higher education: Towards a new academic and professional culture. *Journal of Cleaner Production* 14(9):1028–1038. doi: 10.1016/j.jclepro.2005.11.049.

Juárez-Nájera, Margarita, Martha Magdalena, Chávez Cortés, Enrique Mendieta Márquez, and Octavio Francisco González Castillo. 2010. Characterization of current visions towards sustainability from students to top staff members at a Mexican University. In *14th ERSCP Conference and 6th EMSU Conference*, Delft, the Netherlands, 25–29 October.

Matland, Richard. 1995. Synthesizing the implementation literature: The ambiguity-conflict model of policy implementation. *Journal of Public Administration Research and Theory* 5(2):145.

Meppem, Tony and Roderic Gill. 1998. Planning for sustainability as a learning concept. *Ecological Economics* 26(2):121–137. doi: 10.1016/S0921-8009(97)00117-1.

Mercado, Bravo and María Teresa. 2014. Introducción. In *La planeación ambiental en México. Experiencias desde las instituciones de educación superior*, edited by María Teresa, Bravo Mercado, and Octavio Francisco González Castillo, pp. 3–13. México: UAM-UNAM.

Rule, Peter and Vaughn Mitchell John. 2015. A necessary dialogue: Theory in case study research. *International Journal of Qualitative Methods* 14(4). doi: 10.1177/1609406915611575.

Sabatier, Paul A. 1986. Top-down and bottom-up approaches to implementation research: A critical analysis and suggested synthesis. *Journal of Public Policy* 6(1):21–48. doi: 10.1017/S0143814X00003846.

Sabatier, Paul A. 1998. The advocacy coalition framework: Revisions and relevance for Europe. *Journal of European Public Policy* 5(1):98–130. doi: 10.1080/13501768880000051.

Sabatier, Paul A. and Daniel Mazmanian. 1979. The conditions of effective implementation: A guide to accomplishing policy objectives. *Policy Analysis* 5(4):481.

Trowler, Paul R. 2002. Introduction: Higher education policy, institutional change. In *Higher Education Policy and Institutional Change: Intentions and Outcomes in Turbulent Environments*, edited by Paul R. Trowler, pp. 1–23. Buckingham & Philadelphia: Society for Research into Higher Education and Open University Press.

UAM, Universidad Autónoma Metropolitana. 1981. Reglamento Orgánico de la Universidad Autónoma Metropolitana. UAM, Last modified (10-04-1981).

UAM, Universidad Autónoma Metropolitana. 2006. Plan Institucional hacia la Sustentabilidad de la Universidad Autónoma Metropolitana (iniciativa del Rector General). UAM. www.vinculacion.uam.mx/index.php/uam-sustentable/pihasu-de-la-uam.

UAM, Universidad Autónoma Metropolitana. 2017. Reportes. www.transparencia.uam.mx/inforganos/rg/anterior.html.

UN, United Nations General Assembly. 1987. Our Common Future: Report of the World Commission on Environment and Development. United Nations General Assembly.

UN, United Nations General Assembly. 2015. Transforming Our World: the 2030 Agenda for Sustainable Development. United Nations General Assembly, Last modified Resolution adopted 25 September 2015. https://sustainabledevelopment.un.org/post2015/transformingourworld.

Vaughter, Philip, Marcia McKenzie, Lauri Lidstone, and Tarah Wright. 2016. Campus sustainability governance in Canada: A content analysis of post-secondary institutions' sustainability policies. *International Journal of Sustainability in Higher Education* 17(1):16–39. doi: 10.1108/IJSHE-05-2014-0075.

van Weenen, Hans. 2000. Towards a vision of a sustainable university. *International Journal of Sustainability in Higher Education* 1(1):20–34. doi: 10.1108/1467630010307075.

Weible, Christopher M. 2007. An advocacy coalition framework approach to stakeholder analysis: Understanding the political context of california marine protected area policy. *Journal of Public Administration Research and Theory* 17(1):95–117. doi: 10.1093/jopart/muj015.

Weible, Christopher M. and Hank C. Jenkins-Smith. 2016. The advocacy coalition framework: An approach for the comparative analysis of contentious policy issues. In *Contemporary Approaches to Public Policy*, edited by Guy Peters and Philippe Zittoun, pp. 15–34. Basingstoke: Palgrave Macmillan.

Weible, Christopher M. and Paul A. Sabatier. 2007. A guide to the advocacy coalition framework. In *Handbook of Public Policy Analysis: Theory, Politics, and Methods*, edited by Frank Fischer, Gerald Miller, and Mara S. Sidney, pp. 123–136. Boca Raton, FL: Taylor & Francis.

chapter four

Sustainability in higher education

Implementation in Bulgaria

Kathy Bui and Will O'Brien
Clark University

Contents

4.1 Introduction

It is known that humans are the dominating force of change to our earth and the people who inhabit it. Despite our efforts to conserve the environment, it is being diminished at a faster rate than ever before. "We are severely disrupting the stability of the climate, and there are huge social, economic, and public health challenges worldwide. This is happening with 25% of the world's population consuming 70%–80% of the world's resources."[1] "Current trends and patterns of resource use, coupled with a rapidly changing; increasingly unequal, complex, and interconnected societal structure; and rapid technological change, are impacting human–environment interactions in critical and unsustainable ways."[2] However, imagine if that same human force conducted change for good, change that could restore biodiversity and ecosystems? With the increasing urgency to confront these challenges, many opportunities arise for different stakeholders to find solutions to the issues.

A key stakeholder well positioned to achieve change are higher education institutions (HEIs). "Universities and colleges have been at the forefront of creating as well as deconstructing paradigms. They have led social change through scientific breakthroughs but also through the education of intellectuals, leaders, and future makers."[3] Since its introduction on the international stage, many intuitions have signed on and made efforts to incorporate sustainability into their operations, research, curriculum, and more.

For purposes of this chapter, we will use the definition of "sustainability" provided by the Dow Jones Sustainability Index: "Corporate sustainability is a business approach that creates long-term shareholder value by embracing opportunities and managing risks deriving from economic, environmental, and social developments." Through various literature and anecdotes, the goal of this chapter is to serve as a reference for HEIs in Bulgaria, should they want to incorporate sustainability into their institutions through examples of best practices in sustainability integration in United States' institutions. This chapter also provides assessment questions to gauge the challenges and opportunities for sustainability integration, with respect to variation in culture and context. This chapter argues that HEIs can serve as prominent change agents to address sustainability issues; furthermore, application methods can be applicable across cultures if the assessment of challenges and opportunities are conducted.

1 Hattan, Feder, Naik, Murphy, Davis, Esiet, Vithlani, Rigaud, Advancing Education for Sustainability: Teaching the Concepts of Sustainable Building to All Students (2010) 38.

2 Graham, Hernandez, Stephens, Scholz, Higher education as a change agent for sustainability in different cultures and contexts (2008) 318.

3 Tilbury, Higher Education for Sustainability: A Global Overview of Commitment and Progress (2011) 1.

4.2 Background

Since the Stockholm Conference in 1972, the important role of education in environmental protection and conservation was formally introduced at the international level.[4] Its role was further recognized by the UNESCO–UNEP International Environmental Education Programme by the United Nations Educational, Scientific and Cultural Organization (UNESCO) and the United Nations Environmental Programme (UNEP) in 1975. Afterwards, declarations pertaining to sustainability in HEIs both nationally and internationally were created, gaining acceptance in many higher education communities. The declarations consisted of topics related to sustainable operations, research, public outreach, partnerships with government and non-government organizations, as well as developing new leaders and thinkers to lead a more sustainable life.[5] The Talloires Declaration of 1990 was the first official statement made by presidents and administrators of universities demonstrating their commitment to sustainability in HEIs by drafting a plan for incorporating teaching, research, operations, and outreach of sustainability for universities and college.[6]

Other notable declarations were the Co-operation Program in Europe for Research on Nature and Industry Coordinated University Studies (COPERNICUS), established to promote a better understanding of the interaction between man and the environment and to collaborate on common environmental issues.[7] Since its introduction, colleges and universities across the globe were looked upon to be the leaders in "socializing students to contribute to social progress and the advancement of knowledge. Figure 4.1 provides a historical summary of initiatives taken in society, education and higher education to foster sustainable development.

They have a profound responsibility to impart the moral vision and technical knowledge needed to ensure a high quality of life for future generations."[8] To have a sustainable future through the leadership of higher education, there needs to be a shift in the systematic way things are taught and conditioned. It also required that institutions restructure their missions, courses, research focus, community outreach, and campus operations.

HEI from all over the world came together "to collaboratively champion education, research and actions for sustainable development in the Higher Education Sustainability Initiative (HESI) of the United Nations

[4] Huisingh, Lambrechts, Lozano, Lukman, Declarations for sustainability in higher education: becoming better leaders, through addressing the university system (2011) 2.

[5] Ibid., 2.

[6] Ibid., 4.

[7] Ibid., 4.

[8] Calder, Clugston, Corcoran, Introduction (2002) 10.

History of the initiatives taken in society, education, and higher education to foster sustainable development.

Year	Event/declaration	Level or focus
1972	Stockholm Declaration on the Human Environment, United Nations Conference on the Human Environment, Sweden	Society
1975	The Belgrade Charter, Belgrade Conference on Environmental Education, Yugoslavia	Education
1977	Tbilisi Declaration, Intergovernmental Conference on Environmental Education, Georgia	Education
1987	"Our Common Future", The Brundtland Report	Society
1990	Talloires Declaration, Presidents Conference, France	Higher education
1991	Halifax Declaration, Conference on University Action for Sustainable Development, Canada	Higher education
1992	Report of the United Nations Conference on Environment and Development (Rio Conference): Agenda 21, Chapter 36: Promoting Education, Public Awareness and Training and Chapter 35: Science for Sustainable Development	Society
1992	Association of University Leaders for a Sustainable Future founded, USA	Higher education
1993	Kyoto Declaration, International Association of Universities Ninth Round Table, Japan	Higher education
1993	Swansea Declaration, Association of Commonwealth Universities' Fifteenth Quinquennial Conference, Wales	Higher education
1993	COPERNICUS University Charter, Conference of European Rectors (CRE)	Higher education
1996	Ball State University Greening of the Campus conferences were in 1997, 1999, 2001, 2003, 2005, 2007, and 2009	Higher education
1997	Thessaloniki Declaration, International Conference on Environment and Society: Education and Public Awareness for Sustainability, Greece	Education
1999	Environmental Management for Sustainable Universities (EMSU) conference first held in Sweden. Following conferences in 2002 (South Africa), 2004 (Mexico), 2006 (U.S.A.), 2008 (Spain), and in 2010 in The Netherlands.	Higher education
2000	Millennium Development Goals	Society
2000	The Earth Charter	Society
2000	Global Higher Education for Sustainability Partnership (GHESP)	Higher education
2001	Lüneburg Declaration on Higher Education for Sustainable Development, Germany	Higher education
2002	World Summit on Sustainable Development in Johannesburg, South Africa (Type 1 outcome: Decade of Education for Sustainable Development; Civil Society outcome: the Ubuntu Declaration)	Society
2004	Declaration of Barcelona	Higher education
2005	Start of the UN Decade of Education for Sustainable Development (DESD)	Education
2005	Graz Declaration on Committing Universities to Sustainable Development, Austria	Higher Education
2009	Abuja Declaration on Sustainable Development in Africa: The role of higher education in SD, Nigeria	Higher Education
2009	Torino (Turin) Declaration on Education and Research for Sustainable and Responsible Development, Italy	Higher Education

Source: Adapted and updated from Calder and Clugston (2003) and Wright (2004).

Figure 4.1 History of initiatives taken in society, education, and higher education to foster sustainable development.

Conference on Sustainable Development (Rio+20)."[9] Over 272 organizations from 47 different countries committed to building a more sustainable society.[10] "The importance of reorienting existing education programs to incorporate sustainability-related principles, knowledge, skills, perspectives, and values has been emphasized further by the United Nations Decade of Education for Sustainable Development 2005–2014."[11] The goal of this decade-long declaration was to integrate sustainable development into every aspect of learning in hopes of transforming the view of a sustainable society. The United Nations signed on to mobilize the international community towards sustainability in higher educational institutions and pushed for UN-supported frameworks such as Principles for Responsible Management Education (PRME) and United Nations Global Compact and Principles of Responsible Investing (PRI).[12]

Sustainable development in this context meant "utilizing all aspects of public awareness, education, and training to create and enhance the understanding of the linkages among the diverse issues of sustainable development, of which the objective is to develop the knowledge, skills, perspectives, and values that will empower people of all ages to assume responsibility for creating and enjoying a sustainable future."[13] This idea gained acceptance from many leaders, universities, and colleges across the globe. The declarations fostered a wider consensus that HEIs can be a key player in the sustainable development process and the advancement of the ideas of a more sustainable society. Nevertheless, while these declarations provide commitment and road maps to sustainability, it does not guarantee that institutions will take initiative into implementing more sustainable education.[14] For commitment and ideas to proceed, universities and other HEI must play an active role in the solving of sustainable issues.

4.3 *HEIs as the driver of sustainability*

HEIs are places of critical knowledge production and perpetuation.[15] They hold positions that can encourage critical thinking and understanding to enhance critical methods of change. "The primary role of institutions of

[9] Simon, Haertle, Rio 20 Higher Education Sustainability Initiative (HESI)1 Commitments - A Review of Progress. Report (2014) 4.

[10] Ibid., 4.

[11] Frantzekaki, Mino, Scholz, Yarime, Establishing sustainability science in higher education institutions: Towards an integration of academic development, institutionalization, and stakeholder collaborations (2012) 105.

[12] Ibid., 4.

[13] Frantzekaki, Mino, Scholz, Yarime, Establishing sustainability science in higher education institutions, 106.

[14] Ibid., 106.

[15] Graham, Hernandez, Stephens, Scholz, Higher education as a change agent, 322.

higher education can be viewed in two ways: universities can be perceived as an institution that needs to be changed or universities can be perceived as a potential change agent."[16] There are various perceptions on the roles, values, and potential of universities and how universities can be change agents.

These perceptions vary across different cultures and context. There are a few primary perceptions of how HEIs can play a role in the transition towards sustainability.[17] The first perception is that HEI can model sustainable practices society can follow. The second is that HEI can teach students critical skills to deal with complex problems. Third, HEI can conduct research based on real-world sustainability problems. Lastly, it can help engage stakeholders within and outside the institutions.[18] However, "the potential for higher education to be a change agent accelerating a transition towards sustainability is dependent on a variety of factors including the current position, structure, and arrangement of higher education within its society as well as the location-specific sustainability challenges and opportunities facing a given community or region."[19] This will be further explored later in the chapter. Regardless of perspective, it can be concluded that HEIs are critical stakeholders in solving sustainability issues.

4.4 *Current sustainability integration in Bulgarian universities and higher education*

In 2007, Bulgaria joined the European Union (EU) and began transitioning towards a market economy. The changes that followed the economic and political shift negatively impacted HEIs in Bulgaria.[20] There was a lack of interest from young people, lack of research demand by industries, and lack of investments. While facing the political and economic transition, Bulgarian universities also had to restructure to meet requirements for the Bologna process and to join the European Research Area (ERA).[21] "Opening up of the national system and creating synergies with other educational systems in Europe provided big challenges for Bulgaria, however, also a one-off chance for integration in the European knowledge system."[22] Along with the adjustments, Bulgarian universities

[16] Ibid., 322.
[17] Ibid., 323.
[18] Ibid., 324.
[19] Ibid., 324.
[20] Gourova, Gourvov, Todorova, Skills for future engineers: Challenges for universities in Bulgaria (2009) 387.
[21] Ibid., 387.
[22] Ibid., 387.

incorporated the three-degree education system to include Bachelors (BS), Masters (MS), and Post-Graduate (Ph.D.) degrees into their schools.[23]

These changes were accompanied by new academic curriculums that supported the market shift.[24] "However, the lack of strong university–industry collaboration in developing curricula and new programmes could be considered as a serious drawback influencing the relevance of higher education to economic and societal needs."[25] The lack of state funding also caused a domino effect on demotivating professors from research, decreased teaching quality, and course options.[26] Nevertheless, rather than seeing these barriers as challenges, university presidents and administrators could recognize them and seek different opportunities to advance their university and develop programs and leaders who will be more equipped to create change that will meet societal needs.

4.5 U.S. best practices and examples of sustainability in HEIs

Since the initiation of sustainability on university and college campuses, many North American universities, especially ones in the United States have been making leaps towards more sustainable practices, research, operations, and curriculums. "Students and staff at hundreds of campuses are engaged in sustainability committees and actions, including the following: learning to focus on acquiring sustainability knowledge and application skills; sustainability-oriented film festivals, speakers, and other campus events; socially and environmentally responsible criteria for purchasing and endowments; infusion of sustainability into the general education core requirements, courses, disciplines, whole colleges, and specialized degrees…"[27]

The United States and its "business, architecture, and engineering schools are in the forefront of sustainability education."[28] Many universities in the United States have core requirements within their curriculum that include components related to sustainability, even when sustainability-specific majors are available. Other universities also restructured their teaching and created majors that aim to produce sustainable leaders for the future. A few best practices of these universities pertaining to sustainability will be explored in the following section. The best practices highlighted will be within various academic fields such as

[23] Ibid., 387.
[24] Ibid., 387.
[25] Ibid., 387.
[26] Ibid., 387.
[27] Rowe, Education for a sustainable future (2007) 323.
[28] Ibid., 323.

engineering, sciences, architecture, management, law, and more. Often, the sustainability classes are woven into the program rather than leaving them as options for electives.[29]

4.6 Best practices in building and architecture

4.6.1 Duke University, North Carolina

Duke University is a private research university who signed onto the American College and University President's Climate Commitment in 2007. Undergraduate students engaged in a research-based approach to sustainable living. "This program serves as a test bed for solutions that could be implemented in homes in the larger community."[30] It also provided students with the opportunity to research different Leadership in Energy and Environmental Design (LEED) certification systems such as the water catchment and purification systems and geothermal pumps.[31] The project served as a model for financial security and profitability. Students were continually working towards making homes more efficient, while reducing the costs associated with energy systems through increased efficiency levels.[32] Moreover, students learnt about different green building certifications; the program worked to help shape students to become educated citizens who will influence the future workforce and climate.[33]

4.6.2 Georgia Institute of Technology, Georgia

Georgia's Brook Byers Institute for Sustainable Systems (ISS) developed programs "where green living and sustainability become an integral part of the residential college structure and the academic departments, including informational sessions on "green" resource use, recycling competitions, energy conservation workshops, and orientation sessions for new students and staff members."[34] Georgia Tech sees sustainability as a system-based approach and implemented programs such as Sustainable Energy Systems, Sustainable Enterprise and Sustainable Urban Systems. These programs stressed "understanding the interdependence of the natural and the built environment, as well as environmental values

[29] Ibid., 323.
[30] Hattan, Feder, Naik, Murphy, Davis, Esiet, Vithlani, Rigaud, Advancing Education for Sustainability: Teaching the Concepts of Sustainable Building to All Students (2010) 38.
[31] Ibid., 39.
[32] Ibid., 39.
[33] Ibid., 39.
[34] Ibid., 40.

and ethics."[35] The programs taught students the interconnectedness of the environment, society, finances, politics, and building using renewable sources.

4.6.3 Unity College, Maine

Unity College's educational experience includes the Environmental Stewardship Curriculum, which is designed to incorporate traditional courses with various environmental fields. A prime example of this is their Environmental Citizen curriculum. Students, faculty, and members of the community would come together to build a barn either for the school or the community, allowing students to gain first-hand experience with green building.[36] Although students were not always able to complete a barn in 1 year, the project would continue until it was finished. The hands-on learning let students go beyond theory work and showed them the relationship between architecture and the natural environment.[37] "This educational experience has proved crucial in teaching students about holistic systems thinking and the interconnectedness of systems such as the building infrastructure and agriculture, as well as community development."[38]

4.7 Best practices in management

4.7.1 Georgetown University (GU), Washington D.C.

Georgetown University (GU) is "committed to engaging sustainability issues, creating real-world solutions and using the campus as a living laboratory to develop a long-term strategy."[39] Through many collaborations between corporations and the university, GU is addressing current sustainability challenges. An example of this collaboration is the Global Social Enterprise Initiative (GSEI) at GU's McDonough School of Business.[40] The initiative aims to prepare leaders to make responsible management decisions that are both economically and socially responsible. The initiative focuses on issues relating to "Global Health & Well-Being, Clean-Tech Energy & Environment, Responsible Investing, Economic Growth & Financial Security, and International Development."[41]

[35] Ibid., 40.
[36] Ibid., 37.
[37] Ibid., 37.
[38] Ibid., 37.
[39] International Sustainable Campus Network, Demonstrating Sustainable Development in Higher Education (2016) 11.
[40] Ibid., 11.
[41] Ibid., 11.

Some notable achievements from the initiative are the partnerships with Bank of America's Charitable Foundation. GU awards "fellowship and internship opportunities to GU students, hosts annual leadership series events, and brings the bank's experts in environmental sustainability, impact investing and corporate social responsibility to guest lecture in academic courses."[42] Some students worked with the World Business Council for Sustainable Development's marine plastics initiative and green investments. Additionally, GU collaborated with Phillips to conduct roundtable discussions about the challenges and action steps needed for people to change well.[43]

4.7.2 Yale University, Connecticut

Yale University's School of Management is a member of the Global Network for Advanced Management, a group of 28 business schools aimed at connecting business students around the world. Yale University developed a project to survey how business students felt about climate change and how they expect businesses to respond to the issues.[44] "Students, faculty, and staff collaborated to design and execute this research, generating conversations across Yale and Global Network schools about environmental sustainability, climate change, and MBAs' expectations from business schools and prospective employers in this space."[45] The report received press attention from across the world.[46] It also ignited dialogue with regard to the need for companies, business schools, and governments to lead the effort in environmental sustainability solutions. It was also a call to business schools to switch up their curriculum and resources to allow exploration into solutions for environmental challenges.[47]

4.8 Best practices in health and science

4.8.1 Harvard University, Massachusetts

Harvard University's School of Public Health created the Sustainability and Health Initiative for NetPositive Enterprise (SHINE) is dedicated to working with different companies in various industries to motivate positive change. Their goal is to focus on "raising population well-being through healthy workplaces and meaningful work and raising the health of the planet through the conservation and regeneration of natural

[42] Ibid., 11.
[43] Ibid., 11.
[44] Ibid., 9.
[45] Ibid., 9.
[46] Ibid., 9.
[47] Ibid., 9.

resources."[48] Using the talent of faculty, students, and companies, SHINE helped "develop the framework, metrics, methodology, tools, training, and infrastructure for companies seeking to measure and document the amount of good that they are already doing for human health and the environment and assess where else they can invest their efforts."[49]

SHINE participants worked with investors to establish performance indicators when evaluating health impacts from responsible investment portfolios and SHINE has also partnered with Dassault Systems to conduct research and more.[50] They seek to apply different methods to define and shape the standards for reporting on sustainability and health targets, and "apply rigorous metrics to demonstrate success and results of sustainability and health initiatives and articulate the value of sustainability and health initiatives to corporate leadership."[51]

4.8.2 Princeton University, New Jersey

Princeton University's professor Paul Chirik, along with his research group worked with Momentive Performance Materials to discover a process to replace rare metals found in silicone manufacturing with an abundant metal.[52] Many household items, medical devices, and tires are prepared by an industrial process using rare platinum catalysts.[53] Not only is this costly, "price volatility and toxicity, extraction of such rare elements from the Earth's crust has significant environmental consequences. Obtaining one ounce of a precious metal often requires mining approximately 10 tons of ore, 1 mile deep and as a consequence creates a CO_2 footprint that estimated to be 6,000 times that of iron." [54]

Professor Chirik and his research team "discovered a new class of catalysts based on earth's abundant transition metals such as iron and cobalt that have superior performance to existing platinum catalysts, generate less waste, and require fewer processing steps."[55] Their discovery will likely impact over five major production lines such as agriculture, health care, fiberglass, tires, and more. "This work demonstrates that earth-abundant metals can mimic or even surpass the function of precious elements. These new fundamental chemical concepts change the way chemists approach the periodic table and inspire new approaches to

[48] Ibid., 27.
[49] Ibid., 27.
[50] Ibid., 27.
[51] Ibid., 27.
[52] Ibid., 28.
[53] Ibid., 28.
[54] Ibid., 28.
[55] Ibid., 28.

replacing precious elements in alternative energy, pharmaceutical synthesis, commodity, and fine chemical production."[56]

4.9 Other notable best practices

4.9.1 Carnegie Mellon University, Pennsylvania

Carnegie Mellon University (CMU) held a student-led conference for CMU Sustainability Weekend. The initiative sparked conversations about sustainability issues today. The conference brought together different groups who usually would not converse and facilitated discussion about sustainability issues that are rarely mentioned. The conference had different panels, speakers who spoke about local issues and broad issues such as women's health, environmental justice, and entrepreneurial support and zero waste living.[57]

"The primary intent of the conference was to attract a variety of stakeholders and begin to consider innovation (in the sciences, humanities, art, and business) and the intersections with sustainability. Key points of the educational discussion included climate literacy conversations, evidence-based experience of local and national expert, and city-wide programming and impacts."[58] Students left the conversation with local projects that they could further explore. Although there were no assessment metrics emplaced the first year of the conference, they had ideas on how to improve the conference next year. They succeeded in initiating discussion across departments, with different stakeholders across the sciences, humanities, art, and business, and gave students the tools and agency to succeed.[59]

4.9.2 University of Pennsylvania

The University of Pennsylvania created the Integrating Sustainability Across the Curriculum (ISAC) program, which brought together faculty and undergraduate students to reevaluate current courses or create new courses to incorporate sustainability into the classroom.[60] This program was a part of the University of Pennsylvania's Climate Action Plan, meant to expand climate change and sustainability to the curriculum and educational experience for all students. "ISAC supports this goal by providing an avenue for Penn faculty to incorporate sustainability into new and

[56] Ibid., 28.
[57] International Sustainable Campus Network, Educating for Sustainability (2017) 62.
[58] Ibid., 62.
[59] Ibid., 63.
[60] Ibid., 54.

existing courses across the University."[61] The faculty workshops at the start of the program allowed faculty members to engage in concepts of sustainability across disciplines. The program called in experts to discuss how sustainability was incorporated into their work to spark ideas. The conclusion of the program required students to present their research, and University administrators and guests could come see the presentation.[62]

A few takeaways from this program that other institutions can learn from is providing an outlet for students and faculty to collaborate and work together to find solutions that benefit everyone. During the program, "faculty members receive valuable feedback from student research assistants on what types of readings and assignments are stimulating and interesting. Student research assistants not only become subject-matter experts while researching how to incorporate sustainability into a course but also come to understand the complexity of developing a semester-long course."[63] This method fostered conversations between students, faculty and helps everyone understand the complexity of the topic.[64]

4.10 Assessment and recommendations

The success of institutions of higher education as change agents in the face of climate change and other sustainability issues depend greatly on many factors, such as cultural, societal, and political. Professors from the Clark University, Massachusetts Institute of Technology, National Technical University, Stockholm Environmental Institute, and the Institute for Environmental Decisions describe five specific assessment questions that address critical issues pertaining to the opportunities and challenges of the implementing sustainability in HEIs.[65] "These five critical issues can be explored in the context of any institution or system of higher education throughout the world to assess the potential and limitations for higher education as a change agent. By considering these questions, empirical identification of location-specific characteristics can facilitate the design and implementation of new initiatives and new approaches to maximize the potential for higher education accelerating social change towards sustainability."[66]

The first question is *what are the dominant sustainability challenges?* Sustainability challenges can be location-specific, such as its socio-economic changes, technical, and environmental conditions.[67] "In many

[61] Ibid., 54.
[62] Ibid., 59.
[63] Ibid., 55.
[64] Ibid., 55.
[65] Graham, Hernandez, Stephens, Scholz, Higher education as a change agent, 324.
[66] Ibid., 324.
[67] Ibid., 324.

places throughout the world, low personal income and unequal distribution of wealth exacerbates environmental degradation and limits the capacity for transitioning to alternative pathways."[68] Additionally, local political climates can also affect the feasibility of implementing sustainability. Each region or country holds different cultural attitudes and beliefs that correlate with sustainability challenges and the role of HEIs; therefore, the role of cultural interpretations of sustainability must be recognized. The variation in technological advances is also an important factor to consider. HEI can be a huge benefactor and influence to the advancement of technology based on the country or region's current condition. These factors are relative to Bulgaria; therefore, the assessment of challenges is essential to implementing changes.

The next question is *how is the education financed?* The way the organization is financed has deep influences on the potential of sustainability implementation by a university. There is a global shift away from government and public funding for HEIs to more private.[69] This shift has enabled different stakeholders and interests to emerge onto the education scene and challenging previous notions of what education meant. The reliance on more private funds has opportunities and challenges. It could mean more reliance on private interest, but it can also mean more direct influence on the private industries as private investors want to be more involved in the institution's activities.[70] The shift from public to private could also mean more funding from more sources. Additionally, "with corruption and inefficiencies in some publicly funded educational systems, external support could have a strong benefit, increasing support of critical sustainability programs and allowing higher education to engage in new and different ways as a change agent in society. In some instances, where corruption is widespread in the public system, universities may have more freedom and support to maintain their position as an honest broker of information analysis and dissemination…"[71]. External funding can be an asset in generating more public awareness to sustainability challenges if partnerships with private businesses are there. It can also be beneficial if the government is not leading in sustainability initiatives.

The third question is *what are the organization and structure of the higher education system?* Often, HEIs are traditional, and knowledge exchange tends to happen strictly within the classroom. This narrows the promotion of free thinking and "rewards a narrow disciplinary focus and incentivizes the dissemination of research results primarily through publication in academic journals. The current academic system in most places does not

[68] Ibid., 324.
[69] Ibid., 325.
[70] Ibid., 325.
[71] Ibid., 327.

reward public engagement nor does it create time for academic researchers to reach out to non-academic stakeholders."[72] For Bulgaria, there are very limited interactions between external stakeholders and even foreign experts.[73] With more interactions between non-academic stakeholder and academics, there is potential for diversification of knowledge and skills that can contribute positively to sustainability as seen in some of the best practices mentioned.

Next is *how strong are democratic processes?* Within this question, two components must be considered, "accessibility and rights to obtain higher education; and transparency and neutrality of higher education."[74] This is especially important in places with low college accessibility as education may seem elitist. "With this view, the potential for higher education to have a positive and effective external influence or impact outside the campus borders may be reduced. For promoting a transition towards societal sustainability, broadening the scope and influence of accessibility of higher education could be a critical goal to broaden opportunities for higher education as an agent for social change."[75] Whether education is a right must be determined.

Bulgaria has been adopting educational EU policies like the Strategy for the Development of Higher Education which aims to:

"(1) improving access to higher education and increasing the number of university graduates, (2) increasing the quality of higher education, (3) setting up a sustainable and efficient link between HEIs and the labour market, (4) promoting research, (5) updating the governing system and clearly defining HEIs, (6) increasing the funding of higher education and science, and (7) overcoming the negative trends in the career development of lecturers at HEIs and creating incentives for them."[76]

The adoption of this EU policy can also pose opportunities for universities and colleges in Bulgaria to promote sustainability principles, strategies, and best practices into their curriculums, activities, and research.

Lastly, *what are the major channels for communication and interaction with society?* "This includes mechanisms for communication and dissemination of information both internally, within the higher education system, and externally with non-academic entities in society."[77] Effective communication methods help foster collaborations. If communication is enhanced between institutions, experts from different fields, and members of the community, there could be more opportunities for collaboration on sustainability challenges. "There are wide discrepancies among the type and

72 Ibid., 327.

73 Danchev, Basset, Salmi, Final Report (Washington D.C. 2012) 3.

74 Ibid., 328.

75 Ibid., 328.

76 European Union, Education and Training Monitor 2016 Bulgaria (2016) 9.

77 Ibid., 330.

extent of access to decision makers, policymakers, industry, and community groups that representatives of higher education have in different places. For example, in Sweden, Germany, and many other European countries, there are specific mechanisms such as institutionalized review and consulting processes that require input from academics in the policymaking process."[78] Key stakeholders can share valuable knowledge and collaborate with each other to create a multifaceted approach.

For higher educational institutions to contribute to the societal shift towards sustainability, challenges and opportunities with the country must be assessed, and these five questions are essential in that assessment. "Asking these five questions in specific contexts is likely to highlight issues that span the three levels of transition management: the strategic, the tactical, and the operational. The specific dominant sustainability challenges, the financing structure, the institutional organization, the extent of democratic processes, and the communication and information dissemination situation each have potential relevance to high-level strategic level concerns and decisions, to mid-level tactical decisions, and to more detailed operational-level planning."[79] If after the assessment, it is feasible for a university to advance towards more sustainability teaching, here is one method that an institution implemented to make their Management department more sustainability-oriented.

4.11 *Example of implementation*

Below is one university's approach to integrating sustainability into the institute's curriculum, research, and activities.

4.11.1 *Clark University Graduate School of Management*

Clark University's Graduate School of Management (GSOM) became a signatory of the United Nations PRME program in 2011, upholding the principles of purpose, values, method, research, partnership, and dialogue.[80] To implement sustainability and ethics into the overall curriculum, GSOM followed the following process:

GSOM established a PRME Committee chaired by a senior faculty member with members including faculty, staff, and students. The initial task of the committee was to assess current activities in support of PRME principles related to curriculum, research, and activities. That meant identifying the policies that exist to integrate PRME principles into the

78 Ibid., 331.

79 Ibid., 333.

80 Clark University, Update on the United Nations Principles for Responsible Management Education. Report. Graduate School of Management (2015).

graduate degree programs. "This began with a mapping of the existing curriculum topics, to create a baseline, and to generate discussion on where the principles could be further integrated."[81] The scope of the work included the Master of Business Administration, Master of Science in Accounting, and Master of Science in Finance. "A student member of our PRME Committee met with faculty to understand what additional support would enable each to increase coverage of PRME-related content. It was determined that the best step would be to create a library of resources and relevant content (in the form of articles, cases, papers, and videos) for each program. [The] relevant material was uploaded to a faculty file share drive and organized by topic."[82]

Then, GSOM administrators created a policy and an integration plan on how they will incorporate PRME and their policy commitment into the curriculum. This meant either inviting in a guest speaker to speak on sustainability and social issues or incorporating the topics into lectures and seminars for 1.5h per semester. A notable GSOM course is the Sustainability Consulting Projects Course, which enabled student teams to work with local businesses, non-profits, and more to developed customized sustainable action plans. "The plans recommend and describe initiatives to reduce operating costs and environmental footprint by increasing energy efficiency, reducing water consumption and waste, as well as fostering sustainable behaviour."[83]

GSOM is committed to continuing their efforts. The school produces annual reports to the Dean of GSOM on the progress of the commitment and results. Furthermore, as a PRME signatory, Clark is required to and submits a progress report to the United Nations every 2years. In 2014, to more fully address opportunities to integrate PRME principles into student activities, the PRME Committee partnered with Clark's graduate chapter of Net Impact. "Net Impact is a global community of students and professionals who want to become the most effective change agents they can be. Over 100,000 strong, our emerging leaders take on social challenges, protect the environment, invent new products, and orient business towards the greater good. In short, we help our members turn their passions into a lifetime of world-changing action."

Moreover, students, staff, and faculty members are demonstrating their commitment to PRME and social issues in various ways outside the lecture halls. The Clark Student Sustainability Fund was created to support student-designed or student-led projects that help to improve campus sustainability. Since its initiation in 2012, the fund has funded over $20,000 worth of student projects. Additionally, the Clark Community

[81] Ibid., 6.
[82] Ibid., 6.
[83] Ibid., 11.

Thrift Store is a student-run business that helps to reduce our carbon footprint by passing down their gently used items to each other.[84] GSOM also gives recognition to faculty members and alumnus who demonstrate PRME principles of ethics, social responsibility, and sustainability in their teaching, work, or research. Awards are given annually to GSOM alumni; i.e., the PRME Alumni Award, and to faculty through the PRME Faculty Award. In March 2017, the GSOM faculty approved a revision to its mission statement demonstrating its fundamental commitment to the PRME principles: "Our mission is to engage in consequential research and practice and to prepare students for career and life success through a combination of rigorous academic study and theoretically sound experiential learning with a focus on ethics, social responsibility, and sustainability."

4.12 Conclusion

HEIs are labs for generating knowledge and opportunities for critical engagement as society is rapidly being confronted with climate change, global inequality, social injustices, and loss of biodiversity. For institutions of higher education to lead as change agents in the face of these issues, the challenges and opportunities for the institution must be evaluated in its respective context. The discussion of issues and various best practices from United States' universities are meant to demonstrate that sustainability and social change can occur in many contexts, and there is no one size which fits all. Additionally, with the assessment questions and one example of implementation, this chapter hopes to provide considerations for institutions of higher education in Bulgaria to find methods and practices that are best suited for their culture, roles, and interests. HEIs are capable of spearheading change as they produce citizens and leaders who can persist in the face of existential environmental and social challenges.

References

Best Practice in Campus Sustainability. Report. International Sustainable Campus Network. 2014. 2–57.

Calder, W., and Clugston, R. M. International efforts to promote higher education for sustainable development. *Planning for Higher Education*, 31, (2003), 30–44.

Calder, Wynn, Rick Clugston, and Peter Blaze Corcoran, eds. Introduction. *International Journal of Sustainability in Higher Education* 3, no. 3 (2002), 41–53.

Cortese, Anthony D. The critical role of higher education in creating a sustainable future. *Planning for Higher Education* 31, no. 3 (2003), 15–22.

[84] Ibid., 10.

Danchev, Plamen, Roberta Malee Bassett, Jamil Salmi. 2012. Final Report. Washington, DC: World Bank. http://documents.worldbank.org/curated/en/694441468006586259/Final-report.

Demonstrating Sustainable Development in Higher Education. Report. International Sustainable Campus Network. Davos-Kloster, 2016. 4–51.

Educating for Sustainability. Report. International Sustainable Campus Network. 2017. 4–74.

Education and Training Monitor 2016 Bulgaria. Report. European Commissions, European Union. 2016. 1–12.

Frantzekaki, Niki, Takashi Mino, Roland Scholz, and Masaru Yarime. Establishing sustainability science in higher education institutions: Towards an integration of academic development, institutionalization, and stakeholder collaborations. *Sustainability Science* 7, no. 1 (2012), 101–15.

Hattan, Amy Seif, Julia Feder, Ashka Naik, Kelly Murphy, Nora Davis, Ukeme Esiet, Krupa Vithlani, and Gabrielle Rigaud. Advancing Education for Sustainability: Teaching the Concepts of Sustainable Building to All Students. 2010.

Lozano, Rodrigo, Kim Ceulemans, Mar Alonso-Almeida, Donald Huisingh, Francisco J. Lozano, Tom Waas, Wim Lambrechts, Rebeka Lukman, and Jean Hugé. A review of commitment and implementation of sustainable development in higher education: Results from a worldwide survey. *Journal of Cleaner Production* 108 (2015), 1–18. doi: 10.1016/j.jclepro.2014.09.048.

Lozano, Rodrigo, Rebeka Lukman, Francisco J. Lozano, Donald Huisingh, and Wim Lambrechts. Declarations for sustainability in higher education: Becoming better leaders, through addressing the university system. *Journal of Cleaner Production* 48 (2011), 1–11.

Rowe, Debra. Education for a sustainable future. *Policy Forum* 317 (2007), 323–24.

Simon, Kathleen and Jonas Haertle. Rio 20 Higher Education Sustainability Initiative (HESI)1 Commitments - A Review of Progress. Report. Higher Education Sustainability Initiative (HESI), United Nations. 1–14.

Stephens, Jennie C., Maria E. Hernandez, Mikael Román, Amanda C. Graham, and Roland W. Scholz. Higher education as a change agent for sustainability in different cultures and contexts. *International Journal of Sustainability in Higher Education* 9, no. 3 (2008), 317–38. doi: 10.1108/14676370810885916.

Tilbury, Daniella. *Higher Education for Sustainability: A global Overview of Commitment and Progress*. Barcelona: Global University Network for Innovation, (2011), 1–21.

Update on the United Nations Principles for Responsible Management Education. Report. Graduate School of Management, Clark University. Worcester, MA, 2015. 1–23.

chapter five

The role of universities in sustainable development (SD)

The Spanish framework

Núria Bautista Puig, Elba Mauleón, and Elías Sanz Casado
University Carlos III of Madrid

Contents

5.1 Introduction

5.1.1 Sustainable development (SD) in Higher Education Institutions (HEIs)

Society in the 20th century faced various environmental, social, and economic issues that have become collective problems. This has led to a global concern and awareness about the type of development which was being carried out in each country. This concern started in the 1970s with different Earth Summits/Conferences (United Nations Conference on the Human Environment in Stockholm, 1972). The production model was questioned, and new approaches to sustainability were discussed. After ascertaining the untenable traditional production model, the concept of sustainability, and more precisely, 'sustainable development' (SD), became a central issue and led to a global concern where the progress towards it seemed highly imperative (Waas et al., 2010).

According to the World Commission on Environment and Development in the Brundtland report (Brundtland, 1987), SD can be defined as a production model which aims at better economic results for both humans and the natural environment not only in the present but in the indefinite future as well. This is seen as a way of trying to solve environmental, social, and economic problems and guarantee future generations' needs. In 2000, the conference Rio+20 (Brazil, 2012) was the first formal political endorsement of SD by the global leaders of the world; it led to the creation of a document entitled 'The future we want'. These agreements focused on the adoption of a 10-year plan of new standards of production and sustainable consumption, as well as the launching of a negotiation process to establish the Sustainable Development Goals (SDGs). The result was the formulation of the 17 SDG indicators, based on the achievement of the eight Millennium Development Goals (MDGs), which in turn converge with the development agenda of the United Nations (2015–2030) on the path to sustained economic growth, social development, and environmental protection (United Nations, 2016).

One of the main problems of sustainability arises with the ambivalence of the term. Until the final 1970s, this word was not so frequent, and it had strong connections with the forestry sector, which was considered the origin of the term (Filho, 2000). Its definition depends on the stakeholders involved, and there are also different backgrounds and interpretations of it. Robinson (2004) criticized the vagueness of the concept; Martens (2006) stated there were 'multiple interpretations' of SD, due to the fact it could be interpreted and applied from a variety of perspectives. The first concepts of SD were more environment-related (Sibbel, 2009), but later, economic and social dimensions were incorporated; it is known as the 'triple bottom line' approach (Elkington, 1998) or the 'three-pillar model' (Kastenhofer and Rammel, 2005). Other models incorporate additional aspects to the model, such as institutional (Filho and Pace, 2015), cultural (Axelsson et al., 2013; Filho and Pace, 2015), or spatial (Alshuwaikhat and Abubakar, 2008).

At a European level, SD is of the objectives of the European Union (EU) and has been included in EU policies and regulations. There are significant events related to SD: the creation of the Environment Committee at the European Parliament in 1973 and the launching of the first action program on environment (1973–1976). At the beginning of the 1990s, with the signing of the Maastricht Treaty, community actions linked to the environment gained prominence. They included environmental protection into all EU policies and activities. Other important milestones in the EU entered into effect in 1994 with the creation of the European Environment Agency; in 1997, with the Amsterdam Treaty on Balanced and Sustainable Development; in 2000, with the Lisbon Strategy, and in 2001, with the EU strategy for sustainable development (EU-SDS). Later, in 2010, the Europe 2020 Strategy for smart, sustainable, and inclusive growth was launched with the aim to put EU in a position to lead global SD (European Commission, 2010). Regarding the research activities, the 7th Framework Programme (FP7) (2007–2013) included two thematic fields within the 'Cooperation' programme: one on energy and other on environment (including climate change). Further on, in the H2020 programme (FP8) (2014–2020), in one of the three main pillars, 'Societal Challenge', there are specific calls related to climate, environment, energy, or transport. As a result, it is expected that, at least 60% of the budget for this programme should be related to SD and that climate-related expenditure should exceed 35% of the budget, with measures improving resource efficiency (European Commission, 2018).

Achieving SD can be a challenge, and all societal stakeholders need to be involved (Brown, 2006). On this regard, Higher Education Institutions (HEIs) should play an active and fundamental role in promoting sustainability practices. In the past, universities played a role in transforming societies and serving the greater public good, so there is a societal need for universities to assume responsibility contributing to SD (Waas et al., 2010), especially as agents responsible for knowledge creation and dissemination (Madeira et al., 2011). Also, universities make an important contribution to the development of our society, and they have societal responsibility, not only in training young and future leaders but also stimulating public awareness on sustainability. In this regard, they should be leaders in the search for solutions and alternatives to current environmental problems and agents of change (Hesselbarth and Schaltegger, 2014). Apart from their traditional functions of research and teaching, including this third mission for universities, the transfer knowledge to societies fits with this commitment.

These institutions have made a great effort in integrating sustainability in their actions. This progress is undoubtedly linked to the different conferences held and the declarations and agreements that emerged from them and are exposed below. Universities have signed declarations to show their commitments to sustainability and their number have increased over time. Ever since the Stockholm Conference in 1972, where it was stated that education played an important role in environmental protection and conservation, many declarations, charters, and partnerships have been developed in HEIs. The Stockholm Declaration (1972), the Talloires Declaration (1990), the Copernicus University Charter (1993), the Lüneburg Declaration on Higher Education for Sustainable Development (2001), and the Torino Declaration on Education and Research for Sustainable and Responsible Development (2009) are some examples of it (Lozano et al., 2013). Also, initiatives such as the Higher Education Sustainability Initiative (HESI) or the International Sustainable Campus Network (ISCN) are another proof of it. However, it has been highly debated that this signing commitment does not ensure the implementation of SD into their systems (Grindsted, 2011).

In the last years, many universities have engaged in SD activities, and it is considered to be a big challenge. This challenge is complex and multifaceted, and an involvement of all parties in the university (professors, students, researchers, Research Results Transfer Offices (OTRI), management teams, clerks, and services staff, Green Offices/Environment Offices or services) is required to contribute in substantial ways by working in a third space (Chambers and Walker, 2016). In addition, the concept of SD is being the subject of a debate. It is being questioned if it is a philosophical or an economic concept and how it can be translated into a policy prescription (Meadowcroft, 2007). Several definitions of what sustainability in HEIs is and how to apply it have arisen. According to Chambers and Walker (2016), sustainability must be incorporated into the dimensions of the university: research, teaching and community engagement, and campus operations. Cortese (2003) writes, it is a system that includes education, research, campus operation, and community outreach. Cole (2003) explains a sustainable campus 'acts upon its local and global responsibilities to protect and enhance the health and well-being of humans and ecosystems. It actively engages the knowledge of the university community to address the ecological and social challenges that we face now and in the future'. As stated by Alshuwaikhat and Abubakar (2008), a sustainable university campus should be a 'healthy campus environment' which combines a prosperous economy (energy and resource conservation, waste reduction, etc.) and one that promotes equity and social justice and exports those values to the community. Wals (2014) considers that HEIs are making a systemic change towards sustainability by reorienting their education, research, operations, and community outreach activities, and some of them have converted this sustainability paradigm into a new way of organizing and profiling themselves. According to Ryan et al. (2010), sustainability at HEIs reaches beyond individual curriculum changes and isolated environmental practices/policies; it also requires actions in academic priorities, organizational structures, and financial systems. Regarding the application, none has prescriptions about what needs to be done to contribute to SD; furthermore, no single criteria are provided. While in some universities being considered a 'Sustainable University' is based on having an environmental plan, environmental guidelines, or a statement, others consider declarations, institutional policies, or the implementation of the ISO 140001 standard, among others (Alshuwaikhat and Abubakar, 2008).

A large number of tools are used to assess and rank sustainability actions for HEIs; these allow them to compare their own actions towards sustainability and compare against each other. Some are adaptations of tools like Ecological Footprint or ISOs, while others evaluate campuses (CSAF) or curricula (STAUNCH, CSAF). GreenMetricWorld University Ranking is a global sustainability ranking for universities developed by Universitas Indonesia (UI) since 2010. It assesses the following six categories: Setting and Infrastructures, Energy and Climate Change, Waste, Water, Transportation, and Education. In comparison with the first edition, more indicators were added and verification methods were included to check data validity, among others (Suwartha and Sari, 2013). Some studies have criticized this ranking by arguing its simplicity in terms of categories and indicators in comparison with other systems and that the demands of the data types required are generally low for participants and less empirical than those used in other systems (Lauder et al., 2015).

However, the involvement in SD is voluntary for universities, and many studies have described their resistance. These two issues, the fact that it is voluntary and the rigid structure of the universities, are limiting the expansion of these projects and a greater commitment to SD at the HEIs. Velazquez et al. (2005) underlined the importance of certain problems, such as the lack of awareness, interest, and involvement; the organizational structure; the lack of funding or support from university administrators; the lack of time or training; the lack of data access; the lack of more strict regulations; and the lack of policies to promote sustainability on campus. Another factor, the university conservativism (Lozano et al., 2013) or resistance to change has been highlighted. Fien (2002) highlighted that HEI strategies for advancing sustainability need to be developed by individual systems and institutions because no two institutions are the same. Another argument is the focus of environmental programmes, which has been centred in two issues: reducing energy consumption, waste, and integration into mainstream university operations and 'greening the curriculum' (Roy et al., 2008; Larrán et al., 2015).

Precisely, this has been the reason why universities cannot fully implement SD because this is more than a theory, it is a call for action, a work in progress.

5.1.2 HEIs in Spain and their involvement in sustainability

There is an increasing interest in sustainability in Spain: many institutional statements have emphasized the need to implement SD at HEIs (Larrán et al., 2014). The 2015 University Strategy (Technical Commission of the University Strategy 2015, 2011) gave a great importance to the social responsibility of the university system and highlighted the relationship with the environment. The Spanish Government also fostered Organic Law 4/2007 on Universities, which aimed to incorporate sustainability in areas such as management and accountability, and Law 2/2011 on Sustainable Economy (Larrán et al., 2015). At the research level, the VI National Scientific Research, Development and Technological Innovation Plan (2008–2011), incorporated a strategic action about Energy and Climate Change. In the State Plan (2013–2016), there was a programme called Societal Challenges that included issues such as sustainable transport, action for climate change or energy, secure and efficient energy and clean energy, among others: these sections are also present in the current Innovation Plan (2017–2020).

In Spain, interest in sustainability issues is a recent topic (Larrán et al., 2016). The group of Evaluation of University Sustainability at the Sectoral Commission for Environmental Quality, Sustainable Development and Risk Prevention (CADEP, as per its Spanish acronym) (Conference of Spanish University Rectors, CRUE) was created in 2004 with the aim to increase the incorporation of environment and sustainability concerns in HEIs (Alba, 2007). They established a set of indicators to value the progress of Spanish universities in their path to sustainability. These indicators are grouped into three areas: (1) management, (2) teaching and research, and (3) environmental management. In 2009, they created a sectoral group on sustainability. This was a response by several universities that aimed to 'collect the experience of universities in environmental management, the advances in the environmentalization of the university community and work on risk prevention while promoting cooperation in these areas for the exchange of experiences and the promotion of good practices' (CRUE, 2018). It was made up by nine groups (e.g. Evaluation of University Sustainability, Environmental Improvements in University Buildings, Curriculum Sustainability). This also has led to an autodiagnosis tool, with the participation of 33 universities. In the last report, it was determined that HEIs have improved in these areas and a great effort has been made in environmental aspects; however, curricular sustainability has not been implemented (CRUE, 2018).

Some studies have analysed the sustainability in Spanish HEIs. Larrán et al. (2016) analysed the strategic plans of the universities, and their findings suggest there is a low presence of sustainability strategies at Spanish universities. Alba (2007) stated that all universities have some activity related to sustainability. Some other studies have focused on the learning context or the teachers' competences (Filho, 2000; Albareda-Tiana et al., 2018); others focused on environmental habits (Chuvieco et al., 2018).

Universities are not only a key sector for the analysis of sustainability but also a crucial agent in the generation of knowledge. Between 2012 and 2016, the Spanish university sector (public and private) was responsible for 61% of the Spanish scientific production in the Web of Science (WoS) database. It is the first sector in terms of scientific output, followed by the health sector (28%), mainly hospitals, and the Superior Council of Scientific Research (CSIC), responsible for 16% of the Spanish scientific production in the WoS in that period (Bordons et al., 2017). The Spanish University System (SUE, as per its Spanish acronym) is made up in the term 2017–2018 by 83 universities, 50 of them are public and 33 private (IUNE Observatory, 2018). In this study, all universities have been considered despite the private ones having a greater dedication to teaching, to the detriment of research (Manzano et al., 2016). Teaching and research staff in SUE in 2016 amounted to 64,296.

Despite the growing interest in this subject, we find that there is very little analysis of sustainability at HEIs in Spain from a bibliometric point of view. Some of these studies focused on virtual laboratories (Salmerón-Manzano and Manzano Agugliario, 2018); others analysed the theses and dissertations on sustainability education (Leetch and Hauk, 2017).

5.1.3 Objectives

As a result of the growth and importance of this discipline in the scientific field, WoS created a category called 'Green & Sustainable Science & Technology', leading to the emergence of bibliometric studies on this discipline (Pandiella-Dominique et al., 2018). However, these studies analyse Spanish production as a whole, without differentiating the sectors. The work that we present here analyses the commitment of the university to achieve SD. A thematic analysis through the keywords is carried out, as well as an analysis of the university profiles by specialization through F-measure (Rousseau, 2018). Another added value of this work is the use of different sources of information apart from WoS, such as CORDIS and GreenMetric ranking data.

Bearing in mind the importance of HEIs in the path of sustainability, the main objective of this paper is to evaluate the integration of sustainability efforts into the SUE. According to this objective, the following research questions (RQs) are going to be addressed:

- Is there sustainability research at HEIs in the SUE?
- What are the topics addressed in scientific publications?
- Is there a commitment to this topic at the SUE? Are HEIs participating in European projects?
- Are these universities recognized in the GreenMetric ranking?

The article is structured as follows: After the Introduction, the Sources and Methodology are described, followed by the Results and Discussion sections, where the responses to the RQs and the most prominent findings are discussed. The article ends with Conclusions remarks.

5.2 Sources and methodology

On this framework, this study provides a comprehensive overview about sustainability at HEIs in the SUE. This analysis has been done from different perspectives: (1) Scientific production, (2) Commitment, (3) Internationalization, and (4) Visibility.

The research study was based on the WoS database from 1994 to 2017. The SUE, which is composed of 50 public and 33 private universities, is the case study in this analysis. The bias of this source is a well-known fact (Gómez Caridad and Bordons, 2009), it was used because it stores high-quality documents. The publications were retrieved from the Core Collection: Science Citation Index (SCI), Social Sciences Citation Index (SSCI), and Arts and Humanities Citation Index (A&HCI).

The procedure was the following:

- *Establishment of bibliometric indicators*: The study focused on the following indicators:
    - *Search strategy*. WoS category 'Green & Sustainable Science & Technology' was used, and documents with the involvement of HEIs were selected.
    - *Scientific output retrieval and information processing*: the information was retrieved and exported into a relational database formulated with MySQL.
        a. Research activity.
            - Yearly variation in scientific output.
            - Cumulative average growth rate (CAGR).
            - Output by institution (absolute values and F-measure). The F-measure is defined as follows:

$$\text{F-measure} = \frac{2 * O_{CD}}{O_D + O_C}$$

where O_{CD} is the number of publications in a domain (in this case, in the WoS category) in a country in a specific publication window, O_D is the number of publications in the world in the domain in the same period, and O_C is the number of publications in all domains in the country.
- The citation normalization has been calculated by dividing an HEI citation by the average worldwide citations of the documents in the same area and year.
- Collaboration patterns: type (national/international/without collaboration), countries, and institutions.
- Topic specialization: Identification of clusters based on co-occurrence of article and author keywords, using VosViewer tool (Van Eck and Waltman, 2009).

b. Commitment: This information was collected from the websites of each university. The following indicators were analysed:
   - Inclusion in Strategic Plans/Sustainability Plans.
   - Establishment of a Green Campus.

c. Internationalization: This information was collected using Cordis database.
   - Participation in European projects (FP7, H2020).

d. Visibility: This information was retrieved from the GreenMetric website.
   - Position in GreenMetric ranking.

5.3 *Results*

5.3.1 *Research activity patterns*

This section provides a bibliometric analysis for the publications on sustainability. The SUE published 3,140 documents from 1994 to 2017. The annual distribution is shown in Figure 5.1. Regarding the evolution, the number of documents on this topic shows the growing interest in these issues with a CAGR of 24.09% in the scientific community. This growth is higher than the WoS category (21.14%). The growth implies that this topic is attracting the attention of the academic community. Production on this subject is very recent: 81.18% is concentrated in the last 5 years of the study. Moreover, this is higher than the WoS category during the same period (72.93%).

If we take into consideration the production by HEI in the SUE, Figure 5.2 shows the evolution of the top ten universities by number of documents. The university with a higher number of documents is Polytechnic University of Catalonia (UPC), with 246 documents in the period, followed by Polytechnic University of Madrid (UPM), with

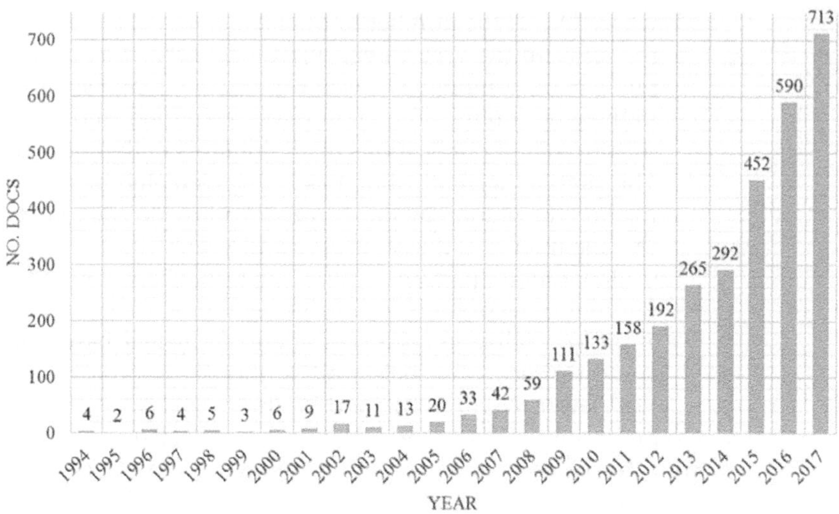

Figure 5.1 Distribution of publications on sustainability from 1994 to 2017.

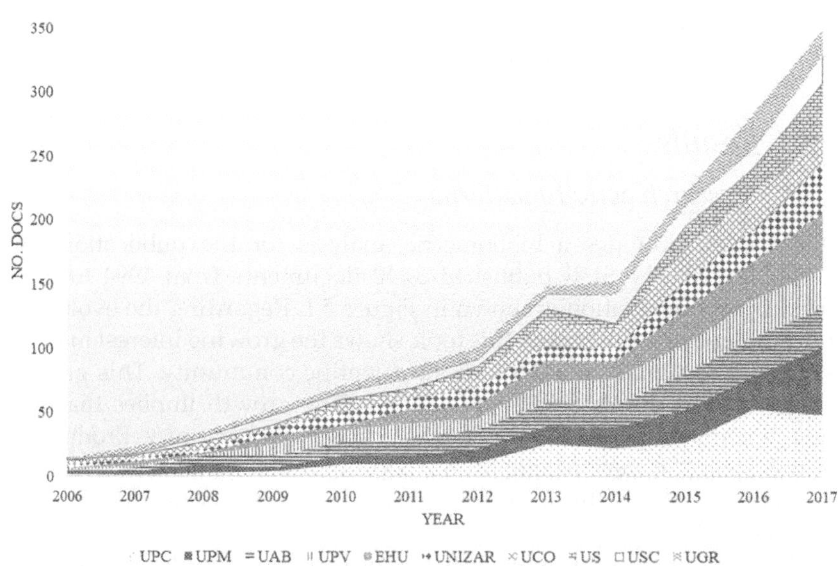

Figure 5.2 Evolution of scientific production by the top ten most productive universities (2006–2017).

227 documents, the Autonomous University of Barcelona (UAB), with 209 documents, and the Polytechnic University of Valencia (UPV), with 207 documents. This proves the leadership of technical universities in the field, representing 21.7% of the scientific production of all the universities. As far as the analysis of the CAGR is concerned, the United States has a higher value (51.08%), followed by the University of Lleida (UdL), University of Vigo (UVIGO), and University of Almería (UAL), with 42.35%, and UPV, with 35.81%. Considering the number of citations of the different topics, MUNI is the university with the highest impact (42.71 citations/paper), followed by the UdL, with 30.7%. Technical universities have a lower impact (13.45 citation/paper for UPC or 12.02% for UPM) (Figure 5.2).

Figure 5.3 shows the F-measure (horizontal axis) and citation normalization (vertical axis) of the top 20 universities with a higher number of documents. The size of the bubbles reflects the scientific output. The profiles of universities are threefold. Group 1: Most specialized universities, according to F-measure, and with a lower impact than average; this group encompasses technical universities (UPC, UPM, UPV) and bigger ones (Autonomous University of Madrid (UAM), University of the Basque Country (EHU)). Notwithstanding, University of Zaragoza (UNIZAR) has a higher impact. Group 2: Universities with higher impact

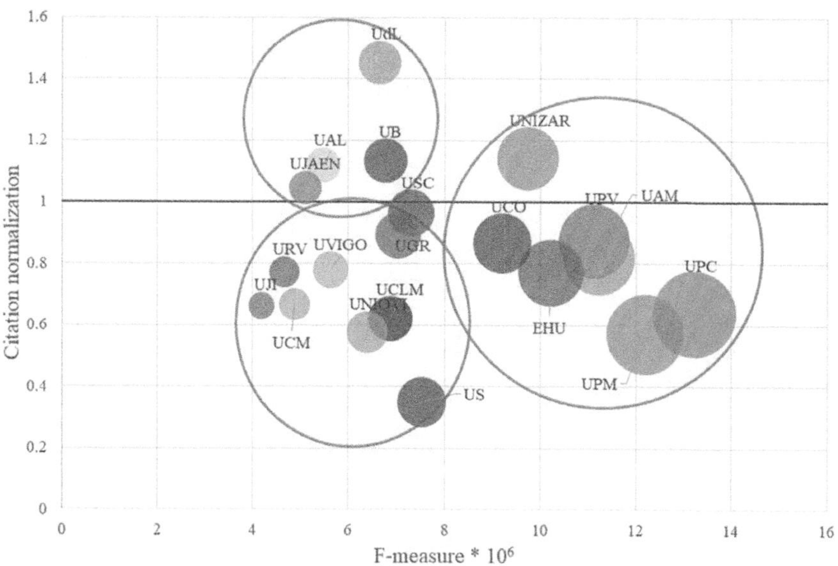

Figure 5.3 F-measure index and citations by universities with a higher scientific production from 1994 to 2017.

but less specialization (University of Barcelona (UB), UdL, UAL, and University of Jaén (UJAÉN)). Group 3: Universities with less production, less specialization, and less impact (Rovira i Virgili University (URV), UVIGO, University of Castilla-La Mancha (UCLM), University of Oviedo (UNIOVI)).

The number of documents with international collaboration adds up to 1,376 (43.8%): 1,181 documents with national collaboration (37.6%) and 1,160 with no collaboration (36.94%). There is a tendency to increase international collaboration over time with a CAGR of 41.30% (Figure 5.3); in 2017, it represents 41.5%. In addition, national collaboration remains stable since 2000. The CAGR is 26.62%, and in 2017 it represents 33%. Documents with no collaboration present a lower CAGR (20.03%) and have decreased over the period from 57% in 2000 to 25.9% in 2017 (Figure 5.4).

Collaboration among countries is shown in Figure 5.5. The highest collaboration is with England (144 documents, 4.59%), United States (118 documents, 3.76%), Italy (116 documents, 3.69%), Portugal (111 documents, 3.54%), France (109 documents, 3.47%), Germany (101 documents, 3.22%), Netherlands (67, 2.13%), Chile (62 documents, 1.97%), Mexico (58 documents, 1.85%), and Brazil (53 documents, 1.69%). Four clusters of collaboration have been identified: (1) Southern European, Central European (England, Portugal, Italy, France, Germany), and South-American countries (Brazil, Chile, Venezuela, Ecuador, etc.); (2) Northern European countries, Northern America (Denmark, Norway, Sweden, Finland, Ireland, and Canada), (3) Northern Africa and Western Asia (Tunisia, Cyprus, Algeria, and United Arab Emirates), and (4) other

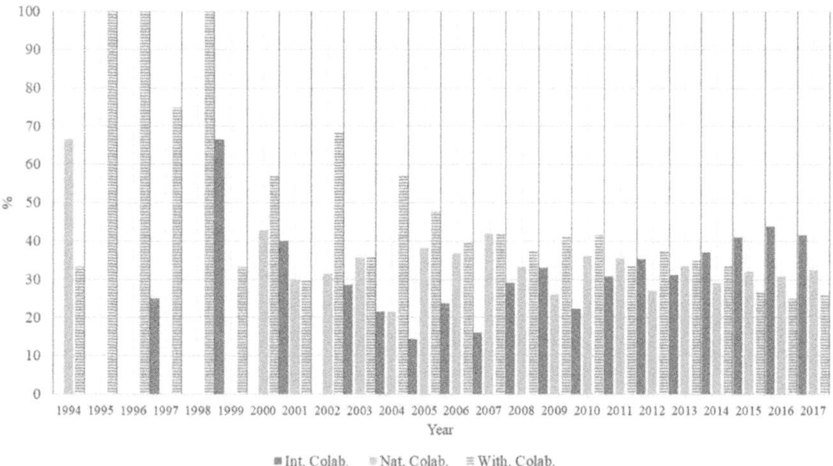

Figure 5.4 Evolution of national and international collaborations and documents without collaborations from 1994 to 2017.

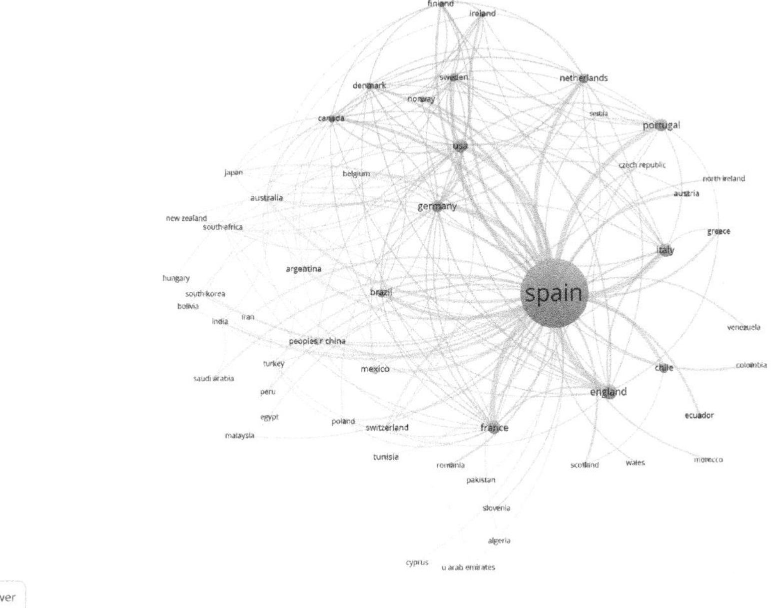

Figure 5.5 Country co-authorship network of sustainability publications from 1994 to 2017.

regions (Argentina, Mexico, China, Australia, and Japan). Considering the institutions, apart from the HEIs involved, other institutions having a strong collaboration with, at least, five documents are research institutes (CSIC, with 78 documents; Research Centre for Energy, Environment and Technology (CIEMAT), with 25 documents; or Catalan Institution for Research and Advanced Studies (ICREA), with 19 documents) and Portuguese universities (Porto, with 26 documents; Lisbon, with 22 documents; or Aveiro, with 21 documents).

With the aim to analyse the topics of this research, a keyword co-occurrence-based clustering was conducted. According to the terms included in each one, the following four clusters have been identified: (1) Technological solutions, (2) renewable energies, (3) life-cycle assessment (LCA), and (4) management and sustainability governance and performance (Figure 5.6).

The two areas studied in the first cluster, '**Technological solutions**' are, on the one hand, the search for scientific and technological solutions capable of mitigating the problems caused by greenhouse emissions. In this sense, terms such as CO_2 capture or biocatalysis appear as a technique for the enzymatic modification of contaminating substances. The second topic is related with the restructure of the current energy system using

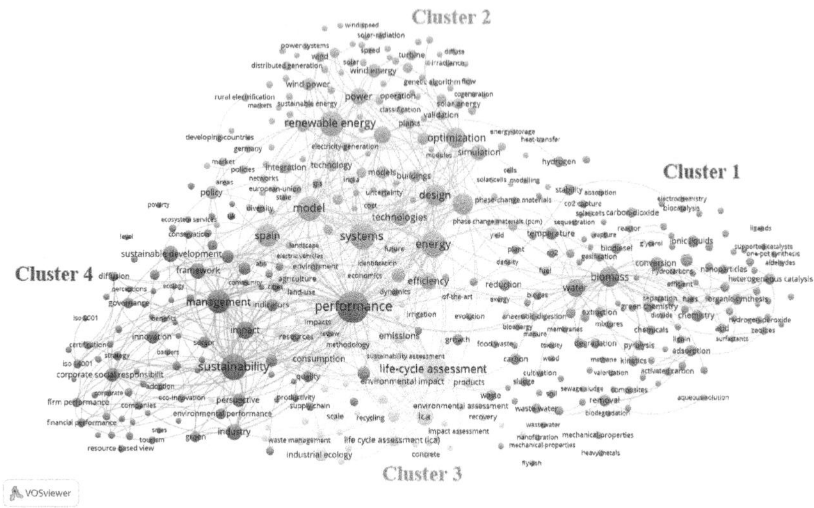

Figure 5.6 Co-occurrence map of subject areas with ≥10 documents on sustainability across the entire period.

biomass as a source of energy to supply energy demands. This cluster is composed of key terms such as 'biomass', 'biodiesel', 'biofuels', or 'bioenergy', among others. Another issue that appears in this cluster is the use of techniques based on the principles of sustainable chemistry, also called 'green chemistry' for the reduction of risks in industrial processes.

The second cluster, **'Renewable energies'**, groups all the works on clean and renewable energy or green energies, which overcome the brown technologies based on the use of fossil fuels (Kim, 2013; Lambertini and Tampieri, 2012). In this sense, there are terms such as 'wind energy', 'solar or renewable energies' that link renewable energies to SD: these are not pollutant.

The third cluster, **'Life-cycle assessment'** (LCA), covers the research that attempts to quantify and characterize the different potential environmental impacts associated with each of the stages of a product's life cycle. This cluster is the smallest within the Spanish university scientific community, which may suggest that these techniques related to the LCA are at an early stage in their development. On the other hand, the keyword map shows that the studies carried out are very partial and applied mostly to construction materials (with terms such as 'construction' and 'residential buildings') and recycling.

The four cluster, **'Management and sustainability governance and performance'**: Research in this cluster puts emphasis on the measures to be adopted by HEIs, governments, stakeholders, corporations, industry, or countries, to achieve a sustainable future. This cluster is about the political,

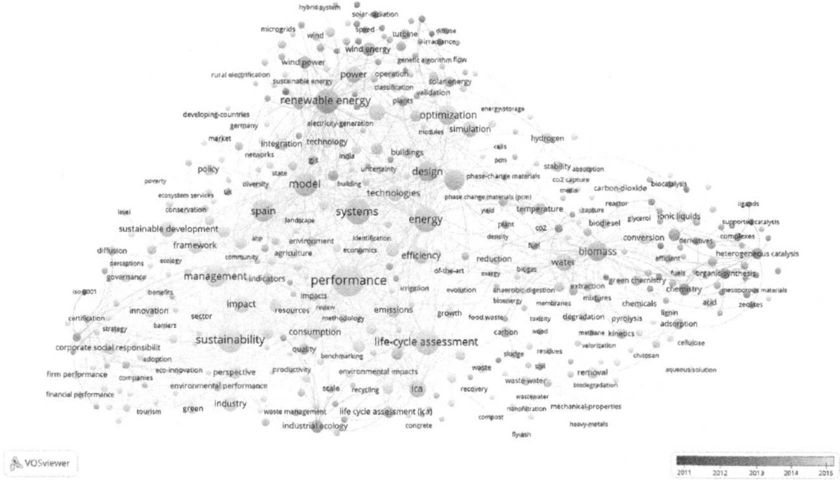

Figure 5.7 Evolution of sustainability research landscape from 2011 to 2015 at universities.

educational, technological, and management measures that would be necessary to establish the bases for a sustainable future. As Vilches and Gil (2003) stated, those would be: education for sustainability (EfS), technologies for sustainability, and universal governance to move towards sustainability. This cluster shows the commitment of these institutions for an education for sustainable, responsible consumption. It is a topic related to working for sustainability, developing technologies for sustainability, with social control (decision-making, participatory processes, corporate social responsibility) and the systematic application of the precautionary principle (empirical analysis).

Figure 5.7 shows the evolution of the keywords in each cluster from the last part of the period analysed (2011–2015). In earlier years (2011–2013), there are topics more related to energies and materials used: chemistry, biomass, or renewables energies. Later, the concepts of LCA or management start to appear. This proves a greater awareness in procedures related to the sustainability of the industry or in consumption.

5.3.2 Commitment

5.3.2.1 Inclusion in strategic plans/sustainability plans

After having analysed the inclusion of sustainability in the Strategic Plans or the Sustainable Plans on each university, the following results have been obtained. Among the 50 public universities analysed in the SUE, the great majority (39 universities, 78%) include sustainability within their strategic plan; only 11 do not include this concept or do not have an online available document. Some of them included this information in the sections devoted to Mission (University of Alicante (UA), University of Murcia (UMU)), Values (University of Burgos (UBU), Miguel Hernández University (UMH), University of Valencia (UV), National Distance Education University (UNED)), or as Strategic Axes (University of Córdoba (UCO), Autonomous University of Barcelona (UAB)). Sustainability is mentioned in different ways: some highlighted the environmental sustainability (University Carlos III of Madrid (UC3M), University of Alcalá (UAH), UCO); others the economic sustainability (Jaume I University (UJI)), and others mentioned the three pillars of sustainability. However, the most mentioned concept is 'environmental sustainability'.

Regarding sustainability plans in public universities, there are Sustainability Plans or Action Sustainability Plans (UAH, University of Salamanca (USAL), UPM, UAB, University of Cádiz (UCA), UAL, UB); Transport/Mobility Plans (University of Granada (UGR), UPV, Public University of Navarra (UNAVARRA)); Energetic Plans (UPC, UNIOVI); Declarations from the deans about this commitment (UC3M) or best practices guide (University Illes Balears (UIB), University Rey Juan Carlos (URJC)). The main focus and actions are related to environmental sustainability.

In private universities, commitment is completely different: only three universities (9.1%) mentioned the sustainability concept in their Strategic Plans. Regarding sustainability plans, only four universities (12.1%) have a document showing this commitment or the actions/plan of action carried out by the institution. Examples of it are the 'Declaration of Environmental Sustainability of the University of Deusto' (DEUSTO), the Framework document for the sustainability and commitment of UIC Barcelona in sustainability policy (International University of Catalonia (UIC)), the Memory for sustainability (Ramon Llull University (URL)), and a Sustainability Plan (University of Vic (UVIC)).

5.3.2.2 *Green campus/green offices*

Another important aspect that shows university commitment is to have a 'green' office or environment office. In public universities that the SUE analysed, 31 of them (61%) have this kind of office. They are called 'Green Office' (UNIZAR, USAL), 'Environment Office' (UABUVIGO), 'Sustainability Office' (UCA, University of La Rioja (UNIRIOJA)). However, 19 of them do not have this kind of office, but some of them have a 'Sustainability Classroom' (University of Huelva (UHU) International University of Andalucía (UIA)). Only three private universities (9.1%) have Green Offices (University of Deusto (DEUSTO), University San Jorge (USJ), UVIC).

5.3.3 *Internationalization: participation in European projects*

Regarding the FP7 on Environment, a total of 97 projects (19.64% of the overall) were carried out with partners from the SUE. Also, projects from other programmes (energy, transport, people, or Social Sciences and Humanities) have been selected for their relation with SD. The Spanish university that participates the most is UAB, with 15 projects, followed by UB, with 11 projects, and UPM, with 10 projects. UAB is also the university with more leaders in the period (three projects). Figure 5.8 shows the main topics, which are related to water or management.

There is a specific programme in H2020 where SD targets are addressed (Societal Challenges). For the purposes of this study, we have selected some of its subprogrammes: 3.2. Food security, sustainable agriculture, and forestry; marine, maritime, and inland water research; and the bioeconomy; 3.3 Secure, clean, and efficient energy; 3.4 Smart, green, and integrated transport; 3.5 Climate action, environment, resource efficiency, and raw materials. Spanish universities took part in 226 projects. The Spanish participation was higher in 3.2 (Food, 37.17%) and 3.3 (Energy, 24.34%) programmes and lower in 3.4 (Transport, 3.4%) programmes. UPM, with 37 projects; UPV, with 26 projects; UPC, with 21 projects; and US, with 17 projects, were the universities with a higher rate of participation. Fourteen projects were led by universities (two by UPM and two by UPC). Topics were related to energies (Figure 5.9). Contrary to what may be expected, there is a remarkable absence of research on SDGs or climate change as a central node.

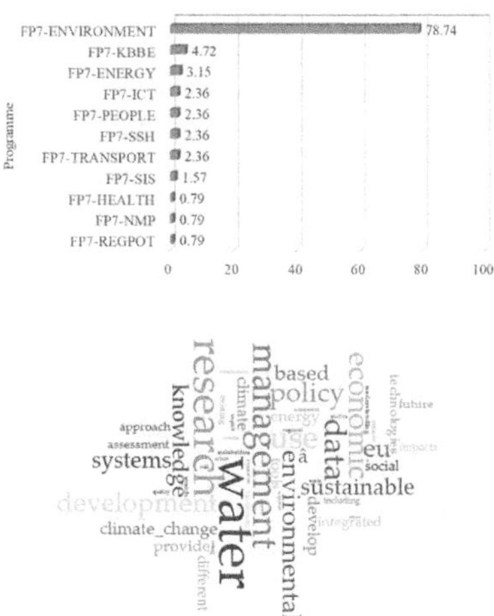

Figure 5.8 Analysis of the abstracts of projects related to SD in FP7.

(a)

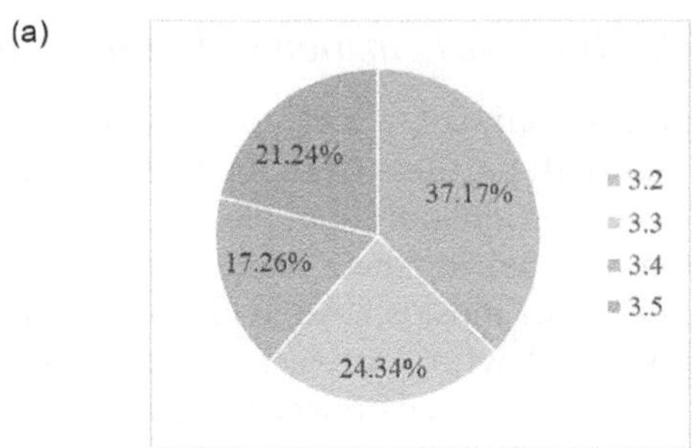

(b)

Figure 5.9 Distribution of projects (a) and abstract analysis (b) of H2020 projects.

5.3.4 Visibility: position in GreenMetric ranking

Overall, Spanish universities' participation has increased, from 95 universities in 2010 to 629 in 2017, with a CAGR of 30.7%. The ranking of Spanish universities in GreenMetric has evolved, from five universities (5.2%) in 2010 to 28 (4.5%) in the last edition, 2017: the first position in all those years was for UAH. A correspondence analysis for the distribution of these universities regarding their score on the different categories of the ranking (Figure 5.10) has been carried out. This resulted in two principal components that explained 57% of the variance in the topic. The figure shows profiles of specialization in the different dimensions related to universities. Regarding the area 'Setting and Infrastructure', UB has a good score, with its 407th position. Not so many universities are well positioned in 'Water', but UNAVARRA and UJI are the best positioned, with the 83rd and 113th positions, respectively. UVIGO, UAH (62nd position), UMH (398th position), and UAB (45th position) are closer to 'Transport'. Universities such as UV, University Santiago de Compostela (USC), and UGR, with positions from 8th, 77th, and 137th, respectively, are closer to 'Education in sustainability'. Regarding 'Waste', UAM is the closest one,

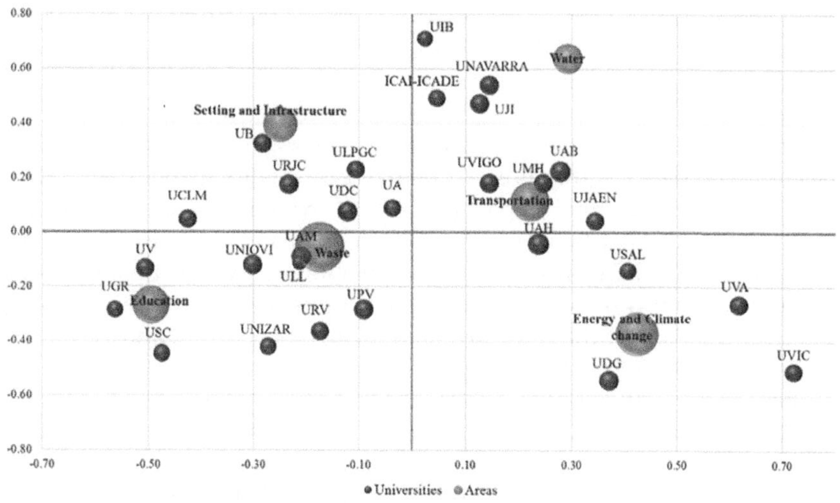

Figure 5.10 Distribution of Spanish universities in 2017 GreenMetric ranking regarding the area of specialization.

with its 153rd position; and in relation to 'Energy and Climate Change', the closest is University of Girona (UDG) (11th position).

Chi = 5153.032 (p=0.000)

5.4 *Discussion and conclusions*

There is a consensus that universities are an important stakeholder, and their researchers have the responsibility to contribute to SD with their research, it being a part of their academic mission (Waas et al., 2010). However, the process of being a sustainable university is still in its infancy stage (Velazquez et al., 2005). In this work, an analysis of the extent of the developments in Spanish HEIs, studying the different dimensions, was carried out.

5.4.1 *RQ1: Is there sustainability research at HEIs in the SUE?*

Regarding the scientific production, although the change in paradigm related to sustainability at HEIs may appear to have a long tradition, previous studies have determined that research on sustainability is not the first priority for many universities and one of the main problems is the lack of interdisciplinary teams (Velazquez et al., 2005; Larrán et al., 2014). Results of this study suggest it is a recent topic on Spanish universities: 81% of the scientific production is concentrated in the last years of the study. This growth continues the findings of the research by Hassan et al.

(2014) and Pulgarín et al. (2015) regarding SD or the study conducted by Nučič (2012) on sustainability science, demonstrating the recent interest on this topic by the research community.

Technical universities (UPC, UPM, UPV) are the main producers. An F-measure analysis, which measures the specialization in a field, showed these universities are the most specialized in this topic; however, other universities have a higher impact (UNIZAR), even if they are less specialized (UB, UdL, UAL, USC). These results are in line with those of the study by Romo- Fernández et al. (2012) on 'Renewable energy, Sustainability and the Environment research' output in the Scopus database, where UPM is also shown as one of the most productive and specialized, and UNIZAR has an impact higher than average.

Regarding collaboration habits, there is a growing rate of international collaboration, which is above 40% in 2017; national collaborations remain stable, around 30% in 2017. The strongest ties, as far as international collaboration is concerned, are with European (England, Italy, Portugal, France, and Germany) and Latin-American countries (Chile and Brazil). This fact may be due to physical proximity and cultural similarities. Universities are collaborating more with research institutes: CSIC, which is the largest public research organization in Spain, and CIEMAT, which is focused on energy and environmental matters, it matches the results of Romo- Fernández et al. (2012).

5.4.2 RQ2: What topics are addressed in scientific publications?

Within the selected WoS category, four clusters have been identified: (1) Technological solutions, (2) renewable energies, (3) LCA, and (4) management and governance of sustainability. Those are mainly related to technology and engineering, which may be the reason why the HEIs with the highest rate of activity in these areas are technical universities.

While it is true that sustainability includes many areas and is a very ambiguous issue (Walker and Shove, 2007), as seen in co-occurrence maps, some topics are not addressed. Social sustainability, economic sustainability, or issues related to the protection of health, or environmental education, are an example of it. That does not mean that these issues are not studied within the SUE. However, when selecting the WoS category, they are left out. Thus, by taking this category into account, it is apparent that the scientific production of the SUE is mainly related to environmental sustainability. Even though sustainability is a transversal issue, when analysing a WoS category, there is less transversality. Another issue that appears not to be as relevant here is that of climate change, which does not appear as a central node. This may be due, in part, to the fact that climate change is usually addressed from a more socioeconomic or healthcare point of view. That is, for now it is seen as a problem whose consequences are to be mitigated, not as a challenge. Other issues that seem to be absent in the network are sustainable mobility and green architecture, which is striking, because cities are making great efforts to meet the challenge of achieving sustainability. This may be a proof that the universities work without taking into account the cities where they are located or, at least in the case of scientific publications, there is no interest in the issue of sustainable cities. Apart from that, there are no nodes on SDGs in the universities' scientific output. This could be related to the fact that it has not been considered as a priority area in research. However, there is a cluster on governance and management (Cluster no. 4, Figure 5.6), indicating the commitment on this study, especially in recent years (Figure 5.7). Results suggest the university carries out basic research.

5.4.3 *RQ3: Is there a commitment in the SUE with this topic? Are HEIs participating in European projects?*

Sustainability assessment and reporting results can help HEIs to focus on coverage and performance weaknesses, thereby highlighting where actions should be taken, as well as better sustainability plans. Another aim is to communicate the university's efforts to its stakeholders and to benchmark against other institutions/companies (Lozano et al., 2013). The findings of our analysis suggest that all universities use their webpages to include this information and disseminate SD practices, as shown in studies of Portuguese universities (Aleixo et al., 2016). The great majority of the universities analysed have shown commitment to SD with its inclusion in strategic plans or sustainability reports. There is a university, the USC, which was one of the first in publishing sustainability reports, in 2003 (Lozano, 2011), followed by UCA and UMH, in 2007 (Zorio-Grima et al., 2018). This increase can also be related to the performance funding system, as stated in Larrán et al. (2014).

The findings of this study suggest that the number of actions/documents by public universities is higher than that by the private ones. This commitment is also quite recent, in line with the awareness on this topic. This was also pointed out by Lozano et al. (2013), who stated there is a tendency of a growing number of HEIs' sustainability reports each year. These documents are mainly related to an environmental perspective, according to similar studies (Velazquez et al., 2006).

Regarding sustainability-related units at the universities, these can be offices (Eco-campus, Green/Environment Offices) or services. Their main functions are waste management, energy efficiency or saving, mobility, and ESD activities (Alba and Blanco, 2008). Thirty-four institutions (40.9%)

included in the present study have these offices; this number has increased from previous studies, where 23 technical units were identified (Alba and Blanco, 2008).

As far as the participation of HEIs in European projects related to SD is concerned, more than 100 projects were identified in FP7 and 226 in H2020. In addition, the structure of the H2020 framework programme indicates that an own line of research has been opened, which reflects the political priorities of the Europe 2020 strategy, where sustainability plays a central role. Results indicate that the universities in the Spanish framework that are taking part in a higher level are UAB, UB, and the technical ones (UPM, UPC, and UPV). These universities have certain common features: they are highly specialized (technical universities) and are located in large cities (UAB, UB) (Manzano et al., 2016).

5.4.4 RQ4: Are these universities recognized in GreenMetric ranking?

Regarding the GreenMetric ranking, which is open to global participation, an increase in the participation of Spanish universities has been observed. It indicates the interest of these HEIs to assess their policies and actions in relation to their efforts on sustainability and green campuses. The universities that have a more active participation in the ranking are mainly those which have already started their path to sustainability (they have included the sustainability on their strategic plan or have sustainability plans). Their ranking is aligned with previous results that suggest Spanish universities have made a greater progress in actions related to waste and teaching and, to a lesser extent, have implemented measures on social responsibility, environmental impact assessment, water, and green purchasing (Hidalgo et al., 2012). This is the reason why the areas 'Water' and 'Energy and Climate Change' have a lower score in Spanish universities.

From all this, it can be inferred that there has been a greater scientific effort by the Spanish universities dedicated to the subject of sustainability during the period studied, where the commitment of the universities has increased through official documents, as well as their participation in projects and visibility in rankings such as GreenMetric. This shows that sustainability crosses all boundaries in universities' activities; however, there should be a greater commitment to SD by university leaders.

Acknowledgments

This research was funded by the Spanish Ministry of Economy and Competitiveness under the Project CSO2014-51916-C2-1-R titled 'Research on energy efficiency and sustainable transport in the urban environment: analysis of the scientific development and the social perception of the topic from the perspective of the metric studies of information' and predoctoral contract BES-2015-075461.

References

Alba, D. (2007). Analysis of management and educational processes for sustainability in public Spanish universities. *Research trends in Environmental Education Relating to Socio-Educational and Community Development*, 197–215.

Alba, D., & Blanco, D. (2008). Indicators for sustainability at Spanish Universities. *Retrieved*, *12*(7), 2014.

Albareda-Tiana, S., Vidal-Raméntol, S., Pujol-Valls, M., & Fernández-Morilla, M. (2018). Holistic approaches to develop sustainability and research competencies in pre-service teacher training. *Sustainability*, *10*(10), 3698.

Aleixo, A. M., Azeiteiro, U. M., & Leal, S. (2016). Toward sustainability through higher education: sustainable development incorporation into Portuguese higher education institutions. In Davim, J., & Leal Filho, W. (eds.) *Challenges in Higher Education for Sustainability* (pp. 159–187). Springer, Cham.

Alshuwaikhat, H. M., & Abubakar, I. (2008). An integrated approach to achieving campus sustainability: Assessment of the current campus environmental management practices. *Journal of Cleaner Production*, *16*(16), 1777–1785. doi: 10.1016/j.jclepro.2007.12.002.

Axelsson, R., Angelstam, P., Degerman, E., Teitelbaum, S., Andersson, K., Elbakidze, M., & Drotz, M. K. (2013). Social and cultural sustainability: Criteria, indicators, verifier variables for measurement and maps for visualization to support planning. *Ambio*, *42*(2), 215–228.

Bordons, M., Morillo, F., Gómez, I., Moreno-Solano, L., Lorenzo, P., Aparicio, J., González-Albo, B. (Coord.). La actividad científica del CSIC a través de indicadores bibliométricos (Web of Science, 2012–2016). *Technical Report*. Madrid: Consejo Superior de Investigaciones científicas (CSIC), 2017.

Brown, L. R. (2006). *Plan B 2.0: Rescuing a Planet Under Stress and a Civilization in Trouble*. W.W. Norton & Company, New York.

Brundtland, G. (1987). Our common future: Report of the 1987 world commission on environment and development. *United Nations, Oslo*, *1*, 59.

Chambers, D. P., & Walker, C. (2016). Sustainability as a catalyst for change in universities: New roles to meet new challenges. In Davim, J., & Leal Filho, W. (eds.) *Challenges in higher education for sustainability* (pp. 1–14). Springer, Cham.

Cole, L. (2003). Assessing sustainability on Canadian University campuses: Development of a campus sustainability assessment framework. *Journal of Environment and Management*, 1–66. doi: 10.1002/cjce.20357.

Cortese, A. (2003). The critical role of higher education in creating a sustainable future: Higher education can serve as a model of sustainability by fully integrating all aspects of campus life. *Planning for Higher Education*, *31*(3), 15–22.

Chuvieco, E., Burgui-Burgui, M., Da Silva, E. V., Hussein, K., & Alkaabi, K. (2018). Factors affecting environmental sustainability habits of university students: Intercomparison analysis in three countries (Spain, Brazil and UAE). *Journal of Cleaner Production*, *198*, 1372–1380.

CRUE (2018). Diagnóstico de la sostenibilidad ambiental en las universidades españolas, Informe 2018. Available at: www.crue.org/Boletin_SG/2018/2018.04.10%20Informe%20Sostenibilidad%20Universidades%20v3.4.pdf.

Elkington, J. (1998). Partnerships from cannibals with forks: The triple bottom line of 21st-century business. *Environmental Quality Management, 8*(1), 37–51.

European Commission (2010). Europe 2020: A Strategy for Smart, Sustainable and Inclusive Growth. Working paper {COM (2010) 2020}.

European Comission (2018). Environment & Climate Action. Available at: http://ec.europa.eu/programmes/horizon2020/en/area/environment-climate-action.

Fien, J. (2002). Advancing sustainability in higher education: Issues and opportunities for research. *Higher Education Policy, 15*(2), 143–152.

Filho, W., Raath, S., Lazzarini, B., Vargas, V. R., de Souza, L., Anholon, R., ... & Orlovic, V. L. (2018). The role of transformation in learning and education for sustainability. *Journal of Cleaner Production, 199*, 286–295.

Filho, W. L. (2000). Dealing with misconceptions on the concept of sustainability. *International Journal of Sustainability in Higher Education*, 1(1), 9–19.

Filho, W. L., & Pace, P. (2015). The future we want: Key issues on sustainable development in higher education after Rio and the UN decade of education for sustainable development. *International Journal of Sustainability in Higher Education*, 16(1), 112–129.

Gómez Caridad, I., & Bordons, M. (2009). Limitaciones en el uso de los indicadores bibliométricos para la evaluación científica.

Grindsted, T. (2011). Sustainable universities–from declarations on sustainability in higher education to national law.

Hassan, S. U., Haddawy, P., & Zhu, J. (2014). A bibliometric study of the world's research activity in sustainable development and its sub-areas using scientific literature. *Scientometrics, 99*(2), 549–579.

Hesselbarth, C., & Schaltegger, S. (2014). Educating change agents for sustainability–learnings from the first sustainability management master of business administration. *Journal of Cleaner Production, 62*, 24–36.

Hidalgo, D. A., Alcántara, R. B., Silva, M. T. B., del Álamo, J. B., Heras, D. B., Antúnez, X. D., ... Comas, P. Y. (2012). Estrategias de sostenibilidad y responsabilidad social en las universidades españolas: Una herramienta para su evaluación. *Profesorado. Revista de Currículum y Formación de Profesorado, 16*(2), 59–75.

IUNE Obervatory (2018). Observatory of Spanish University System (SUE). Available at: www.iune.es/.

Kastenhofer, K., & Rammel, C. (2005). Obstacles to and potentials of the societal implementation of sustainable development: A comparative analysis of two case studies. *Sustainability: Science, Practice and Policy, 1*(2), 5–13. doi: 10.1080/15487733.2005.11907968.

Kim, E. H. (2013). Deregulation and differentiation: Incumbent investment in green technologies. *Strategic Management Journal, 34*(10), 1162–1185.

Lambertini, L., & Tampieri, A. (2012). Vertical differentiation in a Cournot industry: The Porter hypothesis and beyond. *Resource and Energy Economics*, 34(3), 374–38.

Larrán, M., De la Cuesta, M., Fernandez, A., Muñoz, M. J., López, A., Moneva, J. M., … Garde, R., 2014. Analisis del nivel de implantación de políticas de responsabilidad social en las universidades españolas. In: *Conference of Spanish Public Social Councils,* Madrid.

Larrán, J. M., Madueño, J. H., Cejas, M. Y. C., & Peña, F. J. A. (2015). An approach to the implementation of sustainability practices in Spanish universities. *Journal of Cleaner Production, 106,* 34–44.

Larrán, M., Herrera, J., & Andrades, F. J. (2016). Measuring the linkage between strategies on sustainability and institutional forces: An empirical study of Spanish universities. *Journal of Environmental Planning and Management, 59*(6), 967–992. doi: 10.1080/09640568.2015.1050485.

Lauder, A., Sari, R. F., Suwartha, N., & Tjahjono, G. (2015). Critical review of a global campus sustainability ranking: GreenMetric. *Journal of Cleaner Production, 108,* 852–863.

Leetch, A., & Hauk, M. (2017). A decade of earth in the mix: A bibliometric analysis of emergent scholarly research on sustainability education and ecopsychology in higher education. In Leal Filho, W., Mifsud, M., Shiel, C., & Pretorius, R. (eds.) *Handbook of Theory and Practice of Sustainable Development in Higher Education* (pp. 291–306). Springer, Cham.

Lozano, R. (2011). The state of sustainability reporting in universities. *International Journal of Sustainability in Higher Education, 12*(1), 67–78. doi: 10.1108/14676371111098311. Permanent link to this document: https://doi.org/10.1108/14676371111098311.

Lozano, R., Llobet, J., & Tideswell, G. (2013). The process of assessing and reporting sustainability at universities: Preparing the report of the University of Leeds. *Revista Internacional de Tecnología, Sostenibilidad y Humanismo, 8,* 85–112.

Lozano, R., Lukman, R., Lozano, F. J., Huisingh, D., & Lambrechts, W. (2013). Declarations for sustainability in higher education: Becoming better leaders, through addressing the university system. *Journal of Cleaner Production.* doi: 10.1016/j.jclepro.2011.10.006.

Madeira, A. C., Carravilla, M. A., Oliveira, J. F., & Costa, C. A. (2011). A methodology for sustainability evaluation and reporting in higher education institutions. *Higher Education Policy,* 24(4), 459–479

Manzano, J. A., Esteve, A. E., Juan, M. I., & Cano, V. S. (2016). *La universidad española: grupos estratégicos y desempeño.* Fundacion BBVA.

Martens, P. (2006) Sustainability: Science or fiction? *Sustain,* 2(1), 36–41.

Meadowcroft, J. (2007). Who is in charge here? Governance for sustainable development in a complex world. *Journal of Environmental Policy & Planning,* 9(3–4), 299–314.

Nučič, M. (2012). Is sustainability science becoming more interdisciplinary over time? *Acta Geographica Slovenica,* 52(1), 215–236.

Pandiella-Dominique, A., Bautista-Puig, N., & De Filippo, D. (2018). Mapping growth and trends in the category 'Green and Sustainable Science and Technology'. In *23rd International Conference on Science and Technology Indicators (STI 2018),* September 12–14, 2018, Leiden, the Netherlands. Centre for Science and Technology Studies (CWTS).

Pulgarín, A., Eklund, P., Garrote, R., & Escalona-Fernández, M. I. (2015). Evolution and structure of "sustainable development": A bibliometric study. *Brazilian journal of Information Sciecne: Research Trends,* 9(1), 24.

Robinson, J. (2004). Squaring the circle? Some thoughts on the idea of sustainable development. *Ecological Economics, 48*(4), 369–384.

Romo-Fernández, L. M., Guerrero-Bote, V. P., & Moya-Anegón, F. (2012). World scientific production on renewable energy, sustainability and the environment. *Energy for Sustainable Development, 16*(4), 500–508.

Rousseau, R. (2018). The F-measure for research priority. *Journal of Data and Information Science, 3*(1), 1–18.

Roy, R., Potter, S., & Yarrow, K. (2008). Designing low carbon higher education systems: Environmental impacts of campus and distance learning systems. *International Journal of Sustainability in Higher Education, 9*(2), 116–130.

Ryan, A., Tilbury, D., Blaze Corcoran, P., Abe, O., & Nomura, K. (2010). Sustainability in higher education in the Asia-Pacific: Developments, challenges, and prospects. *International Journal of Sustainability in Higher Education, 11*(2), 106–119.

Salmerón-Manzano, E., & Manzano-Agugliaro, F. (2018). The higher education sustainability through virtual laboratories: The Spanish University as case of study. *Sustainability, 10*(11), 4040.

Sibbel, A. (2009). Pathways towards sustainability through higher education. *International Journal of Sustainability in Higher Education, 10*(1), 68–82. doi: 10.1108/14676370910925262.

Suwartha, N., & Sari, R. F. (2013). Evaluating UI GreenMetric as a tool to support green universities development: Assessment of the year 2011 ranking. *Journal of Cleaner Production, 61*, 46–53.

United Nations (2016). The Sustainable Development Goals Report, 2016. Available at: https://unstats.un.org/sdgs/report/2016/The%20Sustainable%20Development%20Goals%20Report%202016.pdf.

van Eck, N., & Waltman, L. (2009). Software survey: VOSviewer, a computer program for bibliometric mapping. *Scientometrics, 84*(2), 523–538.

Velazquez, L., Munguia, N., & Sanchez, M. (2005). Deterring sustainability in higher education institutions: An appraisal of the factors which influence sustainability in higher education institutions. *International Journal of Sustainability in Higher Education, 6*(4), 383–391.

Velazquez, L., Munguia, N., Platt, A., & Taddei, J. (2006). Sustainable university: What can be the matter? *Journal of Cleaner Production, 14*(9–11), 810–819.

Vilches, A., & Gil, D. (2003). Construyamos un futuro sostenible. Diálogos de supervivencia. Madrid: Cambridge University Presss. Capítulo 7.

Waas, T., Verbruggen, A., & Wright, T. (2010). University research for sustainable development: Definition and characteristics explored. *Journal of Cleaner Production, 18*(7), 629–636.

Wals, A. E. (2014). Sustainability in higher education in the context of the UN DESD: A review of learning and institutionalization processes. *Journal of Cleaner Production, 62*, 8–15.

Walker, G., & Shove, E. (2007). Ambivalence, sustainability and the governance of socio-technical transitions. *Journal of Environmental Policy & Planning, 9*(3–4), 213–225.

Zorio-Grima, A., Sierra-García, L., & Garcia-Benau, M. A. (2018). Sustainability reporting experience by universities: A causal configuration approach. *International Journal of Sustainability in Higher Education, 19*(2), 337–352.

chapter six

Application of Interpretive Structural Modelling for analysis of factors influencing Sustainability in Higher Education

S. Vinodh and Rohit Agrawal
National Institute of Technology

Contents

6.1 Introduction

The viewpoint of Sustainability in Higher Education (Sust-HE) is a recent initiative wherein worldwide significant number of universities developed commitment towards sustainability by signing international agreements (Clugston and Calder, 1999). Nowadays, most of the universities have started deploying sustainability in their system for their fitment. Sustainability is the key chain for overall growth of higher education. Higher education contributes a vital role in building the strength of current and further generation and provides an ease with which they can handle and solve real-time problems (Orr, 1995; Rowe, 2002). In line with Sustainable Development Goals (SDGs) which implies that higher education institutions need to focus on deployment of sustainable development across all disciplines and even international journals in the context of Sust-HE were initiated (Svanström, 2008). Many universities and education institutes started assessing their sustainability with the adoption of some tools to increase the sustainability for enriching the student's strength and competitiveness (Azeiteiro et al., 2015). Higher education sector is witnessing several challenges and has to fulfil various stakeholders' requirements. Hence, sustainability of higher education is gaining vital importance. The goal of this study is to recognize the prominent factors which influence Sust-HE. In this viewpoint, this chapter presents the analysis of factors influencing Sust-HE using Interpretive Structural Modelling (ISM) approach. ISM model is being developed based on expert inputs. The derived ISM model indicates the dominant factors influencing Sust-HE. Also, MICMAC (cross-impact matrix multiplication applied to classification) analysis is performed to group the factors in four types. The practical inferences are presented.

6.2 Literature review

Clugston and Calder (1999) showed the growth and future of Sust-HE and mentioned the higher educational institute declarations and commitment for becoming sustainable. They also discussed the critical issues for ensuring success in developing sustainability in Higher Education Institutes (HEIs).

Velazquez et al. (2005) aimed towards finding the barriers that restrict the successful deployment of sustainability action in higher education. They conducted a review on available peer-reviewed papers, books, and conference proceedings. Their finding showed many barriers for successful implementation of sustainability initiatives in institutes. Some of the barriers are lack of funding, lack of training and development and lack of interdisciplinary research and profits mentality.

Beringer et al. (2008) aimed towards finding the condition of Sust-HE in Canada. They sampled all collaborative institutes and universities of Canada to determine sustainable performance. They collected data by making questionnaire related to sustainability assessment. They found that most of the institutes are focusing on their curriculum for becoming sustainable educational institute.

Madeira et al. (2011) presented a methodology for promoting and evaluating Sust-HE. The developed methodology helped in choosing indicators for sustainability assessment, reporting and benchmarking. Through case study, they found current situation of sustainability and established goals for enhancement of current situation.

Wright and Horst (2013) aimed towards finding the initiatives taken by institutes for making it sustainable and also to find the obstacles in developing sustainability in HEI. They collected data from faculties and university leaders through structured interview. They found that financial support and leadership commitment need to be enhanced for becoming sustainable HEI.

Azeiteiro et al. (2015) aimed towards assessing the efficiency of sustainable development education by deploying e-learning in HE. They analysed the competence of students who have registered for environmental/ sustainability science programme. They did a qualitative semi-structured interview for collecting data. The results from the study revealed that, interviewee students have got high motivation towards environment and they came up with competence and good attitude towards environment.

Lozano et al. (2015) analysed the declaration made and initiatives taken to make their institute committed towards sustainable development. They did a survey based on 60 research papers. The study was divided into eight classifications. They got response from 84 experts from 70 institutes across worldwide. The data were analysed by inferential statistics and descriptive analysis. The result revealed a strong connection between declaration sign, initiatives taken and commitment towards sustainability. They found the important factor for signing declaration and initiatives taken for sustainable development is academic leadership's commitment.

Verhulst and Lambrechts (2015) considered the organization management in developing Sust-HE. They considered human factors like communication, organizational culture, resistance and empowerment. They presented a model which integrates human factors in developing sustainability in HEI. The presented model helped in identifying human-related barriers which restricts the inclusion of Sust-HE.

Aleixo et al. (2018) aimed towards finding the awareness on stakeholders regarding Sust-HE. They did a semi-structured interview on 20 stakeholders and found that they are not that much familiar with the sustainability concepts in HEI. They also found that the main barrier which restricts sustainable development in HEI is 'lack of financial support'.

6.3 Methodology

In this chapter, 16 factors are identified belonging to Sust-HE. Then inputs have been collected from educational institute experts in the form of relationship between different factors through email interview. Here, ISM is applied for analysing the factor relationships. ISM methodology is applied because it helps in identifying the relation between factors by showing contextual relationship between factors and also helps in understanding the structure of system (Singh and Khamba, 2011). Then a hierarchical level is developed for structured modelling of factors. The steps involved in ISM are mentioned below.

Step-I: Factors identification and their description

This step is concerned with the identification of factors for which the model has to be made are identified. Factors are collected by reviewing journal papers, conference papers and book chapters.

Step-II: Development of Structural Self-Interaction Matrix (SSIM)

SSIM shows the pair wise connection between factors. V, A, X & O symbols were used in making SSIM, depending on whether factor A depends on factor B, factor B depends on factor A, both factors A & B depend on each other and A & B are independent, respectively.

Step- III: Derivation of initial reachability matrix (RM)

Here, the inputs V, A, X & O of SSIM matrix are changed into binary number i.e., 0 &1 in the RM ((i,j) matrix) (Mandal and Deshmukh, 1994).

V – Signifies entry of 1 from (i,j) column and 0 from (j,i). It means i influences j and j is not influencing i.

A – Signifies entry of 0 from (i,j) column and 1 from (j,i). It means i is not influencing j and j is influencing i.

X – Signifies entry of 1 from (i,j) column and 1 from (j,i). It means both i and j rest on each other.

O – Signifies entry of 0 from (i,j) column and 0 from (j,i). It means both i and j are independent to each other.

Step – IV: Derivation of final RM

Final RM is made by applying transitivity rule in initial RM.

Step – V: Level partition

Level partition is done for making digraph (Sage, 1977). Reachability set and antecedent set are calculated based on final RM. The term reachability, antecedent and intersection set could be referred from Warfield (1974). For those factors, whose reachability set and intersection sets are same, are positioned in level 1 and then those factors are eliminated, same procedure is followed to find levels of all factors (Vinodh and Patil, 2018).

Step – VI: Building ISM

Draw the ISM based on levels of factors achieved in level partition step. Different levels can be connected by arrow based on the relations from final RM.

6.4 Case study

A case study is done by identifying factors influencing Sust-HE. ISM is being applied to derive the hierarchical model of factors. Steps of ISM are mentioned below:

Step-I: Factors and their description

From the literature study and in consultation with institutional experts, 16 factors that are critical for sustainability in HEI have been identified. Description of each factors are given in Table 6.1.

Table 6.1 Factors influencing Sust-HE along with description

Factor	Description	References
Public outreach (F1)	Public outreach means sharing knowledge with others beyond the community. Institutes/universities should have responsibility towards society for public outreach.	Corcoran and Wals (2004)
Develop interdisciplinary curriculum (F2)	By developing interdisciplinary curriculum, students can easily be able to correlate environmental studies with other subjects.	Corcoran and Wals (2004)
Encourage sustainable research (F3)	Most of the universities are encouraging their faculties to do research that will lead to sustainable development.	Corcoran and Wals (2004)
Partnership with government, NGOs and industries (F4)	Partnership with government, NGOs and industries will help to improve education and to make social changes. It will help the underprivileged students.	Corcoran and Wals (2004)
Interuniversity cooperation (F5)	It is the collaboration between two or more universities for exchanging information and sharing knowledge.	Corcoran and Wals (2004)
Interdisciplinary work (F6)	Capability to deal with different discipline work. It is the ability to solve complicated problems from interdisciplinary area.	Stough et al. (2018)
Scientific competence of teachers (F7)	It is the ability to use scientific knowledge to identify problems and draw conclusion based on the evidence which help in making decision.	Azeiteiro et al. (2015)
E-learning platform (F8)	E-learning platform is a set of online services which provide training, learning and various information to enhance education.	Azeiteiro et al. (2015)
Faculty and staff development (F9)	Faculty and staff development is essential to enhance skills and competence, and it leads to enhance sustainability.	Beringer et al. (2008)
Student opportunities (F10)	In most of the educational institute, opportunities for students are the important part in sustainable development.	Beringer et al. (2008)
Enables self-determination in learning (F11)	Self-determination is the ability of an individual that enables a person for goal setting, decision-making and problem solving. Enabling self-determination in learning can be considered as most important practice towards sustainability.	Jucker (2002)

(Continued)

Table 6.1 (Continued) Factors influencing Sust-HE along with description

Factor	Description	References
Organizational structure (F12)	For Sust-HE, an integrated organization structure is needed. Integrated structure will help in addressing problems more efficiently.	Velazquez et al. (2005)
Funding opportunities (F13)	Insufficient funding from top management is a big constraint. It reduces the growth of higher education.	Velazquez et al. (2005)
No resistance to change (F14)	Some people do not need any change in system. They are creating resistance to the growth of Sust-HE.	Velazquez et al. (2005)
Policies to enhance sustainability on campus (F15)	Lack of policies will affect funds and support from higher authorities. It restricts opportunities to the students.	Velazquez et al. (2005)
Institutional commitment (F16)	Lack of institution commitment is a big constraint for Sust-HE. It restricts the growth of Sust-HE.	Mader et al. (2013)

Step-II: Development of SSIM

Data is gathered from educational institute experts to examine all possible pairs of factors.

Sample description of SSIM:

F2-F7 = V. It means F2 influences F7 and F7 is not influencing F2.

F1-F16 = A. It means F1 is not influencing F16 and F16 is influencing F1.

F1-F13 = X. It means both F1 and F13 depend on each other.

F2-F4 = O. It means both F2 and F4 are independent to each other.

SSIM is shown in Table 6.2.

Step-III: Derivation of initial RM

Here the inputs V, A, X & O of SSIM matrix are changed into binary numbers i.e., 0 & 1 in the RM ((i,j) matrix).

Sample description of initial RM:

F2-F7 = V. It signifies F2-F7 = 1 and F7-F2 = 0

F1-F16 = A. It signifies F1-F16 = 0 and F16-F1 = 1

F1-F13 = X. It signifies F1-F13 = 1 and F13-F1 = 1

F2-F4 = O. It signifies F2-F4 = 0 and F4-F2 = 0

Initial RM is shown in Table 6.3.

Step-IV: Derivation of final RM

Final RM is made from initial RM by implementing transitivity rule. Transitivity rule could be referred from Warfield (1974). For transitivity link '0' has to be replaced with '1*'.

Sample description for transitivity rule:

F1-F13 = 1 and F13-F15 = 1, it implies, F1-F15 = 1 and it can be denoted as 1*.

Final RM is depicted in Table 6.4.

Table 6.2 SSIM matrix

	F16	F15	F14	F13	F12	F11	F10	F9	F8	F7	F6	F5	F4	F3	F2	F1
F1	A	A	A	X	A	A	X	A	X	A	A	A	A	A	A	X
F2	A	A	A	X	A	X	X	A	X	V	X	X	O	X	X	
F3	A	A	X	V	A	V	V	V	A	X	X	X	A	X		
F4	A	X	A	X	A	V	V	V	A	V	V	X	X			
F5	A	X	A	X	A	V	X	X	A	X	X	X				
F6	A	X	A	X	A	V	X	X	A	X	X					
F7	A	V	A	A	A	V	O	X	A	X						
F8	A	A	O	V	A	X	V	V	X							
F9	A	A	A	V	A	V	O	X								
F10	A	X	A	A	A	V	X									
F11	A	A	A	A	A	X										
F12	V	V	X	A	X											
F13	A	V	O	X												
F14	X	V	X													
F15	V	X														
F16	X															

Table 6.3 Initial RM

	F16	F15	F14	F13	F12	F11	F10	F9	F8	F7	F6	+F5	F4	F3	F2	F1
F1	0	0	0	1	0	0	1	0	1	0	0	0	0	0	0	1
F2	0	0	0	1	0	1	1	0	1	1	1	1	0	1	1	1
F3	0	0	1	1	0	1	1	1	0	1	1	1	0	1	1	1
F4	0	1	0	1	0	1	1	1	0	1	1	1	1	1	0	1
F5	0	1	0	1	0	1	1	1	0	1	1	1	1	1	1	1
F6	0	1	0	1	0	1	1	1	0	1	1	1	0	1	1	1
F7	0	1	0	0	0	1	0	1	0	1	1	1	0	1	0	1
F8	0	0	0	1	0	1	1	1	1	1	1	1	1	1	1	1
F9	0	0	0	1	0	1	0	1	0	1	1	1	0	0	1	1
F10	0	1	0	0	0	1	1	0	0	0	1	1	0	0	1	1
F11	0	0	0	0	0	1	0	0	1	0	0	0	0	0	1	1
F12	1	1	1	0	1	1	1	1	1	1	1	1	1	1	1	1
F13	0	1	0	1	1	1	1	0	0	1	1	1	1	0	1	1
F14	1	1	1	0	1	1	1	1	0	1	1	1	1	1	1	1
F15	1	1	0	0	0	1	1	1	1	0	1	1	1	1	1	1
F16	1	0	1	1	0	1	1	1	1	1	1	1	1	1	1	1

Table 6.4 Final RM

	F16	F15	F14	F13	F12	F11	F10	F9	F8	F7	F6	F5	F4	F3	F2	F1	Dependence power
F1	0	1*	0	1	0	1*	1	0	1	0	0	0	1*	0	0	1	7
F2	0	1*	0	1	0	1	1	0	1	1	1	1	0	1	1	1	11
F3	0	0	1	1	0	1	1	1	0	1	1	1	0	1	1	1	11
F4	0	1	0	1	0	1	1	1	0	1	1	1	1	1	0	1	11
F5	0	1	0	1	0	1	1	1	0	1	1	1	1	1	1	1	12
F6	0	1	0	1	0	1	1	1	0	1	1	1	0	1	1	1	11
F7	0	1	0	0	0	1	0	1	0	1	1	1	0	1	0	1	8
F8	0	1*	0	1	0	1	1	1	1	1	1	1	1	1	1	1	13
F9	0	0	0	1	0	1	0	1	0	1	1	1	1*	1*	1	1	10
F10	0	1	0	0	0	1	1	0	0	0	1	1	0	0	1	1	7
F11	0	0	0	0	0	1	0	0	1	0	0	0	0	0	1	1	4
F12	1	1	1	0	1	1	1	1	1	1	1	1	1	1	1	1	15
F13	1*	1	0	1	1	1	1	1*	1*	1	1	1	1	1*	1	1	15
F14	1	1	1	0	1	1	1	1	0	1	1	1	1	1	1	1	14
F15	1	1	0	0	0	1	1	1	1	1*	1	1	1	1	1	1	13
F16	1	1*	1	1	1*	1	1	1	1	1	1	1	1	1	1	1	16
Dependence power	5	13	4	10	4	16	13	12	8	13	14	14	10	13	13	16	

Table 6.5 Level partition of RM

Factor	Reachability set	Antecedent set	Intersection	Level
F1	1,4,8,10,11,13,15	1,2,3,4,5,6,7,8,9,10,11, 12,13,14,15,16	1,4,8,10,11,13,15	I
F2	1,2,3,5,6,7,8,10,11,13,15	2,3,5,6,8,9,10,11,12,13, 14,15,16	2,3,5,6,8,10,11, 13,15	III
F3	1,2,3,5,6,7,9,10,11,13,14	2,3,4,5,6,7,8,9,12,13,14, 15,16	2,3,5,6,7,9,13,14	III
F4	1,3,4,5,6,7,9,10,11,13,15	1,4,5,8,9,12,13,14,15,16	1,4,5,9,13,15	IV
F5	1,2,3,4,5,6,7,9,10,11, 13,15	2,3,4,5,6,7,8,9,10,12,13, 14,15,16	2,3,4,5,6,7,9,10, 13,15	II
F6	1,2,3,5,6,7,9,10,11, 13,15	2,3,4,5,6,7,8,9,10,12,13, 14,15,16	2,3,5,6,7,9,10, 13,15	II
F7	1,3,5,6,7,9,11,15	2,3,4,5,6,7,8,9,12,13,14, 15,16	3,5,6,7,9,11	II
F8	1,2,3,4,5,6,7,8,9,10, 11,13,15	1,2,8,11,12,13,15,16	2,8,11,13,15	V
F9	1,2,3,4,5,6,7,9,11,13	3,4,5,6,7,8,9,12,13,14, 15,16	3,4,5,6,7,9,13	IV
F10	1,2,5,6,10,11,15	1,2,3,4,5,6,8,10,12,13, 14,15,16	1,2,5,6,10,15	II
F11	1,2,8,11	1,2,3,4,5,6,7,8,9,10,11, 12,13,14,15,16	1,2,8,11	I
F12	1,2,3,4,5,6,7,8,9,10,11, 12,14,15,16	12,13,14,16	12,14,16	VI
F13	1,2,3,4,5,6,7,8,9,10,11, 12,13,15,16	1,2,3,4,5,6,8,9,13,16	1,2,3,4,5,6,8,9, 13,16	VII
F14	1,2,3,4,5,6,7,9,10,11, 12,14,15,16	3,12,14,16	3,12,14,16	VI
F15	1,2,3,4,5,6,7,8,9,10,11, 15,16	1,2,4,5,6,7,8,10,12,13, 14,15,16	1,2,4,5,6,7,8,10, 15,16	V
F16	1,2,3,4,5,6,7,8,9,10,11, 12,13,14,15,16	12,13,14,15,16	12,13,14,15,16	VI

Step-V: Level partition:

For those factors whose reachability and intersection set are similar and positioned in level 1 and then those factors are eliminated, same procedure is followed for all iterations to find levels of all factors (Vinodh and Patil, 2018). Level partition is shown in Table 6.5. After all iterations, factors with their corresponding levels are depicted in Table 6.6.

Step-VI: Building ISM

Factors are represented in the form of digraph; in digraph, factors having solid relationship are connected through solid lines whereas

Table 6.6 Factors with corresponding level

Factor code	Factor	Level
F1	Public outreach	I
F11	Enables self-determination in learning	I
F5	Interuniversity cooperation	II
F6	Interdisciplinary work	II
F7	Scientific competence of teachers	II
F10	Student opportunities	II
F2	Develop interdisciplinary curriculum	III
F3	Encourage sustainable research	III
F4	Partnership with government, NGOs and industries	IV
F9	Faculty and staff development	IV
F8	E-learning platform	V
F15	Policies to enhance sustainability on campus	V
F12	Organizational structure	VI
F14	No resistance to change	VI
F16	Institutional commitment	VI
F13	Funding opportunities	VII

transitive links are represented as dotted lines. The developed ISM model is depicted in Figure 6.1.

6.5 MICMAC analysis

MICMAC analysis is in line with matrix multiplication properties (Dewangan et al., 2015). MICMAC is performed to find the driving and dependence power of factors. It is done to recognize prominent factors which influence the whole system. Driving power and dependence of each factor are depicted in Table 6.7.

6.6 Results and discussions

ISM is shown in Figure 6.1. It shows the levels of factors affecting Sust-HE. The factors in the top level are those factors which are mostly dependent on other factors. 'Public outreach' and 'Enables self-determination in learning' factors are positioned in top level of ISM. The factors in the bottom level are those factors which are mostly independent and drives all other factors. 'Funding opportunities', 'Organizational structure', 'No resistance to change' and 'Institutional commitment' are positioned at the bottom level of ISM.

MICMAC analysis graphically shows all factors in four different cluster depending on their dependence power and driving power as depicted in Figure 6.2.

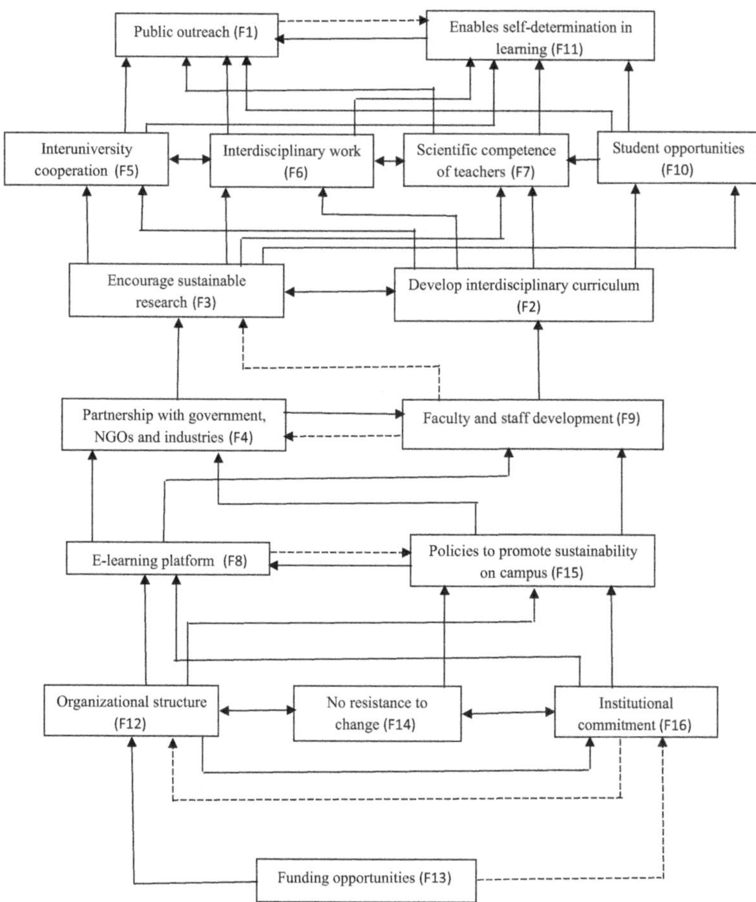

Figure 6.1 ISM model for Sust-HE.

The four clusters are:

Cluster A: Autonomous Factors: It includes the factors which are having low driving as well low dependence power. This type of factor doesn't have that much influence on system. 'Public outreach (F1)' falls under autonomous factor.

Cluster B: Dependent Factors: It includes the factors which are having high dependence power but low driving power. These types of factors are highly reliant on other factors. 'Scientific competence of teachers (F7)', 'Student opportunities (F10)' and 'Enables self-determination in learning (F11)' falls under dependent factor.

Cluster C: Linkage Factors: It includes the factors which are having high dependence and driving power. These types of factors are unstable

Table 6.7 Factors with their driving and dependence power

Factors	Driving power	Dependence power
Public outreach (F1)	7	5
Develop interdisciplinary curriculum (F2)	11	13
Encourage sustainable research (F3)	11	4
Partnership with government, NGOs and industries (F4)	11	10
Interuniversity cooperation (F5)	12	4
Interdisciplinary work (F6)	11	16
Scientific competence of teachers (F7)	8	13
E-learning platform (F8)	13	12
Faculty and staff development (F9)	10	8
Student opportunities (F10)	7	13
Enables self-determination in learning (F11)	4	14
Organizational structure (F12)	15	14
Funding opportunities (F13)	15	10
No resistance to change (F14)	14	13
Policies to enhance sustainability on campus (F15)	13	13
Institutional commitment (F16)	16	16

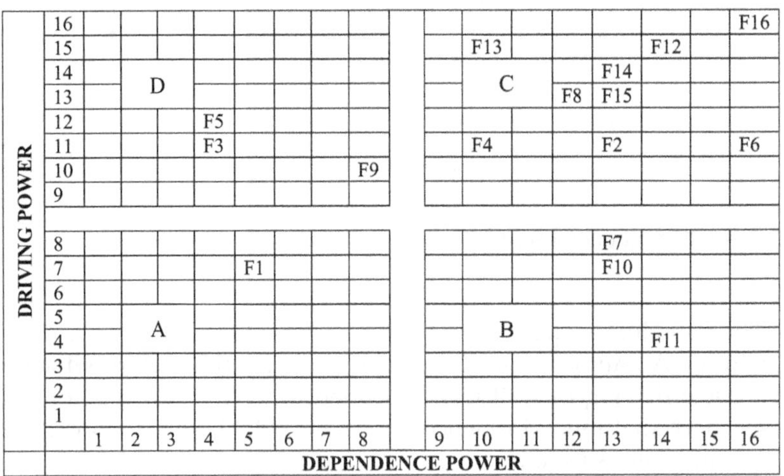

Figure 6.2 MICMAC analysis of factors influencing Sust-HE.

i.e, any impact on these factors will lead to influence the whole system (Jena et al., 2017). 'Develop interdisciplinary curriculum (F2)', 'Partnership with government, NGOs and industries (F4)', 'Interdisciplinary work (F6)', 'E-learning platform (F8)', 'Organizational structure (F12)', 'Funding opportunities (F13)', 'No resistance to change (F14)', 'Policies to enhance sustainability on campus (F15)' and 'Institutional commitment (F16)' are positioned under linkage factors.

Cluster D: Driving Factors: It includes the factors which are having high driving power but low dependence power. 'Encourage sustainable research (F3)', 'Interuniversity cooperation (F5)' and 'Faculty and staff development (F9)' fall under independent factor.

6.7 Conclusions

Higher education has a significant contribution in enabling transition to sustainable society based on knowledge creation and transfer and student preparedness for future societal needs. This aspect necessitates the integration of sustainability into HE system. This chapter presents the analysis of factors influencing Sust-HE. ISM is applied to derive the structural model. Based on the structural model, the factors 'Funding opportunities', 'Organizational structure', 'No resistance to change' and 'Institutional commitment' are found to be dominant. The developed structural model helps the practitioners to analyse and prioritize the most influencing factors. MICMAC analysis indicated that the number of driving, dependent, linkage and autonomous factors to be 3, 3, 9 and 1, respectively. The practical implication of the study is that it presents the systematic analysis of factors and formulation of structural model depicting the dominant factors. The comparison of results with prior research studies is depicted in Table 6.8.

The implications for higher education practitioners are also being presented. The limitation of this study is that only 16 factors are being

Table 6.8 Results from previous studies

Authors	Contributions
Dyehouse et al. (2017)	They aimed towards finding the relationship between 'students' knowledge on environment' and 'resistance towards change'. They found negative correlation between students' environmental knowledge and 'resistance towards change'. They concluded that students with higher resistance towards change were having less environmental knowledge and they won't give much importance to sustainable development.
Hamiti and Wydler (2014)	They aimed towards integrating Sust-HE. They used an assessment tool to recognize the impact of Sust-HE considering lecturers, students and research scholars. The developed spider diagram that focused on lecture content and focuses on priorities of lecturers towards sustainable development.

considered and statistical validation of results have not been reported. Future studies can be done by considering more factors related to Sust-HE to deal with advancements of Sust-HE.

References

Aleixo, A. M., Leal, S., & Azeiteiro, U. M. (2018). Conceptualization of sustainable higher education institutions, roles, barriers, and challenges for sustainability: An exploratory study in Portugal. *Journal of Cleaner Production, 172*, 1664–1673.

Azeiteiro, U. M., Bacelar-Nicolau, P., Caetano, F. J., & Caeiro, S. (2015). Education for sustainable development through e-learning in higher education: Experiences from Portugal. *Journal of Cleaner Production, 106*, 308–319.

Beringer, A., Wright, T., & Malone, L. (2008). Sustainability in higher education in Atlantic Canada. *International Journal of Sustainability in Higher Education, 9*(1), 48–67.

Clugston, R. M., & Calder, W. (1999). Critical dimensions of sustainability in higher education. *Sustainability and University Life, 5*, 31–46.

Corcoran, P. B., & Wals, A. E. (2004). *Higher Education and the Challenge of Sustainability, Vol. 10* (pp. 1–306). Dordrecht: Kluwer Academic Publishers.

Dewangan, D. K., Agrawal, R., & Sharma, V. (2015). Enablers for competitiveness of Indian manufacturing sector: An ISM-fuzzy MICMAC analysis. *Procedia-Social and Behavioral Sciences, 189*, 416–432.

Dyehouse, M., Weber, N., Fang, J., Harris, C., David, R., Hua, I., & Strobel, J. (2017). Examining the relationship between resistance to change and undergraduate engineering students' environmental knowledge and attitudes. *Studies in Higher Education, 42*(2), 390–409.

Hamiti, S. W., & Wydler, H. (2014). Supporting the integration of sustainability into higher education curricula—A case study from Switzerland. *Sustainability, 6*(6), 3291–3300.

Jena, J., Sidharth, S., Thakur, L. S., Kumar Pathak, D., & Pandey, V. C. (2017). Total interpretive structural modeling (TISM): Approach and application. *Journal of Advances in Management Research, 14*(2), 162–181.

Jucker, R. (2002). "Sustainability? Never heard of it!" Some basics we shouldn't ignore when engaging in education for sustainability. *International Journal of Sustainability in Higher Education, 3*(1), 8–18.

Lozano, R., Ceulemans, K., Alonso-Almeida, M., Huisingh, D., Lozano, F. J., Waas, T., … Hugé, J. (2015). A review of commitment and implementation of sustainable development in higher education: results from a worldwide survey. *Journal of Cleaner Production, 108*, 1–18.

Madeira, A. C., Carravilla, M. A., Oliveira, J. F., & Costa, C. A. (2011). A methodology for sustainability evaluation and reporting in higher education institutions. *Higher Education Policy, 24*(4), 459–479.

Mader, C., Scott, G., & Abdul Razak, D. (2013). Effective change management, governance and policy for sustainability transformation in higher education. *Sustainability Accounting, Management and Policy Journal, 4*(3), 264–284.

Mandal, A., & Deshmukh, S. G. (1994). Vendor selection using interpretive structural modelling (ISM). *International Journal of Operations & Production Management, 14*(6), 52–59.

Orr, D. W. (1995). Educating for the environment higher education's challenge of the next century. *Change: The Magazine of Higher Learning, 27*(3), 43–46.

Rowe, D. (2002). Environmental literacy and sustainability as core requirements: Success stories and models. *Teaching Sustainability at Universities,* 79–103.

Sage, A. P. (1977). On interpretive structural models for single-sink digraph trees. In *Proceedings of the IEEE International Conference on Systems, Man, and Cybernetics.*

Singh, H., & Khamba, J. S. (2011). An interpretive structural modelling (ISM) approach for advanced manufacturing technologies (AMTs) utilisation barriers. *International Journal of Mechatronics and Manufacturing Systems, 4*(1), 35–48.

Stough, T., Ceulemans, K., Lambrechts, W., & Cappuyns, V. (2018). Assessing sustainability in higher education curricula: A critical reflection on validity issues. *Journal of Cleaner Production, 172,* 4456–4466.

Svanström, M., Lozano-García, F. J., & Rowe, D. (2008). Learning outcomes for sustainable development in higher education. *International Journal of Sustainability in Higher Education, 9*(3), 339–351.

Velazquez, L., Munguia, N., & Sanchez, M. (2005). Deterring sustainability in higher education institutions: An appraisal of the factors which influence sustainability in higher education institutions. *International Journal of Sustainability in Higher Education, 6*(4), 383–391.

Verhulst, E., & Lambrechts, W. (2015). Fostering the incorporation of sustainable development in higher education. Lessons learned from a change management perspective. *Journal of Cleaner Production, 106,* 189–204.

Vinodh, S., & Patil, D. (2018). Modelling factors influencing lean concept adoption in a food processing SME for ensuring sustainability. In Chakraborty, A., Gouda, S. K., & Gajanand, M. S. (eds.), *Sustainable Operations in India* (pp. 93–111). Singapore: Springer.

Warfield, J. N. (1974). Developing subsystem matrices in structural modeling. *IEEE Transactions on Systems, Man, and Cybernetics, SMC-4*(1), 74–80.

Wright, T., & Horst, N. (2013). Exploring the ambiguity: What faculty leaders really think of sustainability in higher education. *International Journal of Sustainability in Higher Education, 14*(2), 209–227.

chapter seven

Teaching and researching on sustainability

The case of an interdisciplinary initiative at a higher education institution

A. Gomes Martins
University of Coimbra
INESC Coimbra

Contents

7.1 Introduction

Many universities around the world adopt the figure of "initiative" to organize activities which cross borders between scientific disciplines in order to improve their ability to qualified interventions in science and education, or to take action in social responsibility fields, or even to improve the quality of some of their internal operations, always widening their capability to deal with more complex and intricate problems and to contribute more actively to solving contemporary problems. This chapter aims at discussing the role of interdisciplinary initiatives in the context of higher education institutions (HEIs) as instruments of implementing or increasing their contribution to sustainable development and the fulfilment of the Sustainable Development Goals as defined by the United Nations (UN). A brief discussion is made of the possible approaches to organizing interdisciplinary activities in the context of HEI as a framework to a more detailed presentation of a case study at the University of Coimbra, in Portugal, the Energy for Sustainability (EfS) Initiative.

The published literature on sustainable development in HEI follows a common understanding of the characteristics and requirements of institutional policy. For example, UNESCO has been developing valuable guiding documents for educational activities connected with sustainable development. Several examples exist which are dedicated to non-higher education settings [1–3] but also to higher education [4]. In the Guidelines on Sustainability Science in Research and Education, a strong emphasis is given to interdisciplinarity in research and in education as a means of promoting sustainability science and its integration in educational programs at large. For this purpose, encouragement is made to collaboration between researchers, departments and research units within each university and between different HEIs.

Leal Filho [5], in a similar line of thought, points out the need of institutional guidelines to achieve sustainable development and stresses several important components of the sustainability commitment of HEI, namely a strong involvement of senior members as well as the development of new curricula in order to include sustainability.

The International Association of Universities – IAU [6] approved in its annual congress of 2014 the Iquitos Statement on Higher Education for Sustainable Development, where recommendations are found of a very similar nature, with additional emphasis on the need for strong leadership to achieve transdisciplinary approaches to education, research and campus sustainable development and maintenance.

Wu [7] refers a set of aspects that should be present in higher education for sustainable development to conclude that much beyond environmental issues, it must encompass social and economic issues as well, fostering interdisciplinarity.

Many of the aforementioned recommended actions and orientations are currently implemented in the context of the EfS Initiative, as will be seen below.

Examples of interdisciplinary initiatives in HEI around the world could be raised in high numbers. Just to name a few, reference can be made to UC Davis' Initiative to improve graduation rates [8], University of Hartford's The Women's Advancement Initiative [9] or Boston University's Initiative on Cities [10]. On a more focused perspective on energy or sustainability issues, many others exist, some examples following: University of Maine's Green Campus Initiative [11], University of Montana's Smart Buildings Initiative [12], University of Delft's Delft Energy Initiative [13], Cornell University's Sustainable Campus [14], MIT Energy Initiative [15] and Duke University Energy Initiative [16].

The EfS Initiative [17] follows this same logic, and it was, as far as it can be searched, the first of its kind to be created in the Portuguese higher education system. It was created in 2006 at the University of Coimbra, Portugal, by a multidisciplinary group of faculty with a common commitment towards sustainability, by developing scientific activity and cooperation with companies, contributing to sustainability on campus and delivering postgraduate education of interdisciplinary nature. This group of faculty grew slowly and steadily from around 20 in 2006 to more than 100 researchers in 2017, with very diverse backgrounds and scientific skills.

All the examples above, except the case of the EfS Initiative, have the common characteristic of being the result of corporate decisions, following a top-down path towards action. This means that, among other consequences, the initiatives were provided institutional resources to launch their activities and to pursue their mission and objectives.

This was not the case of the EfS Initiative at the University of Coimbra. Since it was created bottom-up by a group of faculty, there was no corporate top-down decision and no resources were initially allocated to it. This had two main immediate consequences: the initiative assumed the singular nature of being driven solely by voluntary efforts of its faculty members and other resources had to be mobilized from somewhere else outside the university, namely from those research units which, although being within the university's boundary, have a private nature.

In spite of its bottom-up nature, after 6 years of activity, the EfS Initiative was officially recognized by the university in late 2012, as a project of strategic value to the institution, although a less than weak top-down commitment to it has been since then the habit of the institution's corporate government. This configures the well-known difficulties associated to the decoupling between the usual silo-oriented organization of universities and the cross-border attitude of interdisciplinary activities and projects [18–20]. In Table 7.1, some characteristics of the initiatives referred above are summarized.

The two most evident differences between the EfS Initiative and the other cases are its bottom-up nature and the absence of regular institutional funding. Several authors have dealt with the top-down versus bottom-up approaches to sustainability in HEI. As an oversimplified synthesis, one can say that successful initiatives require a convergence between the two modes. In fact the levers available to institutional decision makers are essential to implement the necessary measures and programs to a successful sustainable development. On the other hand, the mobilization of the institutional community is indispensable since there will be no actual effects of any measures without understanding and commitment of individual community members or sub-communities associated to organic structures inside the institution, as dormitories, canteens, departments, administrative units, laboratories and so on.

Table 7.1 Comparison of some characteristics of energy and sustainability-related initiatives and of institutional framework in a sample of HEIs

Institution	Initiative designation	Areas of influence HE	R&D	CI	Top-down	Bottom-up	Institutional funding
University of Maine	Green Campus Initiative			✔	✔		✔
University of Montana	Smart Buildings Initiative	✔	✔	✔	✔		✔
University of Delft	Delft Energy Initiative	✔	✔	✔	✔		✔
Cornell University	CU's Sustainable Campus			✔	✔	✔	✔
MIT	MIT Energy Initiative	✔	✔	✔	✔		✔
Duke University	DU Energy Initiative	✔	✔	✔	✔		✔
University of Coimbra	EfS Initiative	✔	✔	✔		✔	(1)

Legend for "Areas of influence": HE, higher education activities; R&D, research and development activities; CI, community involvement.

Note: (1) EfS Initiative is provided a very small fraction of the tuition fees of EfS students, with no fixed rules for each yearly amount.

Pallant [21] supports this perspective from the point of view of overcoming barriers to sustainable development, stressing the need to ensure support from administrators and the involvement of staff, as well as the need to take into account all the dimensions of each problem, from financial to political, moral, physical, educational, etc. Qian [22], focusing on education for sustainable development, suggests a research approach to change as a top-down movement and an educational change as a bottom-up movement namely for the purpose of diffusing sustainability concepts through non-environmental courses. Lozano [23] lists a series of requirements where the support of top-level management is the first, followed by a structured planned approach, education and training of specific actors and the mobilization of individuals. Again, the convergence of top-down and bottom-up approaches to a successful sustainable development.

The EfS Initiative may eventually be considered more a middle-out kind of movement then strictly a bottom-up one. According to Brinkhurst [24], faculty and staff are usually the most important actors of campus sustainability, a role that should be recognized and encouraged by institutional leaders. The official recognition of EfS by the university as a project of strategic value to the institution could be considered one such kind of encouragement. However, the problem remains of lack of actual response of the institution leadership to the proposals made by the EfS Initiative, either of strategic or operational nature, which could contribute to positive change towards sustainability. Even Lozano in Ref. [25] does not account for such a singular form of intervention, when referring examples of forms of implementing sustainable development on campus, since he uses formulations such as "SD groups or networks of staff of different departments working on SD implementation" or "research groups (e.g. on renewable energy in buildings, or in service studies working on SD)" which seem not to be understood as a possible coherent part of a bigger whole integrated approach.

The EfS Initiative is organized to be able to respond to challenges in four main areas:

a. Scientific research and development (R&D), having in mind the cooperation among the university's research units, the participation in inter-institutional networks, the participation in consortiums, the increase of the university's scientific production, the internationalization of R&D, not only in the EU but also worldwide;
b. Advanced education – promotion and support of educational programs, particularly second cycle (MSc) and third cycle (PhD), as well as specialized training courses, but also potentially e-learning courses and training courses customized for companies;
c. Outreach – innovation, knowledge and technology transfer to the economy and the society, through development agreements or contracts established with companies, as well as the promotion of entrepreneurship;
d. Sustainable campus – promoting the development of programs and projects within the university itself, for a sustainable campus approach aiming at the internal application of sustainable management principles of energy resources at several levels – urban, buildings, equipment;

The three first areas are typical of other initiatives [15]. The existence of the fourth area is a consequence of the assumption made by faculty members of their willingness to make their university benefit from their scientific and technical competences towards improving sustainability on campus.

7.2 *Scientific research*

The pillars of the EfS Initiative are 15 scientific research units where its researchers are affiliated (Figure 7.1). Senior researchers may be either faculty members or PhD holders with some kind of formal link to one research unit, leading or collaborating in R&D projects. In many cases,

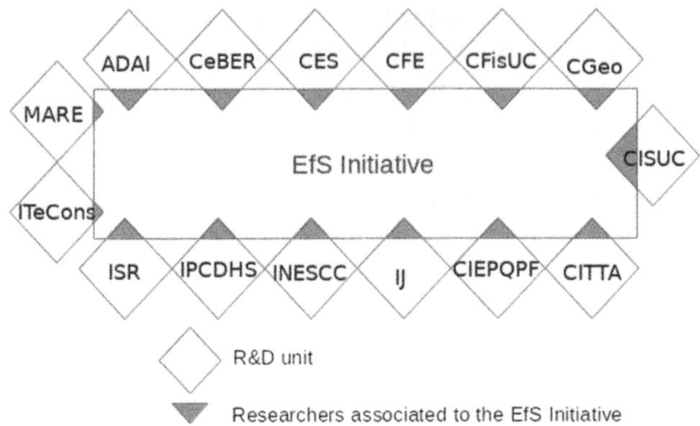

Figure 7.1 The EfS Initiative is made up of researchers from a set of 15 research units.

researchers who are not faculty members obtained their PhD degree through the Sustainable Energy Systems PhD Program, created and managed by the EfS Initiative. Faculty members always have a double affiliation: with a research unit and with a Faculty or a Department of the university. Senior researchers in general have very diverse backgrounds, on domains such as sociology, one of all the classic engineering fields, physics, architecture, behavioural sciences, information and communication technologies, law, economics. A possible illustration of the variety of backgrounds of the faculty involved consists in the identification of their faculties – Faculty of Science and Technology, Faculty of Law, Faculty of Psychology and Educational Sciences, Faculty of Economics – and Departments – Architecture, Earth Sciences, Life Sciences, Physics, Electrical and Computer Engineering, Civil Engineering, Informatics Engineering, Mechanical Engineering, Chemical Engineering.

This diversity and the scientific research-oriented nature of the EfS Initiative have some consequences:

i. The majority of the R&D projects carried out on fields in any way related to Sustainable Development Goals 6, 7, 11, 12 or 13 [26] are based on multidisciplinary teams made up of researchers from more than one research unit, gathering the competences needed to deal with the problem(s) at stake. This procedure is not written in any regulation, it just happens as an acquired orientation resulting from several years of a successful interdisciplinary praxis;
ii. Many students are actively involved in scientific projects, fostering their ability to teamwork, indispensable to interdisciplinary approaches to contemporary problems for which sustainability urgently requires solutions [27];

iii. Research results and achievements translate into up-to-date contents in lecturing, projecting future needs of sustainable development in technology, regulation and policy. There is a permanent concern on making students aware of the importance and virtue of interdisciplinary views and practice on the contribution of progress in energy conversion and use towards the UN Sustainable Development Goals.

Topics of research projects are spread through sustainable energy supply; energy efficiency; industrial ecology; indoor environmental quality; energy demand-side management; bioenergy; sustainable building design construction and retrofit; urban planning; energy economics; energy markets; sustainable mobility; smart cities; smart grids; advanced metering infrastructure; behavioural dimensions of energy demand, energy storage, etc. [17].

Researchers are actively encouraged and technically supported to elaborate applications to calls for projects, in order to increase the number of applications and the number of funded projects (Figure 7.2). This happens synchronously with the announcement of future calls and consists in a selective and systematic dissemination of information on the content of the calls, as well as on the organization of debates among researchers to facilitate the identification of those scientific domains that have potential to cooperate for a particular application.

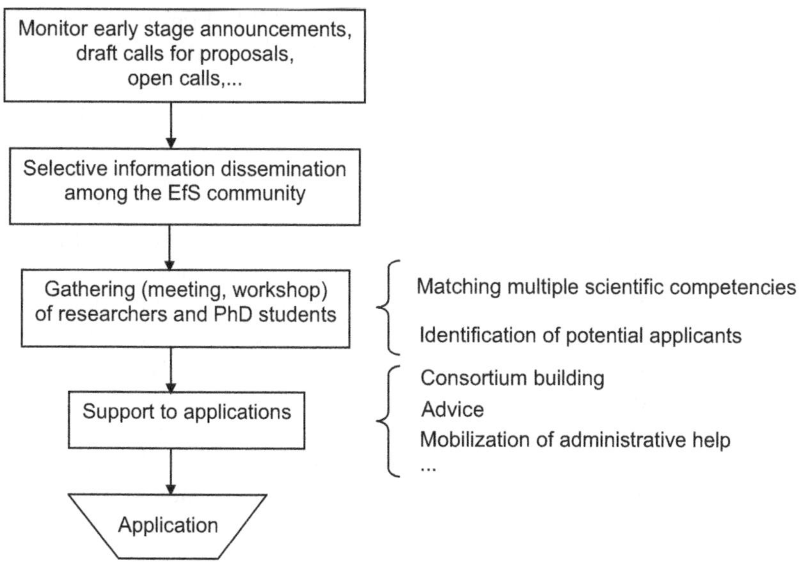

Figure 7.2 Process of fostering competitive funding of scientific interdisciplinary projects.

7.3 Advanced education

The educational programs led by EfS are a PhD on Sustainable Energy Systems, an MSc on EfS and an Advanced Specialization Program which is twin to the MSc. The structure of each program seeks to create a mindset in students towards sustainable development, around a myriad of energy-related topics, configuring many possible approaches to answering the fundamental question: how can energy become a key sustainability driver in the future instead of a sustainability problem?

The design of the educational programs followed a structured approach, linking subjects to competences, which was the case for a multi-authored paper [28]. The average output rates of graduates are six PhD per year and nine MSc per year, from a universe of flowing students with an average size between 40 and 50. Graduates are spread throughout the world in various organizations, including HEIs, scientific research bodies and industrial companies, as well as in the country, serving as professionals at various levels. A few are research leaders or researchers in R&D units at the University of Coimbra.

The PhD Program on Sustainable Energy Systems, organized in a 4-year time span (240 European credit units), has evolved towards a joint program with two other big public universities in the country, University of Lisbon and University of Porto. Students attend courses lectured by a diversity of professors from the three institutions, often by means of videoconference sessions gathering all students in a virtual classroom where interactions are fully operational. A structured path is followed to ensure an effective framework for the students' theses preparation during the first year (Figure 7.3). Students are strongly encouraged to embrace interdisciplinary research themes requiring them to have more than one supervisor. They are also encouraged to have a foreign co-supervisor. As of March 2018, 27 PhD students had graduated in the framework of the EfS Initiative, the first ones in 2013.

The MSc program on EfS is organized in a three-semester time span (90 European credit units), providing three specialization areas: Energy Systems and Policy, Buildings and Urban Environment and Indoor Climate and Comfort. The Advanced Specialization program on EfS (48 European credit units) is a twin program to the MSc, only not requiring the students to develop a thesis and a thesis project thereof.

Both programs, PhD and MSc, have been attracting foreign students (circa 40% in both cases up to now) from 25 countries of all five continents, although predominantly from Europe, Asia, South America and Africa. As of March 2018, 70 MSc students had graduated in the framework of the EfS Initiative, the first ones in 2010.

Students' backgrounds at all levels are strongly diverse, from natural sciences to architecture, economics, management, health, all engineering fields. There is a multicultural atmosphere which requires the extensive use of the English language, together with an intrinsic and spontaneous openness of all agents to interdisciplinarity, not only because it is permanently stimulated but also because of the unavoidable daily acknowledgement of diversity.

The initiative launches at least once a year a call for applications to attend international conferences and to temporary stays at foreign universities or research institutions. This call is directed at PhD students in order to encourage them to present their research work and expose it to criticism and contributions, to facilitate their networking within the international scientific community and, when applicable, to have a stay near their foreign co-supervisors. Successful applications have the right to a subsidy which generally covers a great majority of, if not the whole, expenses involved. This is a very popular initiative of which dozens of students have been benefiting throughout the years, allowing them to merge into the scientific community involved in sustainable development research at large.

The EfS Initiative configures a kind of a pool where students plunge on arrival and become permanently surrounded by references to sustainable development, mainly through energy-related issues, although raising also many aspects which are strongly correlated and very much in tune especially with Sustainable Development Goal number 7, as for example, economic development, human development, energy poverty, public health and others. To this aim, besides other ingredients filling students' everyday life, there are a number of conferences lectured by invited speakers coming either from industry or academia, treating subjects connected to sustainability in general. This happens every now and then during the year, but mandatorily at the beginning of the school year, when an Inaugural Lesson takes place, usually in late September or early October. This lecture is open to all the EfS community and is followed by two welcome sessions specifically dedicated to the PhD students and to the MSc students.

In two other occasions during the year, two events are organized dedicated to PhD first-year students, usually designated seasonal workshops, where all the PhD students are usually present along with many faculty members. The Autumn Workshop, which usually occurs in November, consists of the presentation to the first-year students of the thesis projects developed by their second-year colleagues. The objective is to make the first-year students get acquainted with what a thesis project is, to contact with their older colleagues and faculty members and to have the first contact with scientific debate which usually occurs because the assembly discusses the presentations made by the second-year students. The Spring Workshop, which usually occurs in May, gives the first-year students the first opportunity to present their main ideas on what their

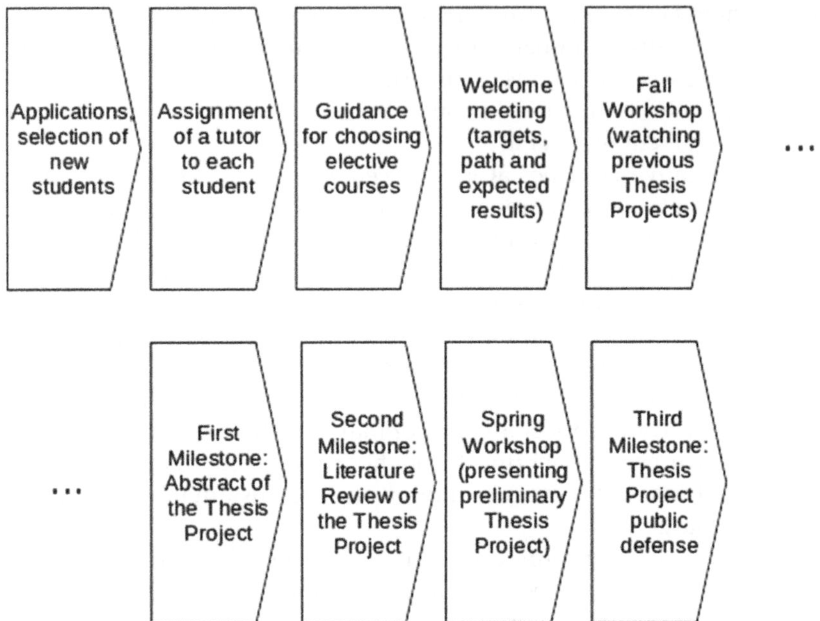

Figure 7.3 Path used to follow first-year PhD students.

thesis projects are going to be, subjecting them to discussion with the audience. Very often these workshops are preceded by an invited lecture provided by a guest from academia.

A recent joint initiative with the Portuguese Open University led to the creation of a distance learning program which aims at spreading knowledge on sustainable development at the local level, mainly directed at qualified personnel of city government administrations: "Local Sustainability – main instruments and practice". This adds to the commitment of the EfS Initiative towards the Sustainable Development Goals.

The EfS Initiative has a responsibility over its educational programs which is equivalent to the responsibility of any university Department or Faculty regarding their own educational programs. However, the virtual nature of the Initiative, which does not correspond to any traditional type of organic unit inside the university, creates a number of difficulties and barriers which are typical in HEIs that are traditionally organized structures specialized in narrow scientific fields [29,30]. Artificial hierarchical dependencies, communication hurdles and other bureaucratic barriers of various sorts create daily difficulties which do not exist in other programs. The Initiative has a specific committee to manage these activities (Figure 7.4) which masters constantly all the aspects of the management of the educational programs, maintaining a close attention to all details

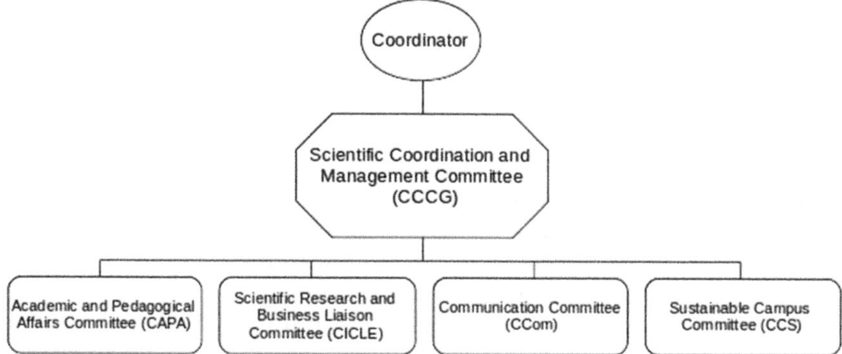

Figure 7.4 EfS Initiative management structure.

and filling gaps and building bridges through permanent contacts with all levels of the university management structure. The EfS Initiative's educational activities are evaluated, as any others at the university, by the national regulatory authority for the higher education system. In this context, some of the roles of the committee are to elaborate the reports, to organize the audit visits and to implement the changes that are required as a consequence of the audits.

7.4 Outreach

A large number of stakeholders are regularly involved in the EfS Initiative activities. As with research projects, there are a large number of bilateral contracts aiming at the development of research or products or new services, not only through contracts with external entities, either companies or public bodies. However, regular contact with the economic tissue is made also through the regular operation of an External Advisory and Assessment Board (EAAB). This structure includes several big energy utility companies, the energy regulator, the national energy agency and several other companies in varied economic fields, including construction and design, vehicles and transportation.

The EAAB meets with students and faculty at least once a year, always at a different location in the university, usually a R&D Unit's facilities, to discuss progress so far and possible future orientation, both in scientific work and in advanced education. It works as an evaluation board, although issuing no formal report, since the recommendations formulated during the meetings are recorded and implemented by the Initiative's directive board. Students are actively involved in these contacts, frequently benefiting from an outside, third-party view on social needs and problems and on both the relevance and gaps of their research.

Frequently, relevant persons from industry, politics or academia provide invited lectures, always open to the university community and the public at large, on themes closely connected with sustainability concerns.

Preceding the meeting, a few weeks before it, a call for posters is launched to students, inviting them to explain their research projects. The posters are organized in an exhibition at the meeting location. The EAAB members are invited to participate in the poll which will result in the choice of the best poster. The chosen poster receives a best poster award which consists of a supplement to the aforementioned subsidy given for presenting papers at international conferences.

EAAB members receive a newsletter detailing the contents of the meeting and the main conclusions thereof, to strengthen the bonds and keep memory of the results and advices.

7.5 Sustainable campus

EfS faculty members have been involved for many years in activities supporting the university needs mainly in the fields of building retrofit and building management, seeking overall efficiency improvements in the use of resources. This translates into two main different types of involvement. In some cases, faculty members involve themselves directly in free consultancy to the university when complex or highly specialized issues are at stake. For example, since the University of Coimbra is one of the few in the world which is in the UNESCO cultural heritage list, there are from time-to-time problems associated to the quality of internal environment of historic buildings or physical preservation issues of those buildings which require very specialized skills that faculty are able to exert. In many other cases, students are mobilized in their dissertation works, or within the scope of specialized courses they attend, to help serving not so difficult issues. In these cases, it is frequent to find energy audit projects, building management systems retrofits and other types of interventions where the guidance of faculty helps students acquire practical competencies in solving problems, at the same time benefiting the university and helping the Facilities Management department.

However, besides this regular type of activity, the EfS Initiative seeks to intervene in the university in a more structured manner, at a higher level of problem hierarchy. In fact, the Initiative has actively participated in the definition of several measures included in the university's strategic plan within the sustainable development dimension of the plan. Besides, it has delivered several proposals regarding the organization of a professional structure dedicated to sustainable campus management. The present university leadership did not respond to these proposals, confirming the traditional barrier of lack of sensitivity of corporate management for sustainable development issues and the classical difficulties related to the needed convergence between top-down and bottom-up approaches to sustainability in HEI. Actually, since the EfS Initiative exists as a bottom-up movement, it lacks the necessary levers to determine decisive practical action. As this configures a stereotype situation [30], the road ahead is also very well known: keep trying to influence decisions, never give up and use all the opportunities possible to demonstrate the need of certain actions.

Additionally, the Initiative is presently planning to organize awareness-raising projects within the university community in order to achieve results in the specific dimension of sustainable campus praxis which corresponds to active dissemination and awareness raising towards sustainability practice and habits and sustainable campus utilization.

7.6 International dissemination and networking

Since 2013, the EfS Initiative has been organizing every 2 years an international conference under the common designation EfS (www.itecons.uc.pt/efs2013/, http://efs2015.uc.pt/index.php?module=sec&id=283&f=1, www.efs2017.uc.pt/index.php?module=sec&id=438&f=1), – see Table 7.2 – which will have its next occurrence in 2019, in Italy, in cooperation with an Italian HEI. Among other positive effects, these conferences have been a proven instrument to attract researchers and foster discussion on several timely topics regarding sustainable development, as well as to stimulate networking and contribute to the powerful stream of scientific contributions to a prosperous sustainable future.

BEHAVE 2016, the fourth European Conference on Behavior and Energy Efficiency (www.uc.pt/en/org/inescc/org_scientific_events/behave2016), which was co-organized by the EfS Initiative, the European Energy Network and the Portuguese Energy Agency, took place in Coimbra, Portugal, on September 2016, being an additional example of the international cooperation for dissemination and networking on sustainability, particularly on the human dimensions of energy use.

Table 7.2 EfS international conferences

Year	General theme	Place	Number of countries of participants
2013	Sustainable cities: designing for people and the planet	Coimbra, Portugal	22 (4 continents)
2015	Sustainable cities: designing for people and the planet	Coimbra, Portugal	20 (5 continents)
2017	Designing cities and communities for the future	Madeira Island, Portugal	22 (4 continents)
2019	Designing a sustainable future	Torino, Italy	na

Table 7.3 Some indicators of the EfS initiative role as a complementary platform to R&D units

Indicator	Yearly average	Obs.
Started R&D and outreach projects	3.3	2007–2017
Concluded PhD theses	5.6	2013–2017
Concluded MSc dissertations	9.0	2010–2017
Competitive funding raised	–	A total of 3.2 million euros (2008–2017)
PhD students funded travels	19.4	2011–2017

The EfS Initiative has also been representing the University of Coimbra at a platform created by the European Universities Association (EUA) designated EUA Energy & Environment Platform which "aims to facilitate the full participation of European universities in energy- and environment-related EU programs and to achieve the creation of the Energy Union and a sustainable energy future for Europe" [31]. The Initiative took part of the inaugural event at the Delft University of Technology in 2012 and has been active since then in the platform, being part of the European Atlas of Universities in Energy Research & Education.

In Table 7.3 some indicators are presented to illustrate the role that the EfS Initiative has been playing as a complementary actor to its associated R&D units.

7.7 Conclusion

The existence of the EfS Initiative is per se a demonstration of the viability of fruitful interdisciplinary work and of structured approaches to make HEIs responsibly acting towards the UN Sustainable Development Goals. It has been active for 12 years in spite of its bottom-up nature, of the scarcity of resources and of the reliance on almost only volunteer effort of a growing number of faculty members. The typical barriers to interdisciplinary education and research apply in this case, stressed by the fact that there is practically no top-down active commitment from the university government. This peculiar resilience is likely due to the fact that faculty and other researchers find more positive value in the advantages of the existence and operation of the Initiative than the negative value of having to deal with the existing institutional, bureaucratic and behavioural barriers.

To the author's knowledge, taking into account the characteristics of the EfS Initiative, it can be said that it does not fit clearly into any of the categories found in the literature, namely in those studies which are based on inquiries launched to HEI. EfS has an intrinsic reluctance to adopt bureaucratic procedures or regulations. The main operational principles it is based on are three simple attitudes: sense, sensibility and mutual trust among all the researchers. It lived and progressed more than 12 years up to now based on these principles. In spite of the limitations that currently exist because of lack of top-down commitment and lack of resources, this track record shows that it is possible to contribute to sustainable development in HEI based on voluntary effort, strong commitment and democratic conviviality.

To deepen the understanding of the relation between bottom-up initiatives and corporate institutional government, as well as to identify the genesis and development of stimuli and detrimental factors to interdisciplinary work, a study is currently being prepared aiming at the definition of policies to foster interdisciplinary education and research on sustainability in HEIs, using the case study of the EfS Initiative.

Acknowledgement

The author acknowledges the support by the Portuguese Foundation for Science and Technology (FCT) under project grant UID/MULTI/00308/2013.

References

1. UNESCO (2012). Exploring Sustainable Development: A Multiple-Perspective Approach. Education for Sustainable Development in Action Learning & Training Tools N°3. UNESCO.
2. UNESCO (2012). Education for Sustainable Development. Source Book. UNESCO. ISBN 978-92-3-001063-8.
3. UNESCO (2017). Education for Sustainable Development Goals. Learning Objectives. Education 2030. United Nations Educational, Scientific and Cultural Organization. France. ISBN 978-92-3-100209-0.
4. UNESCO (2017). Guidelines on Sustainability Science in Research and Education, found online at https://unesdoc.unesco.org/ark:/48223/pf0000260600 on 27/05/2018.
5. Leal Filho W. (2015). Education for sustainable development in higher education: Reviewing needs. In: Leal Filho W. (ed.) *Transformative Approaches to Sustainable Development at Universities*. World Sustainability Series. Cham: Springer. doi: 10.1007/978-3-319-08837-2_1.
6. International Association of Universities - IAU (2014) Iquitos Statement on Higher Education for Sustainable Development, found online at http://iau-hesd.net/sites/default/files/documents/iau_iquitos_statement_on_hesd_2014.pdf on 27/05/2018.
7. Wu, Y-C.J and Shen, J.-P. (2016). Higher education for sustainable development: A systematic review, *International Journal of Sustainability in Higher Education*, 17(5), 633–651. doi: 10.1108/IJSHE-01-2015-0004.
8. UC Davis, www.ucdavis.edu/about/university-initiatives, (visited in 28-03-2018).
9. University of Hartford, www.hartford.edu/aboutuofh/office_pres/womensadvancement/default.aspx, (visited in 28-03-2018).

10. Boston University, www.bu.edu/ioc/, (visited in 28-03-2018).
11. University of Maine Green Campus Initiative, https://umaine.edu/gci/, (visited in 28-03-2018).
12. University of Montana, www.umt.edu/sustainability/operations/Energy/smart%20buildings%20initiative.php, (visited in 28-03-2018)
13. University of Delft, www.tudelft.nl/en/energy/, (visited in 28-03-2018).
14. Cornell University, www.sustainablecampus.cornell.edu/, (visited in 28-03-2018).
15. Massachusetts Institute of Technology, http://energy.mit.edu/, (visited in 28-03-2018).
16. Duke University, https://energy.duke.edu/, (visited in 28-03-2018).
17. University of Coimbra, Energy for Sustainability Initiative, www.uc.pt/en/efs, (visited in 28-03-2018).
18. Morzillo, A.T., Seijo, F., Reddy, S.M., Milder, J.C., Martin, S.L., Kuemerle, T., Rhemtulla, J.M. and Roy, E.D. (2013). The elusive pursuit of interdisciplinarity at the human–environment interface, *BioScience*, 63(9), 745–753, doi: 10.1525/bio.2013.63.9.10.
19. Gaff, J. (1994). Overcoming barriers: Interdisciplinary studies in disciplinary institutions, *Issues in Integrative Studies*, 12, 169–180.
20. Lattuca, L. R. (2001). *Creating Interdisciplinarity: Interdisciplinary Research and Teaching among College and University Faculty*. Nashville, TN: Vanderbilt University Press.
21. Pallant, E., Boulton, K., and McInally, D. (2012). Greening the campus: The economic advantages of research and dialogue. In: Leal Filho W. (ed.) *Sustainable Development at Universities: New Horizons*. Frankfurt: Peter Lang Scientific Publishers; pp. 373–382
22. Qian, W. (2013). Embracing the paradox in educational change for sustainable development: A case of accounting, *Journal of Education for Sustainable Development*, 7(1), 75–93. doi: 10.1177/0973408213495609.
23. Lozano, R. (2006). Incorporation and institutionalization of SD into universities: Breaking through barriers to change, *Journal of Cleaner Production*, 14(9–11), 787–796. doi: 10.1016/j.jclepro.2005.12.010.
24. Brinkhurst, M., Rose, P., Maurice, G., and Ackerman, J. D. (2011). Achieving campus sustainability: Top-down, bottom-up, or neither? *International Journal of Sustainability in Higher Education*, 12(4), 338–354. doi: 10.1108/14676371111168269.
25. Lozano, R., Ceulemans, K., Alonso-Almeida, M., Huisingh, D., Lozano, F., Waas, T., Lambrechts, W., Lukman, R., and Huge, J. (2015). A review of commitment and implementation of sustainable development in higher education: Results from a worldwide survey, *Journal of Cleaner Production*, 108, 1–18.
26. United Nations General Assembly Resolution A/RES/70/1 Transforming our world: the 2030 Agenda for Sustainable Development, 2015.
27. Diab, F. and Molinari, C. (2017). Interdisciplinarity: Practical approach to advancing education for sustainability and for the Sustainable Development Goals, *The International Journal of Management Education*, 15, 73–83.
28. Batterman, S., Martins, A. G., Antunes, C. H., Freire, F., and Silva, M. G. (2011). Development and application of competencies for graduate programs in energy and sustainability, *Journal of Professional Issues in Engineering Education and Practice*. doi: 10.1061/(ASCE)EI.1943-5541.0000069.

29. Ávila, L., Leal Filho, W., Brandli, L., Macgregor, C.J., Molthan-Hill, P., Özuyar, P.G., and Moreira, R.M. (2017). Barriers to innovation and sustainability at universities around the world, *Journal of Cleaner Production*, 164, 1268–1278. doi: 10.1016/j.jclepro.2017.07.025.
30. Filho, W., Wu, Y.C., Brandli, L.L., Avila, L.V., Azeiteiro, U.M., Caeiro, S., and Madruga, L.R. (2017). Identifying and overcoming obstacles to the implementation of sustainable development at universities, *Journal of Integrative Environmental Sciences*, 14(1), 93–108. doi: 10.1080/1943815X.2017.1362007.
31. European Universities Association, http://energy.eua.eu/, (visited in 29-03-2018).

chapter eight

Clean energy and SDGs

Opportunities and challenges for digital transformation of renewable energy

Prakash Rao and Harsha Bake
Symbiosis International (Deemed University)

Contents

8.1 Introduction

In the quest to achieve a 2°C scenario as part of the global climate negotiations process, global efforts are being made to reduce greenhouse gas (GHG) emissions. Various international platforms foster the idea of discussion among developing and developed world to create solutions to the rising issue of global warming. The Paris agreement was one such instrument that is now legally binding, with Europe and India ratifying the agreement in November 2016. Sixty percent of GHG emissions come from power sector, and it is imperative for nations to look at their energy mix. The Conference of Parties at Paris (COP 21) highlighted that climate action was at the centre of the global development goals and also the importance of clean energy sources to provide sustainable energy to all with the aim of keeping the global average temperatures well within 2°C. The agreement was a step forward for a cleaner energy future. It set the stage for international cooperation among the governments and the businesses to adopt low carbon path for development. There are 195 signatories to the Paris Agreement and 169 parties have ratified it. Several countries committed to increase their share of energy consumption through renewables by many folds and drastically reduce their dependence on fossil fuels by 2030.

The global renewable energy demand is expected to grow by 48% (or 825 GW) by 2021. Solar photovoltaic (PV) installations totalling more than 49 GW in 2015 took global PV power-generating capacity to 222 GW by the year end. The total share is 1.1% but has doubled in just 2 years (IEA, Renewable Energy Mid-Term Report, 2016). Three important pillars of a renewable energy system are reliability, affordability and sustainability.

For a reliable energy system, it is of paramount importance that demand and supply are balanced at any moment, and the capacity to transport energy through the system is sufficient. The growth in renewables has posed challenges to demand and supply every time as renewables like wind and solar depend on the availability of wind resource and sunlight, respectively. There is also an increase in the local energy generation technologies across the globe. It is evident that there needs to be a change in the current energy system. The increased use of renewable energy sources and increase in decentralised energy is putting burden on the current energy system. The current energy system was initially designed for the conventional energy sources and passive consumers. Renewables and decentralised energy gives consumers the opportunity to be the energy producers. However, there are challenges in making the electricity grid 100% renewable. Renewable energy is intermittent in nature. This means it is not capable of providing continuous supply of energy to the consumers.

In an Information technology age, the role of digital transformation is one of the major drivers for a sustainable future with societal benefits across organisations, industries, consumers and civil society. The renewable energy sector is currently witnessing a rapid deployment across developed and developing countries as a technology-driven initiative towards (1) providing clean energy access, (2) providing a long-term solution for climate mitigation strategies and (3) a socially relevant technology strategy for emerging developing markets. The importance of digitalisation can play an important part in improving renewable energy systems to make it more reliable, accessible and predictable (IEA,2017). The extremely dynamic nature of the renewable energy systems and processes is adding to this challenge. An overview about the digital technologies that will help transform the electricity grids is described.

8.2 Review of literature

The concept of using digital technologies to develop renewable energy systems has been explored by many authors. A few of the important advances in this area are described below:

8.2.1 Integration of renewable energy sources using artificial intelligent system

Recent studies suggest that artificial intelligence (AI) controlled system is developed to effectively integrate the various sources of renewable energy. The output of this integrated renewable energy system is considered to be efficient and more cost-effective. The voltage and frequency of various loads are mapped, and the sizing of the renewable energy system is decided according to the varying loads. This kind of a system makes better use of the renewable sources of energy with optimised output, thus leading to larger use of renewable sources (Harsh & Singal, 2014).

8.2.2 *Linking renewable energy technologies through data analytics: challenges and opportunities*

The research shows how data analytics workloads are used for supply following mechanisms. The degree of integration of renewable energy sources is also studied. It is found that the supply following scheduling yields a 40%–60% improvement in the integration of the renewable energy. It also tells that the degree of renewable energy integration depends on the variability of the renewable source and intermittence and the scheduling slack in the data analytic workload (Krioukov et al., 2011).

8.2.3 *Data set management in smart grids: concepts, requirements and implementation*

Data management issues are an important element of smart grid processes. It also presents the life cycle of large data sets in smart grid. The implementation of big data analytics is presented with the help of a case of customer data analytics. (Daki, Hannani, Aqqal, Haidine, & Dahbi, 2017).

8.2.4 *Use of machine learning methods for supporting renewable energy generation and integration through surveys*

The authors have illustrated the issues of integrating renewable energy into the power grid. This included conducting a survey on various machine learning techniques which are being used for forecasting and grid management (Perera, Aung, & Woon, 2014).

8.2.5 *Integration of renewable energy sources within the electricity grid*

This section illustrates the issues related to problem of integration of the renewables within the electricity grid. The study explains that in the distribution grid, voltage decreases along the direction of current flow. Rising cable length increases the resistance and inductance of the cable. Renewable power plants are integrated in lower grid levels owing to its less capacity. Therefore, power can flow towards the transformer leading to voltage rise as more distributed generators are connected to the grid. The main problem is the barrier of insufficient grid capacity available for renewable energy. Issues of grid imbalance lead to additional input of fossil fuels in the main grid, thereby increasing the cost of power generation. Decentralised grids, also known as virtual power plants or microgrids, are suggested (Kammer & Kober, 2009).

8.2.6 *Industrial blockchain platforms a case development in the energy industry*

This section provides a practical view on how to develop and to describe blockchain use cases. A provisional use for autonomous machine-to-machine transactions of electricity is described using the case study of a housing sector. This is achieved through an iterative process with stakeholders in the energy industry. It evaluates the concept and its technical specifications against six criteria for a sensible blockchain use case. The study also provides findings and analysis on the use-case development process and its future steps (Mattila et al., 2016).

8.2.7 *Connecting the blockchain through SolarCoin network*

The section explains the SolarCoin network. Mechanisms cited include their structure and process, which is Proof-of-Work (PoW) Phase and Proof-of-Stake-Time (PoST) Phase. It has been shown that SolarCoin uses much less energy compared to BitCoin due to the use of PoST protocol. A solar-powered SolarCoin node has been constructed and tested for a 11-month period in a research lab in Tokyo with long-term benefit. (Johnson, Isam, Gogerty, & Zitoli, 2015).

8.2.8 *Artificial intelligence and the future of renewable energy*

This section describes that by increasingly automating the wind and solar operations with the help of AI, improved forecasting of renewable energy, efficient balancing of the grid and improved analysis of consumers' data can be achieved. It also describes how companies are investing in business intelligence. Various energy platforms that use AI, such as Deep Mind and Nergix, are discussed (Poola, 2017).

8.2.9 *Implementation of blockchain-based energy trading system*

This study has been made to depict the implementation of a system to apply blockchain technology to power trading. The authors have modelled a power trading process by integrating interaction between the administrator, producer and the consumer nodes. Using a blockchain platform known as Multichain, a scenario has been created. This scenario is swift and highly scalable (Oh, Kim, Park, Roh, & Lee, 2017).

8.2.10 *Big data-driven smart energy management: from big data to big insights*

There is large amount of data sets that are currently available in the energy sector through various power systems on the energy consumption and production, energy losses, energy efficiency etc. Smart grids pose a useful opportunity of smart grid management through the use of big data analytics (Shukla, 2017). Characteristics of large data sets and a process model of smart energy management is described. The authors also provide current research status and industrial development of big data-driven smart energy management (Zhou, Fu, & Yang, 2016).

8.2.11 *Issues, challenges, causes, impacts and utilisation of renewable energy sources – grid integration*

There are issues in integration of renewable source of power directly into the grid. The issues of intermittency of solar and wind and their impacts on the traditional grid have been discussed. Low power quality and power fluctuations are the main impact of the grid integration of solar or wind power. Some solutions to overcome these issues have also been discussed (Sandhu & Thakur, 2014).

8.2.12 *Smart grids and renewables – a guide for effective deployment*

This report provides a detailed study on the smart grid technologies such as smart meters, demand response, distribution automation, microgrids and virtual power plants and renewable resource forecasting. It also describes the non-technical barriers to the smart grids such as data ownership and privacy, grid security, control of distributed resources and the need for standards (Kempener, Komor, & Hoke, 2013).

8.3 *Methodology*

The study was conducted based mainly on secondary research. Since the digital technologies such as blockchain are nascent technologies, there is a lack of scientific approval as well as published research papers till date. Sources such as websites of leading companies, start-ups, private sector, news, blogs and articles were taken into consideration for this study. Since these technologies are complex to understand, the analysis is presented in a simple language so that it can be easily understood by those who wish to learn about the application of these technologies in renewable energy sector.

8.3.1 Objective

The objective of the study is to study the recent developments on integration of renewable energy sources into the grid through emerging technologies such as Blockchain, Big Data and AI.

The secondary research considers (1) the challenges posed due to renewable energy integration into the grid, (2) the role of digital technologies in helping the grid to transform it to the next level in line with the SDGs and (3) how can a country like India adopt digital technologies for its renewable energy sector including modernising and strengthening grid structure into a reliable smart grid.

8.4 Results and analysis

A country's development and economic growth is often said to be directly related to energy consumption. With the advent of energy sources, the world relied heavily on the fossil fuels. With more than three decades of industrial revolution, the world has largely been driven by fossil fuel-based economies. This led to exponential increase in the Greenhouse gas levels in the atmosphere thereby increasing the average surface temperature of the planet leading to climate variability. Energy production and use accounts for nearly two-thirds of global greenhouse gas emissions. Therefore, focus on the energy sector is essential for mitigating the adverse effects of climate change. In 2015, the world took an important step in the fight against climate change through the finalisation of the United Nations Framework Convention on Climate Change (UNFCCC)-based Paris Agreement adopted by nearly 200 countries in the world. With energy issues a core topic for discussion, the International Energy Agency proposed to accelerate the investments in renewable energy and technologies. According to Bloomberg New Energy Finance, investments in clean energy exceeded in third quarter of 2017 compared to all previous quarters amounting to $66.9 billion (Figure 8.1).

In 2014, the Government of India set an ambitious target of 175 GW of renewable energy capacity to be installed and generated by 2022. This included, 100 GW from solar energy (60 GW targeted from utility scale solar capacity and 40 GW through rooftop solar capacity), 60 GW from wind energy, 10 GW from biomass-based energy and 5 GW from small hydro power. This would provide immense opportunity to the Indian and foreign investors and the corporate sector to play an important role in the growth of renewable energy in India. Rooftop solar installations in India continue to grow and attract a $23 billion-dollar investment opportunity (Bloomberg New Energy Finance, 2017).

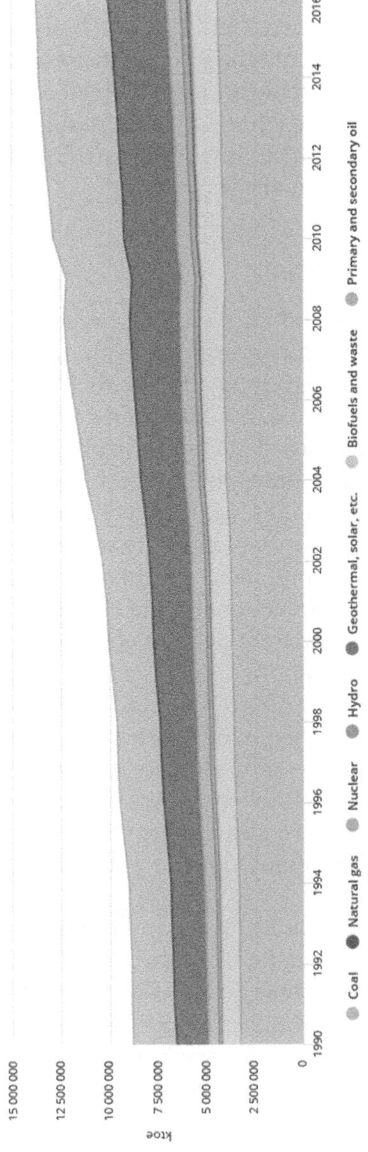

Figure 8.1 World total primary energy supply (TPES) by fuel (1990–2016). (Source: Key World Energy Statistics, International Energy Agency, 2018.)

Renewables are increasingly affordable because of their constant technological development. A renewable energy plant has much lower costs compared to thermal or hydro power plants and generates value immediately at the local level. Renewable energy integration into the traditional grid poses several challenges such as bad power quality due to harmonics and voltage fluctuations, storage issues, protection issues, less availability of transmission lines, among other reasons (Sandhu & Thakur, 2014). In the coming decades, digital technologies are set to make energy systems around the world more connected, intelligent, efficient, reliable and sustainable. Spectacular advances in data, analytics and connectivity are enabling a range of new digital applications such as smart appliances, shared mobility and 3D printing(IEA,2017). Digitalised energy systems in the future may be able to help energy consumers identify the appropriate power sources for their use apart from choosing the reliability, delivery and affordability of the power being generated.

8.4.1 Blockchain

There is a lot of attention being given to the concept of blockchain in recent times (Gupta, 2017). Blockchain first came to be known as a technology behind BitCoins. While the concept had largely been used for financial markets and transactions, its use today could be applied to other sectors as well beyond financial transactions and cryptocurrency and is bound to impact various sectors. Blockchain is defined as a distributed digital ledger, which enables two parties to create secure, verifiable and make transactions without the involvement of an intermediary like a financial institution (Mattila et al., 2016). A ledger is a file where transactions are recorded. A general ledger is centrally owned; blockchain technology is a digital record of ownership distributed across an open network. The information in the distributed ledger is coded with cryptography enabling secure transfer of the assets. It is a giant ledger which is constantly and instantly updated. As soon as the transaction occurs, it gets recorded permanently. Therefore, it eliminates the need of transaction clearing houses. The financial institutions across the world are already using this to fasten up the fund transfers.

8.4.1.1 The present scenario of energy trading

Renewable energy generated can be fed into the grid or can be generated off-grid. Connecting all the energy-consuming devices and its consumers to one national grid is a major challenge. That requires huge transmission and distribution network and huge investments. Hence, decentralised renewable energy sources are gaining a lot of attention especially in rural areas (Kammer & Kober, 2009).

Presently, energy is traded on power exchanges consisting of short-, medium- and long-term trading. Electricity generators sell their electricity to the consumers. Since storage capacity is not developed to a large extent, electricity has to be produced at the time of use. This is the reason why various markets such as the long-term market, the day ahead market and the intraday market have been developed. However, often, the involvement of power utilities between the seller and the buyer leaves little room for tariff that a seller gets paid. Blockchain removes this middle barrier and directly allows the seller to carry the transaction with the buyer.

8.4.1.2 What does blockchain offer in the energy industry?

The use of blockchain as an emerging technology allows peer-to-peer energy trading. In Perth, Australia, there is a peer-to-peer energy trading for residential homes developed by a startup (Power Ledger, 2017). The excess solar power generated by a household can be traded with another household who needs it. In a traditional way, this can be traded on power exchanges. The intermediary power company approves the transaction and pays a small amount to the seller. The power company then sells it to the buyer at a much higher price. This new approach opens the market and removes the power company intermediary. The seller gets the price much higher than the feed-in tariff. The uniqueness of this method allowed the organisation to build a software-based platform to read and analyse the outputs of household electricity meters along with data on the amount of energy consumed or generated. In a short span of time (6 weeks), the organisation raised $34.15 million in Australia's first ever ICO (Initial Coin Offering), which involves a blockchain company selling part of its ledger to investors which gives them access to the function of the blockchain.

Another application which is being explored is to allow multiple suppliers at one connection. In the automobile industry, particularly in the manufacture of electric vehicles (EV) the consumer can choose a supplier for one charging unit and a different supplier for their EVs. Furthermore, it could enable micro-transactions that are selecting the supplier which offers cheapest energy to the consumer.

Blockchain applications need smart meters to work with. This increases the billing process efficiency, provides more accurate energy usage data and gives the users access and visibility of their data. Thus, transparency created is likely to lead to fairer pricing. In a microgrid in Brooklyn, New York, energy trading takes place between the neighbours using the blockchain technology. Direct benefit of this technology is that the buyer and seller trade renewable energy, and at the same time, the seller receives the payment. This reduces the processing time for the payments that would take anywhere between 15 and 40 days in traditional way. SolarCoin (Johnson, Isam, Gogerty, & Zitoli, 2015), a digital currency for renewable energy based on the blockchain technology was an initiative developed in 2014 by a group of volunteers interested in promoting a technology and incentive-based solution aimed to benefit the environment. This was a global payment program for solar electricity generation. One SolarCoin signified 1 MWh of solar electricity generation. SolarCoins are claimed by individuals having solar panels installed on their homes or commercial centres. The goal is to incentivise the solar electricity producers. In many countries, particularly in Europe, the solar subsidies are phasing out or will be phased out in short term. So this kind of personal choices can also help the countries. It will also help reduce the payback time of the solar installation, thereby creating an uptake in the solar equipments and boosting the Renewable Energy Sector jobs. The SolarCoin Foundation as a nodal agency undertook the task of supplying SolarCoins to potential solar electricity generators using verified solar facilities as the "proof of work". The Foundation also maintains a public ledger which tracks each SolarCoin given out to solar electricity generators.

8.4.1.3 Best practices by private sector

The Energy Web Foundation is a global non-profit organisation supporting the growth of blockchain technology in the energy sector. Evolution Energie is building the system on Predix, an industry-based platform for the Industrial Internet. Predix Software gives energy traders real time machine and operations data on how any particular power plant is functioning so that they make better financial decisions. This has helped build one of the first smart solar grid in Carros, in the South of France as well as helped partner with French grid operators to install solar panels in homes and offices, to implement demand response technologies and to create battery storage facility. A Distributed Energy Resource Management (DERM) software manages this information. This helped Carros to build numerous microgrids. People can sell electricity to offices during the day when they are away from homes and buy electricity from those who have stored it in the batteries, electric vehicles or used less energy for the day.

8.4.2 Artificial intelligence in renewable energy industry

To achieve full reliability, increased efficiency and better power quality, efficient algorithms are required (Harsh & Singal, 2014). AI technology can be used to derive past variables, improvise on the present and predict the future variables. AI technology can be used for

- Renewable energy forecasting
- Grid balancing
- Predictive maintenance

AI is in the nascent stages of its implementation but can revolutionise the renewable energy industry. AI is the ability of a machine to perform the same kinds of functions as humans do or humans think (Poola, 2017). This technology will continuously collect the energy and weather data with the help of sensors, synthesise it and give better outputs. This increases the efficiency of predictions and gives better insights to the users of data.

8.4.2.1 What does AI offer for the renewable energy industry?

Renewable energy forecasting – AI machines can gather real-time data from the IoT devices, weather stations, satellites to accurately forecast the solar power generation or wind power generation during different times of the day. This will help in factoring the additional sources of power needed and arranging for the same to cater to the variable loads. Industry players involved in forecasting renewable energy are GE Power, National Center for Atmospheric Research and Xcel Energy (Bullis, 2015).

Grid balancing – Since solar and wind power are intermittent sources, keeping the frequency of the current stable is a problem. Deep learning method can be used to balance the demand and supply side. With the help of smart meters, constant monitoring of demand and supply happen leading to creation of large data sets (Kempener, Komor, & Hoke, 2013). Deep learning algorithms fix any irregularities in large data sets and learn on their own. Along with this, AI can include data about weather, time, region, zones, seasons and provide better clarity on arranging power for the loads in short response time.

Predictive maintenance – It is important that developer or user gets to know when a solar panel or wind turbine would fail so that maintenance can be scheduled to avoid any future losses. This would increase the efficiency of renewable energy (RE) systems. AI offers help in predictive maintenance.

8.4.2.2 AI based Best practices by private sector

Several organisations in the private sector are currently focused on building suitable solutions for blockchain technology across various sectoral disciplines including energy. A global internet-based organisation has applied the AI technology known as 'synchrophasers' that would determine the flow of electricity through the grid in real time. This helps the engineers to avoid any disorders. These sensors communicate with the grid and modify electricity consumption during off-peak times, thereby reducing the workload of the grid and lowering prices for consumers as well as cost savings.

Global IT hardware companies have also developed a tool for solar and wind forecasting technology. The technology is built by utilising dozens of forecasting models and then assimilating a multitude of data sources about the weather, the environment, atmospheric conditions and solar plants and the power grids operating process. The predictions range in availability anywhere between every 15 min up to 30 days in advance.

In the manufacturing industry, organisations have also used these technologies, e.g., installing a digital wind farm in India that leverages AI, IoT and analytics. This has improved the efficiency of the wind turbines through which 30% increase in gross annual energy production over its predecessor is achieved. The technology used has also helped in improved efficiency and reduced costs.

8.4.3 *Opportunities of digital transformation of the RE grid*

Diffusion of RE grids through either the microgrid network systems or processes can be beneficial to large sections of society across the developing world. There are immense opportunities in the integration of such technologies with the goals of SDGs as an income generation activity and also to promote climate action. Other benefits include

1. Transparency and choice for consumers is promoted as the consumers now become prosumers.
2. Decentralised peer-to-peer energy exchanges should be supported by the governments and utilities.
3. The industry must collaborate with the local state units to build smart grid with innovative technologies such as blockchain and AI which will provide real-time data and stabilise the frequency.
4. As the integration of RE sources is improved with innovation, the efficiency of solar power output or wind power output increases. This immensely reduces the cost associated with their operations.
5. As digitalisation creates vast platforms for real-time data sharing, it also calls for cybersecurity and data protection.

8.4.4 *Challenges of digital transformation of the RE grid*

Digitalisation can also bring about challenges in implementation across geographies depending on the scale and nature of the technology to be adopted apart from regulatory barriers. We also believe that

1. Increasing dependency on the digital technologies will further lead to an increase in the energy consumption; if also relied on fossil fuels to satisfy the demand will lead to a rise in the greenhouse gas emissions. To handle the vast amount of data to create and operate the smart grid, additional technology and processing power will be required. Optimisation of energy use may therefore be key to address GHG issues. Therefore, one needs to address this problem and make the smart grid as efficient as possible.

2. There are yet no government regulations towards the use of the SolarCoin which uses blockchain technology. A proper regulation would increase the use of the SolarCoin and increase the penetration of solar energy across economies.
3. With smart grids that has high integration of data flows, security of the grid poses a challenge. Hence, strengthening of the electric grid with advanced firewalls, continuous monitoring and malware protection should be made available.

8.4.5 Blockchain, artificial intelligence and the UN SDGs

In 2015, the United Nations adopted a landmark set of 17 goals to mitigate poverty, protect the planet and ensure prosperity for all as part of a new sustainable development agenda. Each goal has specific targets to be achieved till 2030. These SDGs have been formulated for the welfare of the society and seeks participation from the community, private institutions and governments.

A Blockchain4SDGs event was held in Washington DC that saw various global leaders come together to provide insights and investment advisory to scale up the application of blockchain for achieving the SDGs. This event also saw the applications of blockchain in energy trading. To achieve the targets of clean and affordable energy (SDG 7), scaling up investments in clean energy is essential. With innovation and emergence of digital technologies (SDG 9), use of digital technologies in renewables is fast making progress. This sees a spur in new jobs added to the renewable energy market (SDG 8). Digitalisation is bound to impact each of the 17 SDGs. Smart energy management with the implementation of smart grids and predictive data analytics, billions of MWh could be saved by 2030. This would mean saving million tons of CO_2 emissions in the atmosphere aligning with the climate action goal (SDG 13). Europe has initiated building policies for incorporating the digital technologies into the power sector. India also must follow its leads and partner with the private sector to achieve the SDGs (SDG 17).

8.5 Recommendations

1. Capacity building for the utilities to digitise the contracts, strengthening the smart metering systems and monitoring.
2. Blockchain and AI can prevent blackouts by modernising the grid. It can reroute the power automatically and respond to grid emergencies.
3. Blockchain empowers the people to act as prosumers. Thus, a resilient peer-to-peer power system is built.

4. Renewable energy certificates can be tracked with the help of blockchain enhancing transparency. Since it is driven by smart contracts, it becomes more efficient.
5. Barriers to regulations of peer-to-peer energy trading must be removed.
6. Smart meters deployment should be accelerated, and imposing extra costs on the consumers owning smart meters should be avoided.
7. Local states must guarantee that best cybersecurity standards are put in place to avoid any data loss or theft.
8. Robust regulatory framework has to be built in India to incorporate the applications of blockchain and AI in the energy sector.

8.6 Scope of further research

India's renewable energy market is growing at a rapid pace. The technologies discussed here are not yet mature but growing fast with several start-ups also following suit. UNFCCC recently recognised the potential of blockchain in addressing climate change issues and opportunities that it provides for climate action. Hence, this provides a lot of scope for further research. India has set a roadmap for developing smart cities which would need smart electricity. Microgrids and blockchain could power our energy in future. Few areas of future research could include:

1. Developing and simulating working model for blockchain-based energy trading in India's villages or smart cities.
2. The need of regulatory framework in India to implement the digital technologies discussed in this work, in the renewable energy sector.
3. The need of financial investments for digital transformation of renewable energy sector of India.

8.7 Conclusion

India has a current installed RE capacity close to 79 GW. The installed operational capacity of wind energy is 36 GW, followed by 29 GW of solar energy, 4.6 GW of small hydro power and 9 GW of biomass-based power. While sizeable, this is significantly short for the 175 GW renewable energy capacity target set for 2022. India's renewable dream is growing at a rapid pace. Integrating renewable energy into the power grid is a challenge due to its intermittency that fluctuated the voltage level and frequency causing harmonic distortions. The Government of India has an ambitious plan to develop and install 35 million smart meters by 2019 as part of its campaign towards a digital India. With a target of 175 GW of renewable energy generation projected by 2022, there would be a substantial variability, volume and variety. The National Electric Mobility Mission foresees about 6–7 million electric vehicles in India by 2020, with a chance of overloading the distribution network. Both supply and demand of electricity are on the verge of becoming unpredictable and variable. Therefore, utilities would need automation and machine-to-machine communication systems capable of high reliability operations. Hence, this study gives an overview of emerging digital technologies Blockchain and AI that will help transform the renewable sector. Blockchain applications in energy trading is going live at few places cited here. In India, too trial test versions are taking place. In India, recent government initiatives for fulfilling energy security needs have focused on rural electrification across its many villages. Distributed energy sources and microgrids provide various decentralised options for energy generation in these villages through solar, hydro, wind and biomass-based power generation. However, it is important to mention that, with the convergence of digital technologies in the power sector, the risk of cyberattacks is ever increasing. Thus, utilities will have to embed robustness in their systems. India needs to build a strong regulatory framework for integration of blockchain and AI in power sector. With digital transformation of renewable energy, achieving SDGs 7, 8 and 9 is possible along with India's future vision towards promoting smart clean energy.

References

Bloomberg New Energy Finance. 2017. Accelerating India's Clean Energy Transition.

Bullis, K. 2015. Smart Wind and Solar Power. www.technologyreview.com/s/526541/smart-wind-and-solar-power.

Daki, H., Hannani, A. E., Aqqal, A., Haidine, A., & Dahbi, A. 2017. Big Data management in smart grid: Concepts, requirements and implementation. *Journal of Big Data*, 4(1): 13.

Gupta, V. 2017. A Brief History of Blockchain. https://hbr.org/2017/02/a-brief-history-of-blockchain.

Harsh & Singal, S. K. 2014. Integration of renewable energy sources. *International Journal of Innovative Research in Science, Engineering and Technology*, 3(11): 17291–17305.

International Energy Agency. 2017. *Digitalization & Energy*. Paris: International Energy Agency, 188.

International Energy Agency. 2018. *Key World Energy Statistics*. Paris: International Energy Agency.

IRENA. 2016. Renewable Energy Mid-Term Market Report. p. 12.

Johnson, L., Isam, A., Gogerty, N., & Zitoli, J. 2015. Connecting the Blockchain to the Sun to Save the Planet. https://ssrn.com/abstract=2702639 or doi: 10.2139/ssrn.2702639.

Kammer, P. & Kober, A. 2009. Grid integration of renewable energy sources. *8 EEEIC International Conference on Environment and Electrical Engineering*. Karpacz, Poland: Technical University of Cottbus, Informations-, Kommunikations- und Medienzentrum (IKMZ). pp. 353–356.

Kempener, R., Komor, P., & Hoke, A. 2013. Smart grids and renewables-a guide for effective deployment. *IRENA*. p. 44.

Krioukov, A., Goebely, C., Alspaugh, S., Chen, Y., Culler, D., & Katz, R. 2011. Integrating renewable energy using data analytics systems: Challenges and opportunities. *Bulletin of the IEEE Computer Society Technical Committee on Data Engineering*, *34*(1), pp. 3–11.

Mattila, J., Seppälä, T., Naucler, C., Stahl, R., Tikkanen, M., Bådenlid, A., & Seppälä, J. 2016. Industrial Blockchain Platforms: An Exercise in Use Case Development in the Energy Industry. *ETLA Working Papers*, No 43. http://pub.etla.fi/ETLA-Working-Papers-43.pdf. p. 20.

Oh, S.-C., Kim, M.-S., Park, Y., Roh, G.-T., & Lee, C.-W. 2017. Implementation of blockchain-based energy trading system. *Asia Pacific Journal of Innovation and Entrepreneurship*, 11(3): 322–334.

Perera, K. S., Aung, Z., & Woon, W. L. 2014. Machine learning techniques for supporting renewable energy generation and integration: A survey. In *Data Anlaytics for Renewable Energy Integration* pp. 81–96. Nancy: Springer.

Poola, I. 2017. Artificial intelligence and the future of renewable energy. *International Advanced Research Journal in Science, Engineering and Technology*, 4(11): 216–219.

Power Ledger. 2017. Media Release: Power Ledger. https://web.powerledger.io/mediarelease/.

Sandhu, M. & Thakur, T. 2014. Issues, challenges, causes, impacts and utilization of renewable energy sources - grid integration. *International Journal of Engineering Research and Applications*, 4(3): 636–643.

Shukla, P. 2017. Big data analysis is used in renewable energy power generation. *International Journal of Computer Applications* 174(2): 37–39.

Zhou, K., Fu, C., & Yang, S. 2016. Big data driven smart energy management: From big data. *Renewable and Sustainable Energy Reviews*, 56: 215–225.

chapter nine

Paying or saving? The Greek drama that leads to achievement of Sustainable Development Goals (SDGs)

John Gelegenis
University of West Attica

Evanthie Michalena
University of the Sunshine Coast

Contents

9.1 Introduction

Global financial crises negatively affect most countries' economies and societies, delaying their economic development plans and deteriorating living standards. This was the case of the recent 2005–2008 financial global crisis, too. Greece was probably one of the most severely affected countries. In Greece, crisis and economic recession mainly lasted from 2009 to 2016 (Kindreich, 2017), which was a longer time period, due to inherent problems of the economy and the limited effectiveness of the government in the implementation of financial reforms (Artelaris, 2017; Kindreich, 2017). The country has not yet recovered, and the evolution of gross domestic product (GDP) from 2000 to 2017 (Figure 9.1) indicatively depicts the path of financial crisis in Greece.

Transitions regarding energy sustainability started getting formulated in Greece well before the recent financial crisis, in response to climate change and oil prices volatility and certainly in accordance to European energy policy. A long listing of respective legislation had been enforced to this aim, and many institutional interventions have taken place (Frantzeskaki et al., 2008). However, the initial optimistic indications related to energy sustainability faded, for several administrating and other reasons (Michalena and Frantzeskaki, 2013). State efforts towards sustainability ignored the self-generated completion of sustainability goals because of existing present or future financial forces. Indeed, authors note barriers towards sustainability transition, not only of technical and institutional nature (such as the environmental impacts

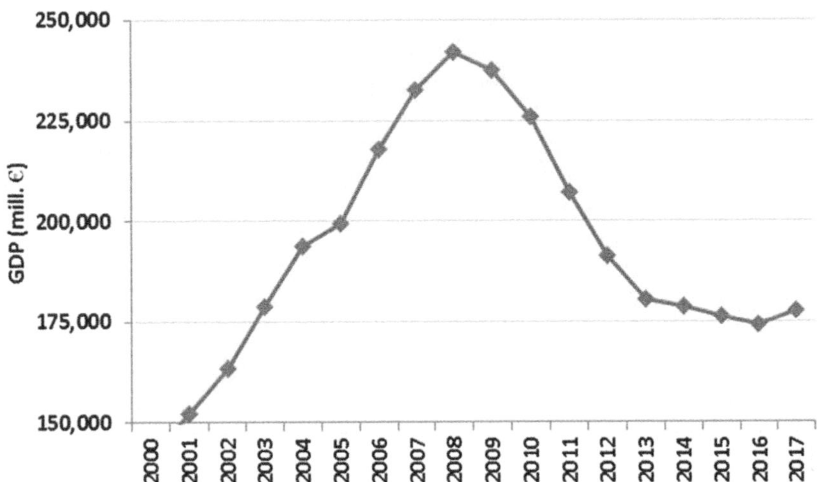

Figure 9.1 The evolution of GDP in Greece from 2000 to 2017. (Data from ELSTAT, 2018.)

of renewable energy (RE) but also the weak integration of technical, political and social factors (Michalena and Frantzeskaki, 2013).

The financial crisis in Greece is still having a significant impact on family income, which resulted in the reduction of energy consumption. Financial crisis leads, in general, to expenses decrease due to family income reduction; in this context, consumption of goods is also decreased. Exemption to this constitutes the "inferior" products' category (Ferguson, 2001), without energy belonging to this category, however. In Greece, total fuel consumption dropped dramatically (–20% since 2005) to 1.5 tonnes of oil equivalent/capita in 2015, getting the second lowest value among international energy agency (IEA) countries (IEA, 2017a). This phenomenon presents similarities with the oil price crises back in 1970s where attention was analogously drawn to the execution of energy preservation and management measures (Azam et al., 2016).

Economic recession affects national efforts to sustainable goals but not always negatively (Zarotiadis and Michalena, 2010). Focusing on energy, general public behaviours and consuming practices may change, with most of them deteriorating but some others surprisingly improving. United nations development program is using Sustainable Development Indicators and Goals to measure achievements related to sustainable development (UN, 2018a). In this work, sustainable development goal (SDG) 7 (related to energy)[1] and SDG 13 (related to climate change)[2] are mostly of interest. An interesting question to be posed here would be how a financial crisis can generate forces and reactions that can lead to continuous efforts and permanent good sustainability indicators. In other words, we attempt to explore the prerequisites through which energy and climate SDGs can be achieved in Greece.

Although focusing on Greece, the topic may be interesting even for the wealthier European countries. According to official statistics, nearly 17% of the European Union (EU)-28 population is at risk of income poverty while almost 11% of the EU population is unable to afford a proper indoor thermal comfort (BPIE, 2015). Shrinkage of disposable income may lead to cancellation of funding towards energy efficiency projects, but at the same time, it provokes energy efficiency as a *necessity*, as far as low or no cost "housekeeping" measures are concerned. Improvement in energy efficiency (which is suggested as the "first fuel", IEA, 2017b) can probably alleviate the problem.

Consequently, the aim of this work, is to investigate if economic recession in Greece and the respective decrease on GDP and the families' income, does stimulate the further development of energy consciousness

[1] Description of SDG7: "To Ensure access to affordable, reliable, sustainable and modern energy for all" (UN, 2015).

[2] Description of SDG 13: "Take urgent action to combat climate change and its impacts" (UN, 2015).

in tandem with the necessary cut-offs in energy expenses. The situation is quite complicated to be analysed through an econometric model, because the improvements achieved in energy efficiency are masked by the reduction in energy use due to the economic crisis. Besides, there is lack of detailed data related to fuel poverty in Greece, as elsewhere noticed (Atsalis et al., 2016) while there are no recent data relating to the sustainable goals of the country – the last ones go back to 2007 (HMEPPPW, 2007). However, a qualitative analysis and some rough indicators may still provide insights and reveal potential contribution of financial crisis to the achievement of energy sustainable goals. In the following lines, a set of figures have been used to feed insights and analysis regarding the relationship between financial crisis, energy savings and SDGs. The majority of the figures (unless stated differently) have been elaborated with data from UN[3] and Eurostat[4].

The chapter will develop as follows: Section 9.2 will present the financial situation in Greece and the (in)ability to pay the bills and Section 9.3 presents an outline of the weak energy governance in Greece, the opportunities lost (when it comes to energy efficiency) and the high potential for the situation to improve as long as certain measures are adopted. Then the sections related to discussion and conclusion follow, in which strong arguments are discussed on why financial crisis can be an excellent opportunity to accelerate energy efficiency measures and deepening of energy efficiency mentality.

9.2 The financial situation in Greece and the (in)ability to pay the bills

Financial crisis affected dramatically the life of Greek people. It is estimated that from 2008 until 2017, Greek households lost 25% of their income in 2015 (Eurostat, 2018b), and during the same period, the risk of poverty rate increased to 36% (Papada and Kaliampakos, 2016). Indexes related to health, education and social services have substantially deteriorated, with some returning back to their 1990s level (Artelaris, 2017). Social expenditure has significantly decreased and austerity policy measures imposed considerable cutbacks in regional development policies. The distribution of income has been shifted lower, with a recent profile as depicted in Figure 9.2.

The relative poverty limit (set at 40% of mean personal income) fell from 7,178€ in 2010 to 4,500€ in 2016 (IME GSEVEE, 2018). According to

[3] UN 2018b. Sustainable development goals. SDG Indicators. Area: Greece (available at https://unstats.un.org/sdgs/indicators/database/?area=GRC visited at February 2018).

[4] EUROSTAT 2018a. Main Tables: Energy (available at http://ec.europa.eu/eurostat/web/energy/data/main-tables visited at February 2018).

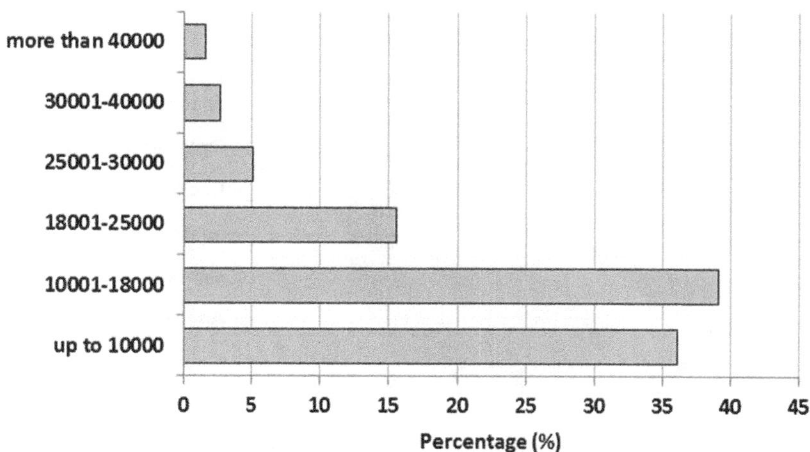

Figure 9.2 Distribution of household income, in 2017 (in €, IME GSEVEE, 2018).

IME GSEVEE (2018) almost 70% of households have at least one member unemployed looking for a job, 47.8% of households delayed or postponed medical care, 40.2% decreased food expenses while 14.6% of households cannot satisfy their primary needs. This means that, according to the poverty limit valid in 2010, almost half of the households are regarded as poor, when applying indicators of the organisation for economic co-operation and development (OECD) (OECD, 2018). Financial crisis has consequently removed society away from the sustainability goals related to poverty, proper nutrition, well-being, employment and social cohesion.

9.3 Using and managing energy in Greece during the recession years

According to a recent research (data 2011–2012), domestic energy consumption in Greece is estimated at 14,000 kWh annually per capita, composed by 10,250 thermal kWh and 3,750 electrical kWh (ELSTAT, 2013), summing a mean energy cost of 1,200–1,500€ annually per household. Although this is an apparently minor share of total mean income, it constitutes a remarkable part of the disposable income left after the subtraction of all principal expenses consisting of food, health, education costs and taxes with the latter getting enormously increased during the same period. According to a recent investigation (IME GSEVEE, 2018), a significant part of households (31%) applied trimming in heating expenses during the last 2 years, 2016–2017.

At the same time, needs for energy are high: Buildings in Greece are mostly old (41.5% of Greek residences have been built before 1979 when

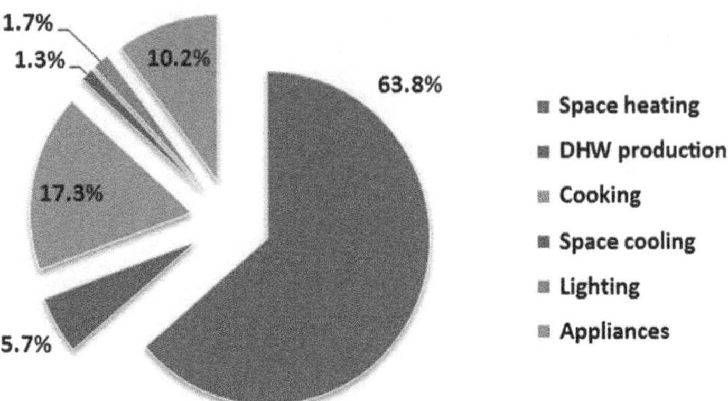

Figure 9.3 Breakdown of end use of energy in Greek households. (Data from ELSTAT, 2013.)

Insulation Regulation was enacted). While data regarding energy consumption in Greek households are not systematically collected, according to results out of a survey of European union statistics on income and living conditions, 37.5% of the respondents in Greece reported leakage, damp walls and mold, which was the higher percentage in the EU28 (Eurostat, 2015). According to another research on the topic which was undertaken by the Hellenic Statistical Authority (abbreviated in Greek as ELSTAT) back in 2011 and so in crisis early years (ELSTAT, 2013), more than 60% of energy end use is consumed in space heating, with the following consumptions accounting for much smaller percentages (see Figure 9.3).

The energy management situation proved dynamic during crisis' years. Significant changes in the energy management environment have come as a result of the general public opting for more economical solutions in their attempt to reduce of energy expenses. For instance, many people turned to electrical heating instead of oil, when the latter reached the price of 1.4€/L (in winter 2013–2014). In this perspective, it is not surprising that the second most popular heating system are now electrical heaters (24% of households, Papada and Kaliambakos, 2016), which come as a second choice behind oil boilers that remain the main system for 42% of Greek households; before crisis, the most popular systems in the country were central oil-fired boiler, unitary gas-fired boilers and unitary heat pumps (Papadopoulos et al., 2008).

9.4 SDGs and energy indicators in Greece

In Greece, a positive association between energy consumption and growth has been identified for the period 1990–2010 (Azam et al., 2016), and so including crisis period, which actually states that decrease of GDP is correlated to the decrease of energy consumption and vice versa. A similar interrelation was identified between residential energy consumption and economic growth for the period 1960–2006, before crisis (Tsani, 2010). However, these outcomes were based on macroscopic econometric analysis; hence, some technical analysis is further required to validate these findings. Some of the energy savings might be the result of energy service poverty,[5] while others might be the result of energy consciousness behaviour[6]and state's initiatives.

Scientific literature demonstrates a lack of consensus when it comes to relating energy saving and living standards in Greece: In a survey on 814 individuals in northern Greece, 80% of the interviewed consumers mentioned that they use less heat than needed to satisfy their needs (Panas, 2012). Atsalis et al. (2016) investigated the correlation between fuel poverty and health impacts; they realized a dramatic increase in fuel poverty after 2010 and attributed to this 1%–2.7% of deaths recorded (due to cardiovascular diseases and respiratory infections). Boemi et al. (2017b) elaborated field studies (in 2014–2015) through questionnaires and interviews in Western Macedonia and concluded, through a sample of 241 households, that almost 58% cannot or rarely afford maintaining the temperature inside the house (which is quiet unusual for Greece, due to its mild climate and the relatively – to other European countries – lower heating energy needs); furthermore, a 32% face mold issues which are associated to illnesses like asthma, influenza, bronchitis, migraine and even with depression and stress (WHO, 2009).

On the other hand, however, there are also findings suggesting that energy savings do not necessarily mean lowering of living standards. For example, in a survey (2011–2012) on 598 households of a wide variety of geographical regions of Greece, it was found that 15% energy savings came as a result of decreasing indoor temperatures and shortening household heating time (Santamouris et al., 2013). In another field targeting Central and Western Macedonia areas, Boemi et al. (2017a) analysed the association between educational level and residential energy behaviour and attitudes during financial crisis, by processing 762 questionnaires collected mainly via interviews. The vast majority declared their dwelling quality at acceptable levels, although 35.2% of the households face issues of mold; 22.6% responded that they are rarely unable to maintain the temperature inside their house, but at the same time, 77.9% answered that

[5] *"Energy service poverty"* term was introduced by Bouzarovski and Petrova (2015) to distinguish the situation from "energy poverty" (where the non-use of electricity is actually due to missing electricity supply infrastructure) and to include all energy sources further to "fuel poverty" term.

[6] *"Energy conscious behavior"* could be used to express utilization of energy in an efficient and rational way; in other words by trying first to reduce energy needs/demand and then to serve/cover them by using the less possible amount of energy.

they do not heat their whole dwelling, implementing in this way some kind of energy management. Additionally, some sustainability indicators have improved due to recession. One example is the reduction of regional inequalities (related to SDG10), a phenomenon which can be interpreted by the growth pole theory of Perroux (1970). Actually, this theory supports that economic development that takes place around specific poles leads to non-uniform development in favour of the poles; in this sense, the opposite is also reasonable, namely recession limiting such poles' benefits, relieving in a way the unbalanced development. Reduction of inequalities as a result of recession was indeed confirmed in the Greek case, by Petrakos and Saratsis (2000), although this fact can certainly not be regarded as an achievement. Another example is the reduction of CO_2 emissions (related to SDG 13) which is a direct outcome of the reduction of electricity consumption and energy intensity in Greece. Indeed, CO_2 emissions dropped dramatically during the years of the crisis, as it is depicted in Figure 9.4, probably because of energy savings taking place or RE increase or the increase of use of NG (natural gas) – to be discussed in following sections-.

Due to recession which led to the awakening of energy consciousness and a better energy management, electricity consumers reduced expenditures (IME GSEVEE, 2018). More in particular, at the beginning of the crisis, electrical energy demand was expected to reach 80,000 GWh and peak load was expected to reach 16,000 MW by 2020 from 10,600 MW being in 2007 (NBG, 2010). However, total electricity consumption in 2015 has already been 28% less than what was expected to be, and the same happened with the residential use of electricity, as shown in Figure 9.5.

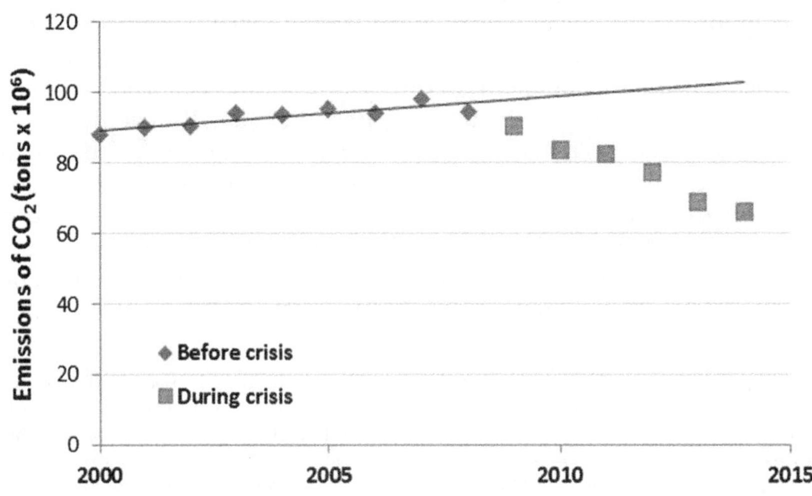

Figure 9.4 Evolution of CO_2 emissions from 2000 to 2014. (Data from UN, 2018a.)

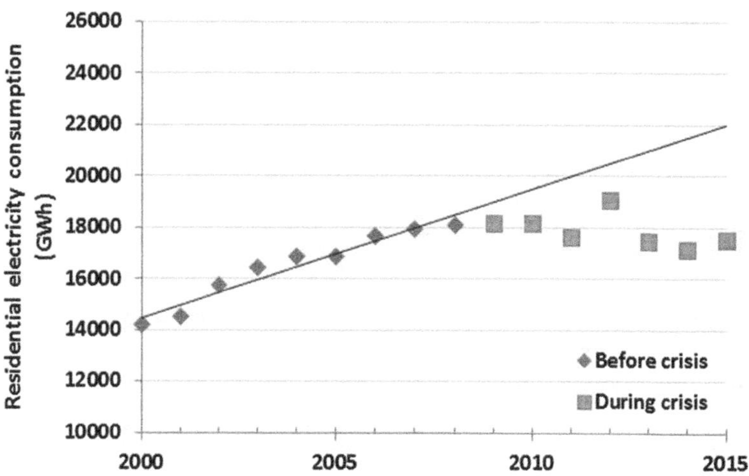

Figure 9.5 Electrical energy consumption in the residential sector of Greece. (Data from IEA, 2018.)

The steadily increasing electricity consumption till 2008 stabilized later on, during the crisis period, where some substitution of heating oil with electricity was partly applied (explained later in the text). In the same context, electrical energy production and energy intensity in Greece have also been reduced, as a result of electricity consumption shrinkage (Figures 9.6 and 9.7).

The SDG 7, however, related to sustainable energy, needs more discussion. According to research observations Boemi et al (2017b), to deal

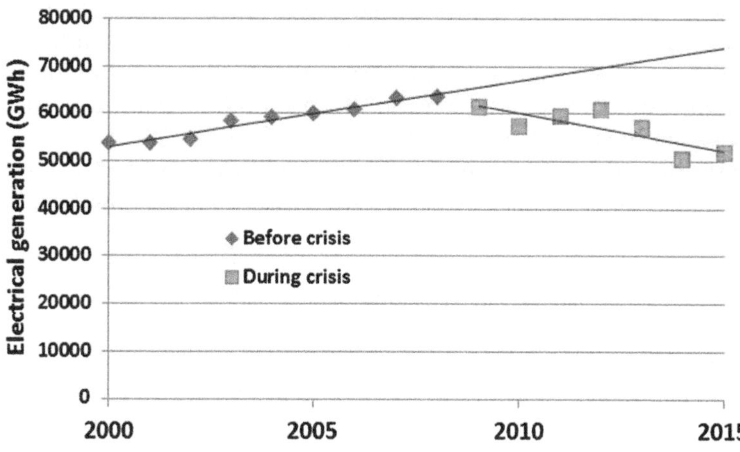

Figure 9.6 Evolution of total gross electricity production in Greece from 2000 to 2015. (Data from IEA, 2018.)

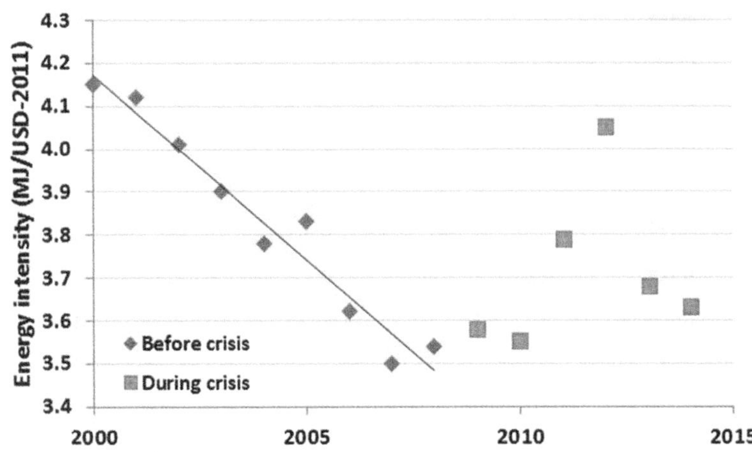

Figure 9.7 Evolution of energy intensity from 2000 to 2014. (Data from UN, 2018a.)

with impacts of financial crisis (income shrinkage), exogenous parameters (increase of energy prices – especially heating oil) and endemic features of Greece (low performance buildings), 57% of Greek energy users have adopted energy conservation measures like the shutdown of heating, cooling and shut off of unnecessary lighting in order to lower their energy costs. In addition, 19% of electricity users started heating only the rooms that they were using instead of the whole house (74%). In another research elaborated by Boemi at al. (2017a), it was found that 43.3% of electricity consumers overall have adopted energy conservation measures. However, while reduction of electricity coming from solid fuels can lead to sustainable energy supply, clean growth, sustainable cities, sustainable industrialisation etc. (Rosen, 2017). In Greece, cuts in electricity consumption from solid fuels and the progress in the increase of RE (discussed in following paragraphs) do not necessarily mean that all conditions towards sustainable energy transition are satisfied.

Typically, in Greece, the indicators that politically define "sustainability of energy sector" are (as described in HMEPPPW (2007) and explicitly reported in UN, (2018c)):

- Evolution of primary energy demand
- Evolution of final energy consumption
- Per capita energy consumption and CO_2 emissions
- Contribution of the energy sector to total CO_2emissions
- Decoupling the economy from energy demand and pollutants emissions
- Energy intensity
- Relative evolution of energy and electricity demand

- Composition of the electricity production mix
- Participation of RE sources (RES) in electricity production
- Electricity production from RES excluding large hydroelectric facilities
- Installation of solar collectors for water heating.

However, according to scientific literature, the transformation of wide-scale socio-technical systems – such as the development of sustainable energy systems – involves co-evolutionary changes between technologies, infrastructures, institutions and people (Grin et al., 2010; Elzen et al., 2004; Smith et al., 2005; Michalena and Frantzeskaki, 2013).

Therefore, the role of civil society is substantial in every sustainable energy transition. Is the Greek civil society participating adequately in the transition towards energy sustainability? The answer can be illustrated through the example of RE increase: Before the Greek financial crisis, the relevant situation regarding RE was rather discouraging with RE share increasing slowly or even dropping (Frantzeskaki et al., 2008). More recently, however, RE share has been steadily increased (see Figures 9.8 and 9.9) except from the period 2007 to 2009 when production has been temporarily reduced (this was due to decrease in hydro production that recovered in the coming years, as also shown in the same figure). Regarding wind and solar photovoltaic generation, the increment between 2010 and 2015 was indeed impressive reaching 300% (IEA, 2017b); total RE increased by almost 68% in the same period, but this increase is not sufficient to interpret increase of RE share in electricity production from 7.5% to 16.1%.

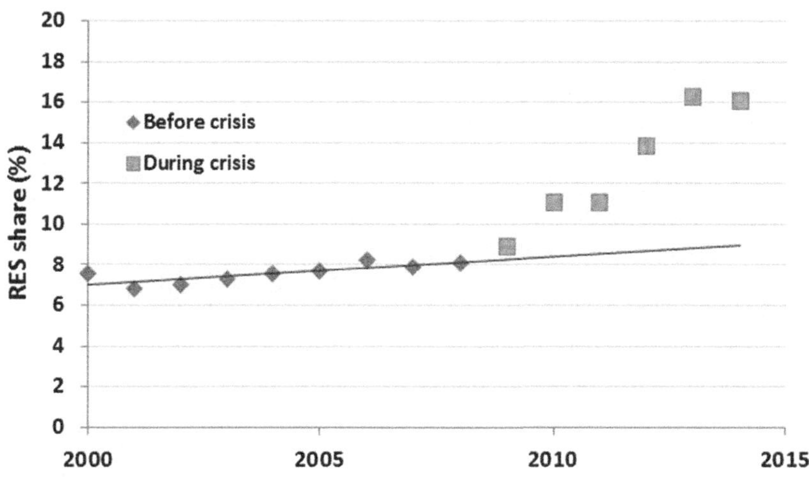

Figure 9.8 Evolution of RE share in final electricity consumption in Greece from 2000 to 2014. (Data from UN, 2018a.)

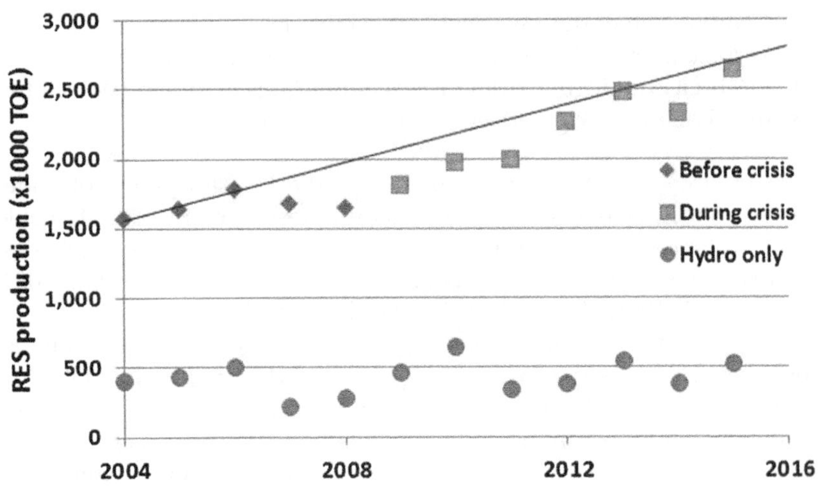

Figure 9.9 Primary production of RE in Greece from 2000 to 2014. (Data from EUROSTAT, 2018a.)

However, as consumers in Greece are not yet free to choose between RE and solid fuels, in the sense that they ignore which mix of energy fuels their electricity providers offer in their supply, the continuation of RE increase independently from their choice between solid and renewable fuels. This makes it clear that the increase does not really mean a "transition towards sustainable energy" but an outcome that has come as a result of the political will of the Greek state to coordinate well with representatives of the private RE industry. In this case, we cannot really talk for "transition towards sustainable energy" since civil society does not consciously participate in this transition. In fact, what is really happening, is that consumers are not happy with RE presence in Greece as RE contribution meant higher electricity prices for them. In fact, since the beginning of the crisis, electrical energy prices have been growing slightly but steadily and only recently seemed to get stabilized as shown in Figure 9.10, where taxes including prices taxes are depicted.

The reason for this increase was certainly not the increase of oil price at the international wholesale market which has actually dropped for the same period (as shown in Figure 9.11) but mostly amounts increase in taxation and that electricity were called to pay so as to subsidize RE penetration in the energy mix of the country.

In other words, the notable changes in the energy mix in households (Figure 9.12) and the cuts in electricity consumption look like energy sustainability indicators are positive; however, the driving force behind those moves is the high price of electricity (probably due to RE existence) to be paid rather than anything else to do with sustainability.

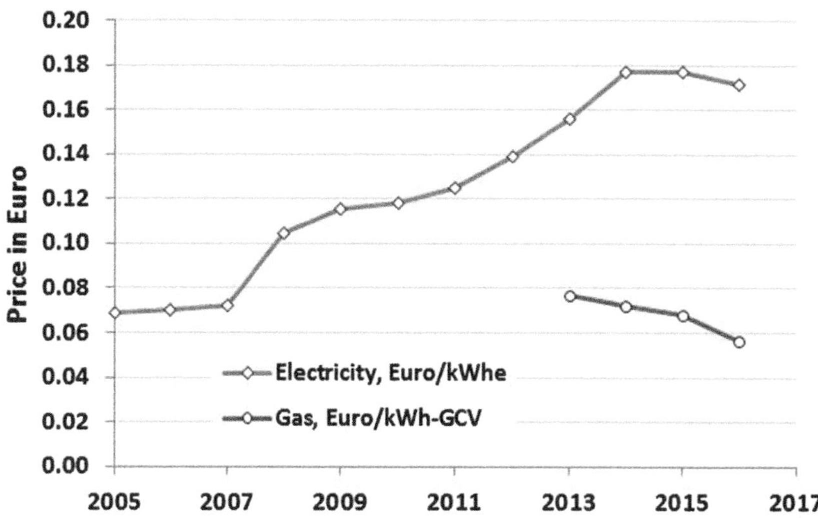

Figure 9.10 Changes in energy prices in Greece. (Data from EUROSTAT, 2018a.)

It is therefore important not only to note the existence of the improvement of some sustainability indicators (like per capita energy consumption and CO_2 emissions, participation of RES in electricity production etc.) but also to investigate the driving forces behind this "improvement".

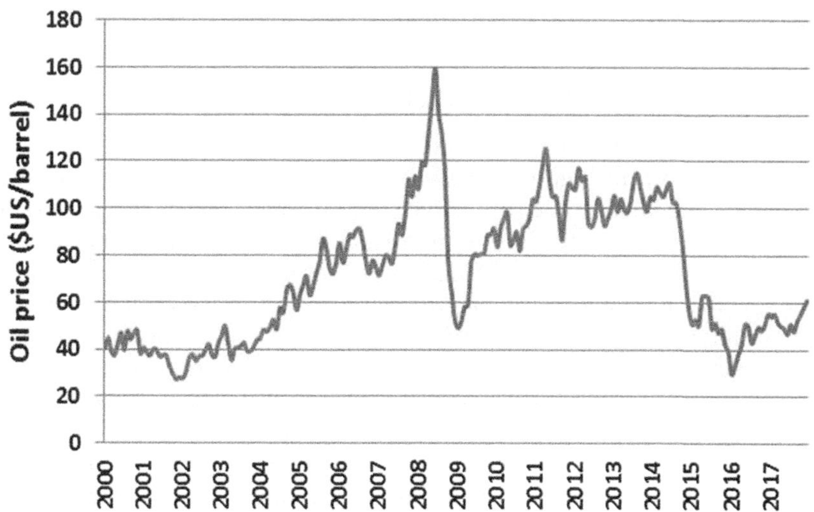

Figure 9.11 Oil real prices, as adjusted for inflation. (Data from Macrotrends, 2018.)

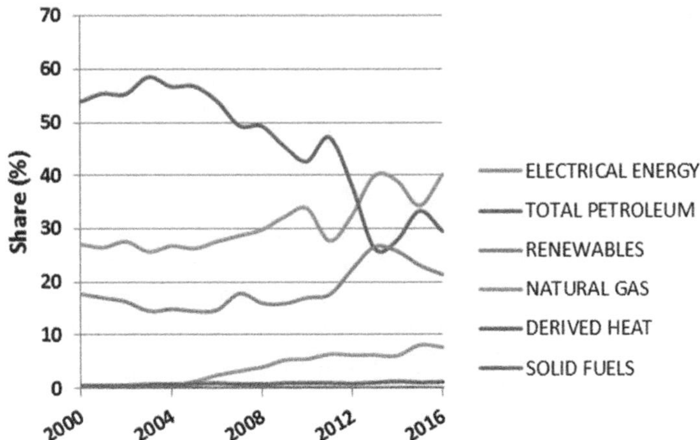

Figure 9.12 Changes in the energy mix used in households. (Data from EUROSTAT, 2018a.)

9.5 *Further weaknesses of the energy system governance (as worked till now) and further potential for the crisis to constitute a good friend towards Greek energy sustainability*

Due to weaknesses of the Greek energy system governance nowadays, the Greek case study encapsulates a high potential for improvement of energy indicators towards sustainability in all energy-intensive sectors.

9.5.1 *Energy surveys and energy audits*

The concept behind energy management is monitoring and comparing energy consumption against a base line that it is desirable not to get surpassed (as specified in the respective standard ISO 50001). The process of extracting such a base line is dynamic, since any interventions or even tenants' behavioural changes may lead to a different base line. This saying, a base line that has been developed under energy conscious behaviour (as it is expected to happen in crisis periods) sets lower consumption standards than a base line developed under relaxed conditions, contributing in this way to greater wakefulness regarding energy utilization. In this sense, *crisis may constitute an appropriate opportunity to extract most effective base lines, to compare and evaluate energy consumptions according to the standard.*

Energy surveys and audits are the most valuable and effective tool to save energy in buildings and the industry. Substantial energy savings can be achieved by investing in highly efficient equipment and building materials on the one hand and good energy management on the other. The Greek state has circulated several leaflets and guidelines towards this aim, and organizations have offered support to general public during crisis period towards energy saving in households (e.g. WWF, 2012). The Greek state has typically started to get engaged to energy classification of buildings and the respective energy surveys in accordance to the 91/2002 EU directive, while the whole framework of energy surveying and classification of buildings was finally set in operation in late 2010, in the middle of the crisis. All accompanying information and support, however, are of limited value to the end-user, who is not usually in a position to rank it and select the most cost-effective solution for each case. In fact, this effort was ungracefully implemented and communicated to the people as the draft law introduced initially a relatively high cost for the edition of an energy certificate which provoked the opposition of home owners association. The state finally reduced this amount in an attempt to finally focus on the incurred cost for the energy survey and edition of the energy certificate instead of the potential value of this certificate. Since then, a great number of certificates have been edited, but from the first year of implementation, it was realized that this process has rendered to a just bureaucratic procedure (Gelegenis et al., 2014) with reduced value for the users. This is because, often, energy certificates constitute just an additional document required in selling a house while the buyer knows little on how to read it or to value it. To make full use of the value of energy audit, the services of an energy manager/consultant might be proven necessary, not however without cost for the energy end-users.

The simultaneous monitoring of consumption and internal temperature conditions would only allow distinguishing between the two above, and simultaneously to extract a convenient base line for comparison and evaluation of energy consumptions, as per ISO 50001 standards. Energy audits place the context for the verification of savings and third-party financing of relevant initiatives. The obligation for energy audits has only been legislated in Greece in 2017. This approach is additionally expected to stimulate larger projects, dealing for instance with entire buildings instead of single apartments. In spite of this seemingly arising market, the initiative by the state to allow the hydronic isolation of apartments from a preexisting central heating system (explained further in the next paragraph) practically is restricting, in this way, the applicability of such initiatives.

9.5.2 Space heating

In Greece, energy intensity for space heating decreased from 0.43 GJ/m^2-year in 2010 to 0.24 GJ/m^2-year in 2014 (IEA, 2017b), as a combined result of energy efficiency policy (regulations, incentives), economic crisis, better management and utilization of energy. The vast majority (98.9%) of the dwellings in the country has a central heating system, 63.8% have a central heating system using oil and almost 30% have an auxiliary heating system (Boemi et al., 2017b). Operation of these systems requires the fuel purchase commonly by the occupants of a multi-floor building. As a consequence of the financial crisis however, more and more renters/owners of apartments are not in a position to pay their share in common expenses related to heating. Another problem was that in the cases that the oil-fueled central heating systems are not operating, many people run room air-condition units for their heating needs (Boemi et al., 2017a).

To cope with this problem, in 2017, the state allowed the hydronic isolation of an apartment from the building's central heating system, provided that it would run by a highly effective autonomous modern heating system like an NG-fueled condensing boiler or highly performing heat pumps. Nevertheless, this option enlarged a pre-existing problem namely the inefficient operation of central heating systems at low part loads. The situation could be alleviated if an exchange of problems and solutions would be set between users and the state on how to operate heating systems in a more efficient way (Gelegenis et al., 2015); which did not happen. Therefore, a good chance to improve the overall efficiency of already badly designed heating systems was lost.

Due to the availability of local resources, and of course due to the economic crisis, fuel wood consumption (biomass) increased extensively – e.g. in Northern Greece, it was rocketed from 12,000 t/year in 2008 to almost 90,000 t/year in 2012 (Slini et al., 2015) – and this happened especially in the rural areas (Azam et al., 2016). Obviously, the use of biofuels in the form of chips or pellets –instead of wood logs – has been considered both economical and environmentally friendly (Tsoutsos et al., 2007). What has happened however is that appropriate quality wood logs was not always used, e.g. it was also utilized waste painted wood which is impregnated with chemicals and releases toxic substances when burnt. Additionally, burnt wood in urban environment is causing suffocating conditions exceeding threshold values (e.g. benzene, PN10). Scientists even suggest the complete ban of burning wood not only in urban but in rural settings too (Elafros, 2018). The state should have ex ante undertaken initiatives to discourage the construction and operation of fireplaces within the big urban centres.

9.5.3 *Cooking*

Electrically driven cookers constitute almost 90% of the total, with the minority (~10%) been driven by liquefied petroleum or NG. Hence, electricity is mainly used to serve this secondly important consumption (after space heating), and probable substitution with fuels (LPG or NG) could be an interesting alternative. Indeed, people turned from LPG in the past to electrical cooking for various reasons (operability, safety); these reasons seem to be still valid as people still prefer electrical cooking than NG, as revealed by the very low penetration of NG cooking (~0.4%) in spite of its remarkable share 8.7% in heating. The use of wood stove or the fireplace is another option for electricity substitution in cooking, but it is strongly season dependent, having also an impact on the air quality (as already explained). Besides, clean cooking (by the use of electricity, LPG or NG) is an indicator of progress by IEA, therefore, such a solution is not recommendable. Other solutions such as pressure cookers and microwave ovens have been considered by consumers during financial crisis for alternative cooking with a goal to save energy and money. However, although consumers have entered the procedure of thinking alternatively regarding energy sustainability solutions, and although the use of appliances such as microwave ovens and induction cookers may indeed lead to significant electrical energy savings all year round, consumers are not adequately informed on the crucial issues related to those appliances (e.g. safety and health issues). Therefore, although those appliances are significantly merchandized, this commercialization is not followed by information for appropriate and safe use, not excluding the responsibility of the state for the imported and distributed equipment. For instance, accidents out of appliances have been broadly reported but without supplying specific details on the nature of those accidents, which has led to the discouragement and disbelief from the behalf of consumers for the use of such appliances (Lampropoulos, 2016). So, despite existing possibilities and materials in the market, energy intensity in cooking has not been reduced (as achieved with space heating), but on the contrary increased slightly, from 3.1 (in 2010) to 3.3 GJ per dwelling in 2014 (IEA, 2017b). Taking into consideration the above lines, potential energy saving interventions in cooking should probably focus on applying energy saving practices (e.g. use of suitable for the put hot plate, avoid boiling with uncovered pot, cook electricity non-demanding dishes etc.) with an informed combination of using pressure cookers, microwave ovens and induction cookers.

9.5.4 *Other consumptions*

There is also a slight increase in consumption with *appliances* (from 7.7 to 8.3 GJ per dwelling), like happened with cooking, but this has been partly attributed to the reduced ability for Greek households to purchase more new efficient appliances (IEA, 2017b). Regarding *lighting*, consumers change their lamps into highly efficient fluorescent or – more recently – LED lamps; however, this change in behaviour came as a result of the obligation for sellers to only sell energy-efficient lamps rather than the traditional ones. Therefore, consumers have been "obliged" to follow the relevant EU guidelines, rather than choosing themselves this "sustainable" path. No matter which was the motivation in this case, it has to be noted that many energy-efficient lamps which promise to last many times more than the traditional ones (and, hence, they are for one more reason "cost-effective") may last much less than expected (Electronics Weekly, 2009; Mail Online, 2014). In those cases, the concept of "sustainability" collapses; therefore, it is a state's job to support and show the way to consumers on how to report or/and replace defected lamps.

Important energy savings – within the same period 2000–2014 – have been achieved in domestic hot water production, as respective energy intensity has decreased from 3.4 GJ/dw to 2.8 GJ/dw (IEA, 2017b), and this is obviously due to the use of solar collectors. Indeed, during last 10 years (period 2006–2015), there is an almost constant rate of newly installed capacity with a mean value of 22.9 m^2/1,000 capita (ESTIF, 2018). Nevertheless, the capacity in operation has not equally increased (e.g. increased at a mean rate of 11.3 m^2/1,000 capita only, see Figure 9.13) which means that these installations refer at a great part to replacement of existing units (e.g. by more efficient ones with selective painting).

The state offers some funds – even for the replacement of old units – which are mainly addressed to low-income households, however. Conclusively, these incentives proved insufficient to stimulate the desired penetration of this technology, and the needs for other more effective form of incentives (like tax exemptions, as has also been successfully applied in the past in the country) become obvious (Maniatis, 2016).

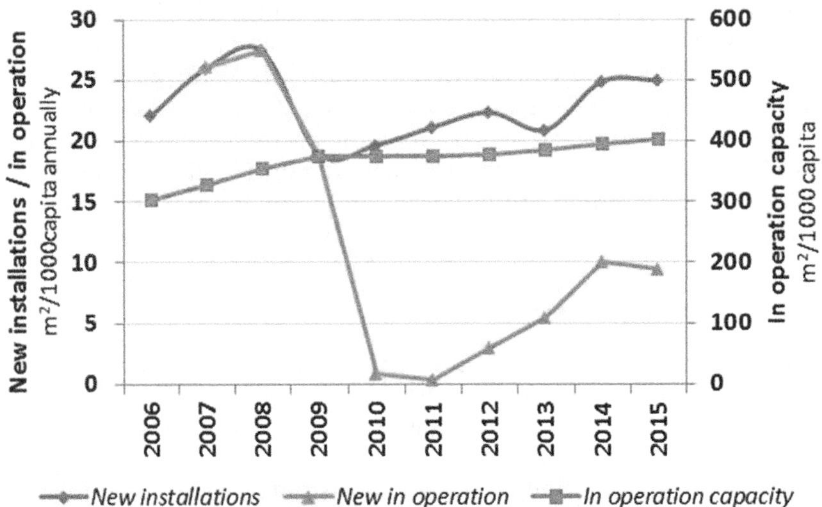

Figure 9.13 Newly installed capacity and in operation capacity of solar collectors in Greece, expressed in m^2/1,000 capita. (Data from ESTIF, 2018.)

9.5.5 Energy consumption in passengers' transportation

Transport has the largest share (35%) in total fuel consumption in Greece with oil accounting for 97% of energy used in transports. Potential of energy saving in this sector is very high. According to IEA, cars in IEA countries in 2014 used more energy than the entire residential sector (IEA, 2017b). In Greece, financial crisis affected energy expenses for transportation, however, it fell dramatically (by 30%) after a peak in 2009. The decrease is partly due to less utilization of passengers' cars. Indeed, cars and light tracks consumption decreased from 144.8 PJ in 2010 to 109.5 PJ in 2014 (–25%), while at the same time, total passengers' consumption decreased from 174.8 PJ to 133.7 PJ. In addition, there was a remarkable improvement in energy intensity of passengers' cars, as it reduced from 1.9 MJ per passenger-kilometer (pkm) in 2000 to 1.1 MJ/pkm in 2014, which could be attributed to a more rational utilization.

The Greek massive transportation fleet is not so old (mean age 13 years), while the country has one of the biggest "green" fleet in Europe (560 buses are fueled with NG). People in Greece used more of the public transportation means in the beginning of crisis, as it was realized a shift from passenger vehicles to public transport, as people couldn't afford fuel expenses. At the same time; however, this shift was initially supported by new infrastructure and relevant (eco-driving training) programs (IEA, 2017b). Unfortunately, the full potential to satisfactorily serve this shift to massive transportation has not been fulfilled. The whole fleet (including diesel-fueled buses) was not well maintained; as a result, almost 60% of them became decommissioned in the depot. The replacement of deferred spare parts with used ones (from the decommissioned buses) has resulted to additional defects when the buses were on the road and hence to additional waiting time at the stops. This situation finally discouraged passengers from using massive transportation means. Only recently, the state has announced that intends to purchase 3,000–4,000 modern technology buses (electrical or NG fueled), but unfortunately, an ex-Minister (Mr. Y. Maniatis) announced at the first Conference for Eco Mobility (2018) that during the period 2014–2017, a 50% reduction in public means utilization has been noted (Insider, 2018).

Many vehicle owners attempted fuel substitution, opting for a lower cost fuel like LPG. It is estimated that about 200,000 private passengers' vehicles operate now on LPG, and this shift took place during the crisis period. The diesel-fueled cars' share reached 80% of cars' sales in crisis period, but this was additionally due to the fact that only recently (2011) was allowed their use in the big urban centres of the country (Athens and Salonica) provided they fulfil Euro V standards (Directive EC/715/2007) or better (Euro VI). Unfortunately, the infrastructure of NG supply is still insufficient; in the same context, the utilization of biogas in cars is not applied in Greece.

9.5.6 Vulnerable social groups

The Greek Government undertook a few important initiatives to support vulnerable groups, dealing with several social issues including energy supply. For instance, the application of social electricity tariff (discount up to 40%), the offer of 0.125€/L allowance on heating oil price etc. were a few of the energy-related measures. In addition, the Greek authorities have been commended for the significant efforts to the energy sector reforms in the past years including the crisis years (IEA, 2017a). This notwithstanding, the Greek authorities seem not having exploited fully the opportunities to promote energy efficiency in a sustainable way and on a long-term basis. Authors (Papada and Kaliambakos, 2016) have noted this insufficiency from the side of the Greek Government and have suggested the provision of real incentives to support low-income households (e.g. instead of short-term financial support of them or by subsidizing electricity prices (IEA, 2017b). What is more to be noted, however, is that those incentives should be given timely and properly communicated and followed-up for all end users without exemptions.

9.6 Discussion

The prolonged financial crisis of the period from 2009 and which lasts to the present manifested through salaries and pensions reduction, heavy taxation, capital controls and high unemployment rate has threatened the economic and social cohesion of the country (Artelaris, 2017). Paradoxically, the strenuous financial situation in the country has led to the improvement of some energy and climate change indicators (such as reduction per capita energy consumption, energy saving, increase of RE, reduction of CO_2 emissions).

The question, however, is whether the above improvements constitute a "transition to sustainable energy" which is meant to last and improve quality of life for energy users and result in social and environmental protection? For example, will energy-saving practices continue, if the Greek economy recovers? In Figure 9.14, the progress in the relationship between oil consumption and GDP is presented. We distinguish three components, the before crisis ascending set of data, a descending set during crisis and a stagnation (regarding outcome) set, obviously uprising to higher consumption values, but it is unknown where exactly the price level is going to be stabilized. In other words, the question arises if any of these savings will be permanent or if they will be totally lost together with the recession.

Further, what if the state decides to turn energy policy push and energy market pull tools (Michalena and Hills, 2016; Hills and Michalena, 2017) to solid fuels again? Or what will happen if Greek electricity consumers protest because of the high price of their electricity consumed due

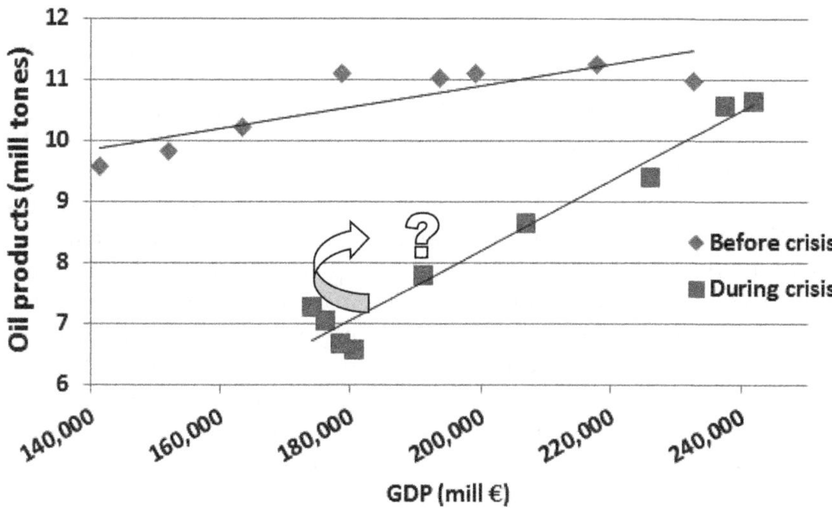

Figure 9.14 Variation of oil products consumption with GDP. (Data from EUROSTAT, 2018a.)

to RE? Or, what if the lack of information to electricity end users results to unsafe or unhealthy choices? Especially the power of the electricity consumers can be decisive when it comes to state policy as the energy consumption in both the transports and the residential sector accounts for 65% of the total energy consumption in the country (Figure 9.15).

At this perspective, studying thoroughly the case of Greece may constitute a valuable real-field experiment on how the state's energy

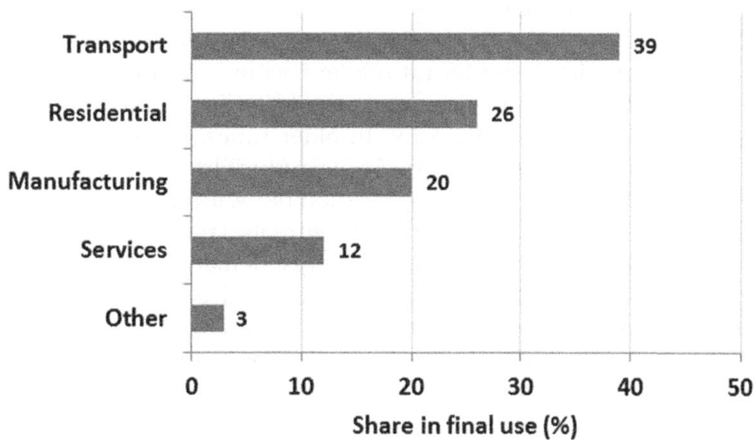

Figure 9.15 Distribution of energy consumption to the various sectors of Greece. (Data from IEA, 2017b.)

policy and general public's energy behaviour and practices can interact under financial crisis conditions and how those interfere with the sustainable goals.

For example, according to authors (Boemi et al., 2017a), there is still room for low-cost energy conservation measures which can alleviate the problem and have to be implemented (like improving the air tightness of windows and balcony doors by sealing their joints, insulating the distribution piping of the heating systems etc.). Though, the role of state in disseminating information regarding energy-saving approaches through respective ministries (e.g. for energy, environment, climate change etc.), agencies (such as energy agencies) and non-governmental organizations should be strengthened. Not only regarding the safety and health process of using appliances but also regarding programs and actions which are funded by international or European organizations towards attaining sustainability. Examples of those programs – though addressing to collective institutions – are the Intelligent Energy Europe program and programs such as Energy Cities, Manage Energy and Covenant of Mayors did also contribute to the same aim (Berardi, 2017). The tendency for the state to turn initially smart and daring approaches into bureaucratic procedures (such as the case of energy certificates) must stop. The collaboration of relevant associations of consumers is essential towards this end.

Technically speaking, energy management practices could be fully automated (e.g. as it happens with "smart" buildings), but unfortunately, this will not make huge difference in countries such as Greece where the majority of buildings are traditional and conventional. Electricity consumption reduction may be either the result of better energy management or of improper conditioning.

Researches have demonstrated that education and dissemination of information, not only on technical aspects but also on behavioural components, can play a significant role in turning consumers into users with energy efficiency practices and mindset (Zografakis et al., 2008; Michalena, 2009; Ntona et al., 2015). In older times, European countries (such as United Kingdom) have tried to introduce lessons on energy efficiency to primary and secondary education but with limited success due to lack of cohesion in the relevant teaching material (Newborough et al., 1991). Nowadays, however, increasing energy demands not only require governments to take action on education around energy efficiency but also make them working on energy efficiency initiatives in the education sector too; such as ones related to the design of energy-efficient schools (Naji et al., 2015). Still, although education tools (such as training and certifications) on sustainable energy approaches have been much discussed and suggested (Tsoutsos et al., 2013), there is still the observation that much has to be done for younger generations (in particular) to change their energy-related behaviours (Ntona et al., 2015).

9.7 Conclusions

The financial crisis in Greece caused lower family income, poverty and unemployment which resulted, among others, to a substantially lower energy use per capita. The situation stimulated reconsideration and re-prioritization of usual behaviours and practices concerning energy utilization, in a way probably similar to the public response to the preceded oil price crises back in the 1970s. In this framework, economic recession stimulated some good practices in linearization with sustainability principles, and this has been reflected to respective energy sustainability indicators quiet impressively. At the same time, the state undertook important initiatives to support vulnerable groups and combat energy service poverty. This notwithstanding, the state could have undertaken additional initiatives to support general public with better energy management, and on the other hand, the state again could have postponed or entirely avoided to legislate costly, bureaucratic and other processes that finally damaged energy-sustainability efforts. Greece could provide an important case study of approaching sustainable energy goals in times of crisis but rather limited progress has happened to this direction. In any case, the few energy savings benefits witnessed during the crisis in Greece, could be scientifically documented and communicated in a way to become lessons out an unwished – still very useful – consumers approach towards energy efficiency. Then the state should use those lessons learnt and all tools available to create an environment where "temporary" crisis outcomes would turn into "permanent" national energy efficiency achievements and even achievement of SDGs. Education and the dissemination of information is key to the creation of this much wanted environment.

Acknowledgements

Authors would like to thank Dr. Jeremy Maxwell Hills for his insights in this study.

References

Artelaris, P. Geographies of crisis in Greece: A social well-being approach. *Geoforum* 84 (2017) 59–69.

Atsalis, A., Mirasgedis, S., Tourkolias, C., Diakoulaki, D. Fuel poverty in Greece: Quantitative analysis and implications for policy. *Energy and Buildings* 131 (2016) 87–98.

Azam, M, Khan, A.Q., Zafeiriou, E., Arabatzis, G. Socio-economic determinants of energy consumption: An empirical survey for Greece. *Renewable and Sustainable Energy Reviews* 57 (2016) 1556–1567.

Berardi, U. A cross-country comparison of the building energy consumptions and their trends. *Resources, Conservation and Recycling* 123 (2017) 230–241.

Boemi, S.N., Avdimiotis, S., Papadopoulos, A.M. Domestic energy deprivation in Greece: A field study. *Energy and Buildings* 144 (2017a) 167–174.

Boemi, S.-N., Panaras, G., Papadopoulos, A.M. Residential heating under energy poverty conditions: A field study. *Procedia Environmental Sciences* 38 (2017b) 867–874.

Bouzarovski, S., Petrova, S. A global perspective on domestic energy deprivation: Overcoming the energy poverty–fuel poverty binary. *Energy Research & Social Science* 10 (2015) 31–40.

BPIE, 2015. *Alleviating Fuel Poverty in the EU. Investigating in Home Renovation a Sustainable and Inclusive Solution*. Brussels: Buildings Performance Institute Europe.

Elafros, J. Fire-places and wood ovens "choke" cities. Gazette "Kathimerini", daily issue of January 21, 2018.

Electronics Weekly. LED life expectancy. Uploaded in February 4, 2009 (available at the web at www.electronicsweekly.com/blogs/led-luminaries/led-life-expectancy-2009-02/).

ELSTAT, 2013 (Hellenic Statistical Authority). Research on energy consumption in households (period 2011–2012) Press Release dated October 29, 2013.

ELSTAT, 2018 (Hellenic Statistical Authority). Gross Domestic Product, time series 1995–2017 in xls file (available at www.statistics.gr/el/statistics/-/publication/SEL15/-).

Elzen, B., Geels, F., Green, K. (Eds.), 2004. *System Innovation and the Transition to Sustainability: Theory, Evidence and Policy*. Camberley: Edward Elgar.

ESTIF (European Solar Thermal Industry Federation), 2018. Solar Heat Europe. Solar Heat interactive statistics (available at http://solarheateurope.eu/publications/market-statistics/interactive-statistic/).

EUROSTAT, 2015. EU statistics on income and living conditions (EU-SILC) methodology – economic strain (available at http:/ec.europa.eu/eurostat/statistics-explained/index.php/EU_statistics_on_income_and_living_conditions_(EU-SILC)_methodology_-_economic_strain#Main_tables).

EUROSTAT, 2018a. Eurostat databases (available at http://ec.europa.eu/eurostat/data/database).

EUROSTAT, 2018b. Unemployment statistics at regional level (available at the web at http://ec.europa.eu/eurostat/statistics-explained/index.php/Unemployment_statistics_at_regional_level).

Ferguson, K., 2001. *Essential Economics*. Basingstoke: Palgrave Macmillan.

Frantzeskaki, N., Michalena, E., Angeon, V., Van Daalen, E. The on-going Greek energy transition to sustainability. In *Conference Proceedings, International Conference Protection and Restoration of the Environment XI*, 29 June – 3 July 2008, Kefalonia, Greece, 2008.

Gelegenis, J., Diakoulaki, D., Lampropoulou, H., Giannakidis, G., Samarakou, M., Plytas, N. Perspectives of energy efficient technologies penetration in the Greek domestic sector, through the analysis of energy performance certificates. *Energy Policy* 67 (2014) 56–67.

Gelegenis, J., Harris, D., Diakoulaki, D., Lampropoulou, H., Giannakidis, G. Determination of fixed expenses in central heating costs allocation: An arising issue of dispute. *Management of Environmental Quality: An International Journal* 26 (2015) 810–825.

Grin, J., Rotmans, J., Schot, J., 2010. *Transitions to Sustainable Development: New Directions in the Study of Long Term Transformative Change*. New York: Routledge.

Hills, J. M., Michalena, E. Renewable energy "pioneers" are threatened by EU policy reform. *Renewable Energy* 108 (2017) 26–36.

HMEPPPW (Hellenic Ministry for the Environment, Physical Planning and Public Works), 2007. National strategy for sustainable development, 2002 Greece. *Progress Report*.

IEA (International Energy Agency), 2017a. Energy and climate change. *World Energy Outlook Special Report*. IEA Publications, Paris, France (available at https://www.iea.org/weo2017/).

IEA (International Energy Agency), 2017b. *Energy Efficiency Indicators: Highlights (2017 Edition)*. Paris: IEA Publications.

IEA, 2018. Statistics (available at www.iea.org/statistics/).

IME GSEVEE (Hellenic Confederation of Professionals, Craftsmen and Merchants). Annual report on households' income and expenses. February 2018 Insider 2018. Press release: The 1st EcoMobility Conference 2018 has been successfully completed (available at www.insider.gr/epiheiriseis/aytokinito/79394/me-epityhia-oloklirothike-1st-ecomobility-conference-2018).

Kindreich, A., 2017. The Greek financial crisis (2009–2016) (available via the CFA Institute at the web at https://www.econcrises.org/2017/07/20/the-greek-financial-crisis-2009-2016/).

Lampropoulos, V., 2016. When pressure cookers kill. (published in the electronic version of newspaper "TO VIMA" (in Greek), available at www.tovima.gr/society/article/?aid=778237).

Macrotrends, 2018. Crude oil prices - 70 year historical chart (available at www.macrotrends.net/1369/crude-oil-price-history-chart).

Mail Online, 2014. The great LED lightbulb rip-off: One in four expensive 'long-life' bulbs doesn't last anything like as long as the makers claim. (published in January 26, 2014, available at the web at www.dailymail.co.uk/news/article-2546363/The-great-LED-lightbulb-rip-One-four-expensive-long-life-bulbs-doesnt-like-long-makers-claim.html#ixzz3yvocKGpp).

Maniatis, G. Feasibility and consequences by supplying incentives for the installation of solar thermal systems in dwellings. In *Presented at EBHE General Assembly*, February 5, 2016.

Michalena, E. Methods of promotion of renewable energy among local municipalities of Poland. *Geomatics and Environmental Engineering* 2(2) (2009) 59.

Michalena, E., Frantzeskaki, N. Moving forward or slowing down? Exploring what impedes the Hellenic energy transition to a sustainable energy future. *Technological Forecasting and Social Change* 80 (2013) 977–991.

Michalena, E., Hills, J.M. Stepping up but back: How European policy reform fails to meet the needs of renewable energy actors. *Renewable and Sustainable Energy Reviews* 64 (2016) 716–726.

Naji, M., Salleh, M., Kandar, M.Z., Sakip, S.R.M., Johari, N. Users' perception of energy efficiency in school design. *Procedia - Social and Behavioral Sciences* 170 (2015) 155–164.

NBG (National Bank of Greece), 2010. Renewable energy constitutes a necessary and attractive investment (available at sup.kathimerini.gr/xtra/media/files/meletes/ener/ape_ete.doc, August 30, 2011).

Newborough, M., Getvoldsen, P., Probert, D., Page, P. Primary- and secondary-level energy education in the UK. *Applied Energy* 40(2) (1991) 119–156.

Ntona, E., Arabatzis, G., Kyriakopoulos, G.L. Energy saving: Views and attitudes of students in secondary education. *Renewable and Sustainable Energy Reviews* 46 (2015) 1–15.

OECD, 2018. What are equivalence scales? (available at www.oecd.org/eco/growth/OECD-Note-EquivalenceScales.pdf).

Papada, L., Kaliampakos, D. Measuring energy poverty in Greece. *Energy Policy* 94 (2016) 157–165.

Papadopoulos, A.M., Oxizidis, S., Papandritsas, G. Energy, economic and environmental performance of heating systems in Greek buildings. *Energy and Buildings* 40 (2008) 224–230.

Panas, E., 2012. Research of energy poverty in Greece, Department of Statistics, Athens University of Economics and Business, Athens (in Greek) (available at www.energypoverty.eu/publication/research-energy-poverty-greece visited at 10/03/2018).

Perroux, F., 1970. A note on the concept of growth poles. In: Dean, R., Leahy, W., McKee, D. (Eds.), *Regional Economics Theory and Practice*. New York: Free Press, pp. 93–103.

Petrakos, G., Saratsis, Y. Regional inequalities in Greece. *Papers in Regional Science* 79 (2000) 57–74.

Rosen, M. How can we achieve the UN sustainable development goals? *European Journal of Sustainable Development Research* 1(2) (2017) 06.

Santamouris, M., Paravantis, J.A., Founda, D., Kolokotsa, D., Michalakakou, P., Papadopoulos, A.M., Kontoulis, N., Tzavali, A., Stigka, E.K., Ioannidis, Z., Mehilli, A., Matthiessen, A., Servou, E. Financial crisis and energy consumption: A household survey in Greece. *Energy and Buildings* 65 (2013) 477–487.

Slini, T., Giama, E. Papadopoulos, A. The impact of economic recession on domestic energy consumption. *International Journal of Sustainable Energy* 34 (2015) 259–270.

Smith, A., Stirling, A., Berkhout, F. The governance of sustainable sociotechnical transitions. *Research Policy* 34 (2005) 1491–1510.

Tsani, S. Energy consumption and economic growth: A causality analysis for Greece. *Energy Economics* 32 (2010) 582–590.

Tsoutsos, T., Kouloumpis, V., Kalogerakis, A. Thermal use of biofuels in buildings. Environmental and economic evaluation. *Fresenius Environmental Bulletin* 16 (2007) 735–744.

Tsoutsos, T., Tournaki, S., Gkouskos, Z., Masson, G., Holden, J., Huidobro, A., Stoykova, E., Rata, C., Bacan, A., Maxoulis, C., Charalambous, A. Training and certification of PV installers in Europe: A transnational need for PV industry's competitive growth. *Energy Policy* 55 (2013) 593–601.

UN, 2015. Resolution adopted by the General Assembly on 25 September 2015. A/RES/70/1. Seventh session. United Nations, pp. 14/35 (available at www.un.org/en/development/desa/population/migration/generalassembly/docs/globalcompact/A_RES_70_1_E.pdf).

UN, 2018a. Transforming our world: the 2030 Agenda for Sustainable Development (available at https://sustainabledevelopment.un.org/post2015/transformingourworld).

UN, 2018b. Sustainable development goals. SDG Indicators. Area: Greece (available at https://unstats.un.org/sdgs/indicators/database/?area=GRC).

UN, 2018c. Greece: National indicators of SD (accessible through UN Sustainable development knowledge platform, available at https://sustainabledevelopment.un.org/index.php?page=view&type=6&nr=172&menu=139).

World Health Organization (WHO), 2009. Large analysis and review of European housing and health status (LARES), WHO regional Office for Europe, DK-2100 Copenhagen, Denmark.

WWF, 2012. *Energy Savings Guide*. Athens: Intelligent Energy Europe.

Zarotiadis, G., Michalena, E., 2010. Green Technology: A European way-out of the crisis? In *International Conference "The Economic Crisis and the Process of European Integration"*. Organized by the University of Antwerp in European Parliament, Brussels/Belgium, June 2010.

Zografakis, N., Menegaki, A.N., Tsagarakis, K.P. Effective education for energy efficiency. *Energy Policy* 36(8) (2008) 3226–3232.

chapter ten

A transition to sustainable lifestyles

The role of universities

Bojan Baletić, Rene Lisac, and Morana Pap
University of Zagreb

Contents

10.1 Introduction

This chapter is based on the results from the research project Innovative Green Building Research on the Campus Living Lab (CLL) which was supported by the Croatian Science Foundation (HRZZ). The project builds on the experience of previously developing a master plan for the new sustainable university campus area in Zagreb.

It is expected from the universities to prepare their students, the future professionals, for creating a sustainable society. With a student population of 13.6 million, the European Union (EU) universities seem well positioned to influence, educate, involve and offer alternative lifestyle choices to students on campuses and through them influence the wider society. This challenge should be central to the evolving "third-generation university" (3GU) and a part of the living laboratory practice that is being adopted by a growing number of universities.

The question is how can they advance the process of student transition towards proposed sustainable lifestyles? We believe that the role of student housing is crucial and that it needs to evolve and respond to the future needs. As Moneta points out, despite decades of analysis, it's quite stunning how little coherence there is across campuses and designers about optimal residence halls design considerations (Campus Living Lab (CLL), 2017). The evolving campus culture and on-campus housing could be an instrument for the promotion of new sustainable lifestyles related to living, mobility, social innovation, smart urban solutions and food. Students have great potential to change the community around them through their activities. In the context of the campus living laboratory, part of housing units could have advanced features and thus serve as an educational, research and social tool for trying out advanced energy, technology and living propositions. The research was based on design questions and, therefore, it considers a wide area of interest in order to develop a framework for a different student housing program that should be a stimulus for the student population to transition to sustainable lifestyles and by doing so influence the wider society.

This chapter not only develops the proposition but also discusses the necessary tools for design inspiration and evaluation of sustainable campus developments. All insights and methodologies have been implemented on the actual development, University of Zagreb's Borongaj campus that is presented in the chapter's second part.

10.2 When is our lifestyle sustainable?

One of the most descriptive and motivating indicators for the global impact of mankind on the environment is the number of planets we consume. It is hard to precisely measure this indicator due to complexity of world ecosystem and our influence on it. But without a doubt, it shows that we are rushing towards a climate, resource and pollution "dead end", living on the account of billions of citizens in underdeveloped countries, which are developing fast.

In the attempt to define when our lifestyle becomes sustainable, we identified basically two approaches. One is a top-down and the second a bottom-up approach. One defines a metric for the society to be achieved and engineers the lifestyle, and the other builds on social awareness and sets the individual or specific goals to be reached in building a sustainable society. The first one seems hard to fully implement, and the second doesn't give an indicator whether we have done enough. Here we give an overview and discuss the approaches.

Twenty years ago, the experts from ETH Zurich (UNU, 2009) proposed a strategy for a "2,000-watt society". By this, they tried to address the continuing growth in energy consumption that was due to rising income, consumerism and changes in lifestyle which brought increase in the number of dwellings and energy-consuming appliances. Analysing the average European use of 6,000 W at the time (compared to 12,000 W in the United States, 1,500 W in China, 300 W in Bangladesh), they set the target of 2,000 W per person if we are to achieve a balanced development by the year 2050, without compromising living standards and mobility. The experts were confident that the technological solutions required for changes in the energy demanding building and road transport sector were already present. What was needed was a significant change in human behaviour and a political will. To achieve the goals of the "2,000-watt society", a fundamental change in social norms, values and practices was required, together with an innovation system (research policy, education, standards, incentives etc.) as a part of a national policy on sustainable development. This approach offered a metric based on energy use. At the time, this proposal was considered as a utopian vision although all the predictions about available resources continued to require important changes in our living habits. Since then, the society has adopted the new regulation on building energy use, and the rising interest for electric vehicles is reaffirming a need for a fresh vision of the sustainable society. We strive to design new buildings in accordance with the "nearly zero energy building" (nZEB) standard, though, to precisely define, it is still a complex question. On the energy efficiency side, new houses will be expected soon, to produce and store energy, i.e. to become micro power plants functioning as parts of a smart grid. Arup Forsight (2013) study suggests an even bolder vision for the future, buildings as living organisms.

A more recently developed model for measuring human behaviour and directing the needed changes was proposed in 2012 by a European social platform project: *SPREAD – Sustainable Lifestyles 2050* (SPREAD, 2012). The project states that the researchers have "taken a systemic, human-centred approach and emphasizes the importance of social innovation and behaviour change in order to achieve more sustainable living for all by 2050". The project has identified four key enablers: policy and governance, the economy and the monetary system, social innovation and individual change in behaviour. The project team states that "sustainable living goes beyond the consumption of the most sustainable material goods and/or services, into the re-design of ways of living, feeling, communicating and thinking". To measure sustainable lifestyles, they set the material footprint of a sustainable lifestyle at 8,000 kg per person per annum. As the project describes, "sustainable lifestyle material footprint means the use of renewable and non-renewable material resources (excl. water and air) plus the erosion caused by agriculture and forestry. It covers the whole lifecycle from the extraction of raw materials to the processing industry, distribution, consumption, recycling and disposal. The idea of the material footprint is to provide a comprehensive and understandable tool to reduce different kinds of present and future environmental challenges". The present material footprint of an average European lifestyle is 27,000–40,000 kg per person per year. As the material footprint drops to 8,000 kg per year, the environmental and resulting social impacts of our lifestyles should decrease and change considerably. Consumption will differ based on the values, needs and aspirations of each person. Using forecasting and backcasting methodology, they developed four scenarios for future sustainable living. Different scenarios provide variation, in some cases significant, in values for mobility, housing building, housing electricity, product consumption, leisure time and food. The study opened a complex discourse on sustainability lifestyles and provides an important reference point. Both proposals described, the 2,000-watt society and the 8,000 kg sustainable lifestyle present a base for comprehensive policymaking and need state management.

The other approach to promoting sustainable lifestyles takes advantage of technological advancement and promotes social innovation as the necessary complementary part. The social challenge, supported by new design principles, has been taken on by many different programs and initiatives that advocate a change in present lifestyles. One such is the EU programme Nature-Based Solutions (EC, 2018). As defined, "nature-based solutions are actions that are inspired by, supported by or copied from nature. They have tremendous potential to be energy and resource-efficient and resilient to change, but to be successful, they must be adapted to local conditions." It is expected that the nature-based solutions will result in multiple co-benefits for health, the economy, society and the environment, and thus, they can represent more efficient and cost-effective solutions than more traditional approaches. The other interesting example is the "Living Building Challenge" (ILFI, 2018a) that tries to promote a more holistic approach to sustainable living. It provides tools for creating a "symbiotic relationship between people and all aspects of the built environment". Their goal is to make the buildings self-sufficient and limit their resource usage to the resources available on-site. In their next step, the "Living Community Challenge" framework (ILFI, 2018b), they focus on communities designed using multipurpose elements, which are net positive when it comes to water and energy and which promote healthy lifestyles for everyone. They are walkable, bikeable and have affordable public transport. The challenge presented in these initiatives is to develop a lifestyle model and make it appealing for citizens to make the choice to adopt sustainable lifestyles.

The most recent attempt to change the present social and economic practices towards a more sustainable society comes with full formal political support. In 2015, the United Nations (UN) put forward the new sustainable agenda to guide the strategies and policies of member countries for the next 15 years. The set of 17 goals to end poverty, protect the planet and ensure prosperity for all, known as the Sustainable Development Goals (SDGs), were adopted by all countries. Unlike the Millennium Development Goals for 2015, the new SDGs represent a much more holistic approach. They include two specific focuses: raising the Human Development Index (HDI) of developing countries and lowering the ecological footprint of developed countries.

What is not so obvious is that the goals could be used as a methodological framework to work out the policies and strategies. Thus, the 17 goals have 169 sub-goals with 126 targets and 43 means of implementation. There are also 232 indicators to verify the progress of implementation. At first glance, this seems rather overwhelming as a method. Some graphical representations of the goals' relationship and impact provide a more practical way of understanding the necessary tasks. The SDG Compass provides guidance for companies on how they can align their strategies to the SDG and measure and manage their contribution. More useful is the SDG ring (UN, 2015) that can be used to measure the distance to the SDG targets for countries, cities, companies, universities, projects, etc. If we put goal no.4: Quality education in the centre of the SDG ring, we can analyse its relation to other goals and their influence. Goal number 11: Sustainable cities and communities brings us back to sustainable living, but its relationship to other goals is important and provides the base for qualitative analysis for future sustainable policies. It could be expected that the public sector, as well as private, would consider from now on their investments with more social awareness. The SDGs do not set a numerical target, it would be impossible to do so for a global audience, but they direct the discussion and focus towards awareness for what constitutes our sustainable existence.

10.3 Sustainable campus planning as a learning process for society

Educating future students about sustainability, moreover bringing them up into future sustainable citizens, has been recognized by universities worldwide as one of the main paths towards a sustainable society. This importance of the universities is further powered by their other three main objectives: scientific research, professional work and outreach activities. When they are intelligently combined on a campus, they form a living laboratory for the society in general. Since the early 1990s, through several conventions and declarations, university leaders have committed their institutions to the integration of sustainable development into all parts of their institutions by dedication to follow a number of guidelines (Baletić, Lisac, and Vdović, 2015).

What makes a university campus a sustainability living laboratory environment? Interdisciplinary research teams are developing and testing inventions and models in the spirit of holistic approach. New technologies and social solutions are being implemented and evaluated on-site. Students, the new generations of professionals, are raised in the context of new trends and sustainability, and they take part in the processes of implementation. Creative and interdisciplinary environment combined with concentration of university infrastructure and knowledge opens the opportunities for start-ups and spin-offs, for new economic activities. Campuses connect with local communities, disseminate experiences in the scientific community and public, cooperate with the government in policymaking, and so on. Although universities live in different sociocultural contexts and their activities take various shapes, they have influential roles in the urbanity of the surrounding city and a potential to act as cradles for sustainable future (Baletić, Lisac, and Vdović, 2017).

In this perspective, planning a sustainable university campus represents an important and even more complex task that raises numerous questions, some of them are yet to be answered in the future. But first, before designing elements of new campus, we have to give a holistic answer what makes this urban development sustainable, a representative self-developing role model place for future society? What are the criteria that can inspire or describe all possible efforts to make a campus sustainable? In 2010, the International Sustainable Campus Network (ISCN) Guidelines document (ISCN-GULF, 2010) was released to describe this phenomenon. In collaboration with University of Zagreb research team, the criteria list was further structured through theoretical insights in sustainability principles and campus structure. Within the framework of the CLL project, the list has been finalized through additional systematization, case studies testing and fine tuning. The present criteria (70) tend to represent all sustainable campus activities, and they are structured in four main categories: environmental criteria, social processes criteria, architecture and urban planning criteria and academic activity criteria.

10.3.1 Sustainable Campus Knowledge Base

Worldwide universities are investing efforts to empower sustainable lifestyles trough different activities on their campuses. To get a wide view on possible campus strategies and acquire inspiration, in the framework of the CLL research project Sustainable Campus Knowledge Base was developed (Baletić, Lisac, and Vdović, 2017). The knowledge base is based on an integral set of criteria that could fully cover campus sustainability and at the same time describe each possible activity performed on campus. In terms of Borongaj campus planning, it facilitates decision-making process, and it provides inspiration from worldwide successful practices (Figure 10.1). Worldwide examples of sustainable campus activities are basic units in the knowledge base. They are linked to the context of each university on the one hand and on the other linked through sustainable campus criteria that are awarded to each activity. The knowledge base is expected to grow with the input and support of ISCN and UNICA Green members and to become a contemporary tool for worldwide university collaboration – sharing cutting edge experience in sustainability on their campuses; a global living laboratory tool for emerging of new and unexpected sustainability activities and solutions.

In 2017, in line with contemporary UN recommendations, the Sustainable Campus Knowledge Base incorporated the SDGs as tags to be cross referenced in the examples of sustainability practices. Overall,

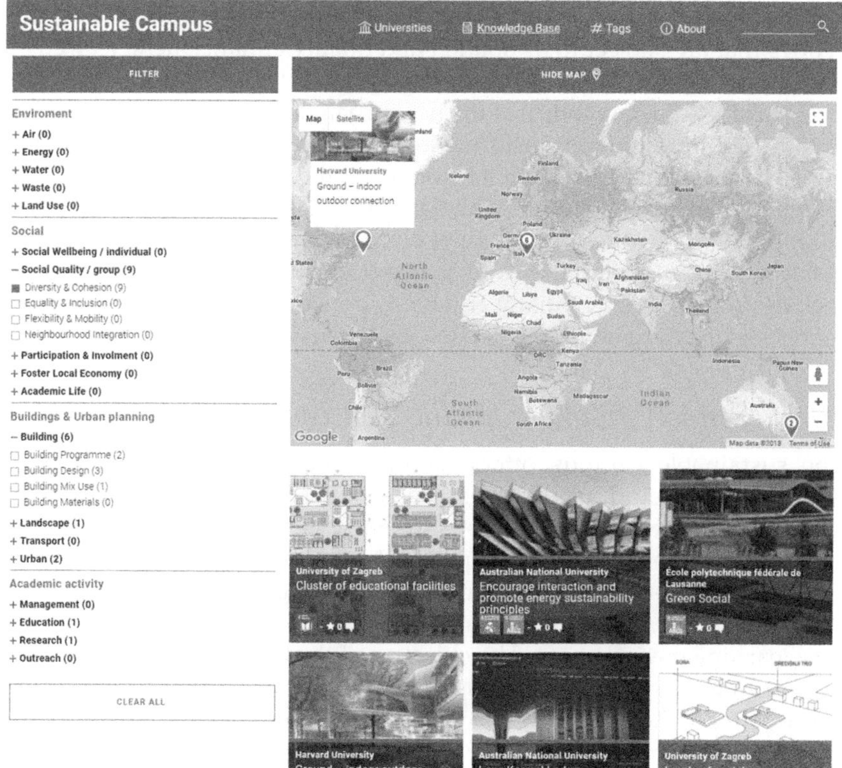

Figure 10.1 Sustainable Campus Knowledge Base. (Source: CLL project 2014–2017, http://www.sc-kb.net.)

both SDGs and Sustainable Campus Knowledge Base and its criteria represent tools for inspiration, discovery, planning and managing a sustainable campus.

10.3.2 Evaluating the sustainable campus

For sustainability evaluation of campus urban planning and architectural design, in the framework of the CLL research project, we focused on the *Eco-systemic urbanism certification*, developed by the Directorate-General for Land and Urban Policies, Department of Housing and Urban Interventions – Ministry of Development, Government of Spain (UEAB, 2012). Although Leadership in Energy and Environmental Design (LEED), Building Research Establishment Environmental Assessment Method (BREAM), Comprehensive Assessment System for Built Environment Efficiency (CASBEE), Minergie label and similar systems are widely accepted, there was an evident need for a more comprehensive tool in urban development planning. The development of more elaborate guides was motivated by the lack of connections between the sustainability categories that are subject to evaluation and quality enhancement, as well as social sustainability criteria in existing systems. Its aim is to more objectively evaluate sustainability of the new urban development and the existing urban environment transformations. Due to the level of detail in this guide, it represents very elaborate guidelines and recommendations linked to examples of urban solutions with their influences on sustainability (Lisac, 2012). Recognizing the complexity of planning, importance of connections between separate criteria, as well as integral planning which intelligently includes solutions for as many criteria as possible in only one measure, it represents the most advanced guidebook for sustainable urban planning.

10.4 Student housing for the 21st century

The sustainable campus offers an understanding of what our built environment should aspire to. It also stimulates the occupants, present and future professionals, to adopt the lifestyles that are crucial for the sustainable development of the planet. In this context, all student housing on campus should have a role of promoting sustainable living and social initiatives. Student residences need to be regarded as part of the new learning environment. The objective of the future designs needs not only to conceive them as "lodging places" but also as communities in the broad sense of the word, where a web of human relationships that support learning experiences will thrive the transitory nature of student living needs to be countered by the sense of permanency that a good design offers. Students must feel that their room is their dwelling unit and part of a neighbourhood.

Student residences are currently going through dynamic shifts in design and functionality. Where once a student dormitory merely required beds, appropriate kitchen facilities and reliable Wi-Fi connections (in recent years), student housing today is being refashioned around the concept of "community" which blends living, leisure, socializing and studying functions. The student housing has slowly evolved during the 20th century. In 1900, the only concern was to provide shelter, study and economics, but at the end of the century, as Friedman shows, concerns have shifted to the following factors: students, shelter/study and economics/social/technology/environment (Friedman, 2016).

Having this overview in mind, we decided to extend our research to the future (2020–2030–2050). Our analysis (Pap, 2019) showed that student residences will need to be a community in the broad sense (the first universities were conceptualized as societies rather than physical spaces). Student residences will need to be collaborative, aiming at bringing students together. They should also be heterogeneous and multifunctional and intelligent in the sense of smart living (Big Data and intelligent technologies). It will be important to encourage design concepts that conserve natural resources and energy and to offer more flexible spaces to accommodate growth and change (easy alteration or expansion). Thus, the role of student residences will be further expanded to encompass the teaching instrument role covering future housing and living in general. For example, student housing can encourage sustainable lifestyle or be a backdrop for social learning incubators for new ideas. New factors, which expand the role of a mere shelter, could be prioritized: student needs and aspirations, health & well-being, shelter, entrepreneurship (making & sharing), social, technology and environment (Pap, 2019).

To be able to design and develop the environment where students will learn informally and interact, as well as experiment with new living habits, we would need to know their present preferences, their future needs and their willingness to adapt to the new ways. The annual *Eurostudent* survey gives us only the statistical insight into the present state of student demographic and situation in the EU. Informal surveys and studies have proved to be far more revealing for the understanding what the situation might be in the next 10 years. The challenge is not only to change the habits but also to enable students to acquire new skill sets for future social involvement and preparedness for the workplace.

10.4.1 *Forming the future citizen*

Many factors could be influential on the education of students in the future. We choose three factors which we consider to be the most important for our research: environmental awareness, online culture and future workspace needs.

It is well known that environmental awareness is important to students. It changes their habits and their value systems. The sharing culture is making an impact on the cities. We see more initiatives for urban gardening, cycling and green activism on campuses. Social innovation initiatives are gaining presence and attracting more students. An important area of student engagement is food, which is a sensitive part of lifestyle with serious impact on sustainability. One such example is Green Monday (Green Monday, 2016), a student initiative from Hong Kong, advocating that students and teachers observe one vegetarian day in a week. This project has been very successful. Their analysis showed that going green would save 1,611 L of water per capita per day, reduce CO_2 emissions from livestock and alleviate food insecurity. Also, by not funneling large quantities of food to livestock, instead of feeding starving humans, we could impact the world hunger situation. Their message is simple: the impactful action everybody can take starts from their dining plates. The initiative now continues from mindful eating to mindful living, emphasizing the need for the change of lifestyle.

Regarding the online culture, in 2014, The Student Room, a UK network boasting of the largest membership among students worldwide, presented their findings in the report: *The Future Student: What to expect in the next ten years* (The Student Room Group, 2014). The study made a projection for 2024 by exploring current significant trends. We will mention some of their projections here. As students are readily embracing today's new technologies, the future student (Generation Z) will learn and communicate in a way never quite seen before. The students will tend to choose and design their own unique higher education portfolio. The educational institutions will cater to all kinds of educational formats. Expectations from on-campus living experience will be important. The importance of well-rounded skill sets will have an impact on teaching and recognition. There will be private initiatives like start-ups and maker hubs offering the very latest technologies for outsourcing to both universities and students. Innovators are setting up businesses in university incubators and student residences. There will be a growing demand for "Earn as you learn". Students will be more enabled and empowered. The authors are cautious as they conclude that future could be unimaginably progressive, but it could also be rather dysfunctional.

The future workspace challenges to education were recently analysed by the Economist Intelligence Unit (EIU). Their research was summed up in the 2016 study: *Worldwide Educating for the Future Index* (The Economist Group, 2018) in which they listed the skills they considered crucial for the future workplace: leadership, entrepreneurial, interdisciplinary, digital and technical, creative and analytical, global awareness and civil education. The study asked the question whether the education systems were equipped to teach these skills. The EIU developed indicators to evaluate the higher education area. In their conclusions, along with better policies and funding, there were three recommendations which pertain to our research: reforming systems should improve global citizenship skills and focus on project-based learning; learning both inside and outside of the classroom, as well as across disciplines, should be fostered and encouraged; and efforts should be made to foster societal openness and tolerance as these are linked to future-oriented skills.

It is important to acknowledge these trends when programming a new sustainable campus and, especially, student housing. For student housing will be the stage for more individual and informal forming of students that will complement their professional development.

10.4.2 Student housing development – factors and attributes for the future

The housing aspect is an important part of the living/learning/exploring experience. It raises an interesting question about how we perceive students and their part in society. It would be tempting to provide an intense environment where students live, learn and play 24/7 and thus graduate faster and perform better. But where does one learn best, with whom and how? The library, lecture hall, classroom, laboratory or a studio are necessary parts, but learning also happens outside of these places in the physical and virtual world. As Baletić and Holms Samsoe (2013) point out, the physical density and sociocultural pattern of a city is interpreted and transformed into various scales in university planning ranging from the interior of, to the space between the buildings. We talk about "city halls", "knowledge squares", meeting places and cafes, transparent grounds, urban gardens, overlapping and commercial programmes and double use. Housing on campus could, from a financial point of view, contribute to an intense and effective learning experience where more students get through university in shorter time. From an urban planning point of view, it can be seen as a way of making non-functional university areas livable also after classes. From a research and development point of view, it can be a living laboratory facility for housing in general.

To approach the topic of student housing, practice in a systematic way we developed three tools: a database with examples from the 20th century, a graphical representation of characteristic spatial–functional schemes and a systemized table of the design criteria.

As the design of new housing should be informed by past traditions, it was necessary to look at the existing examples of student residences. Therefore, the examples of international and Croatian student residences from the 20th century were analysed. The criteria used in the database (Pap, 2012a,b) were the location of student residences (city or campus); development of standard, function and programme and the spatial organization of the building. Thus, created database represents a repository which stimulates the discussion and generation of new student housing proposals (Figure 10.2). What the databases show is that the programme of student residence has grown but has not necessarily brought about a new typology. The organization reinterprets the spatial characteristics of the existing typologies (Pap, 2019). The historical development related to spatial organization is linked more to the increase in standards and less to the development of the programme and function of the student residence.

To relate the new programme to the spatial organization, we have developed a graphical representation (Pap, 2019). We have used it to analyse and describe new and interesting examples of student housing. These schemes have been designed in accordance to student experiences when living in a student residence. The scheme can be simple or complex, depending on the location of the student residence (suburbs, city or campus) (Figure 10.3). If the surroundings are better equipped with amenities, then the student residence is simple. If there is a lack of amenities in the surroundings, then the student residence is more complex. The programme for the residences should be emphasizing the student experience and satisfying different students' lifestyles. It will be necessary to provide flexibility, variability and adaptability, regardless of the chosen typology or morphology of the building. The spatial–functional scheme of future student residences will become more complex, both in spatial and programmatic sense, and it will have the ability to be easily transformed.

Spatial organization is followed by design criteria that are organized in four groups. The first group refers to the campus neighbourhood: campus connectivity, productive landscapes in the vicinity, direct contact with green outdoors from the spaces in interior and bioclimatic design. The second group describes the scale of the student residence: fractal distribution of spaces with gradation from private to public spaces; flexibility, adaptability and variability of spaces; informal and specialized work spaces. The third group of design criteria refers to sustainable development and efficiency: zero CO_2, energy plus, water efficiency, zero waste, local economy engagement, health and comfort for residents, enabling social cohesion, diversity and equality and a building that teaches and encourages sustainable practices. Finally, the fourth group covers new technologies such as new materials and constructions, sensors, Internet of Things (IoT), Big Data, artificial intelligence, automation technology (>Intelligent>Smart>Thinking).

From our analysis comes an insight that in addition to the knowledge about the building's physical aspects, the residences can become a backdrop for social learning and incubators for new ideas. If society regards its student population as its future brain trust, developing opportunities for innovative critical thinking outside the classroom can be part of the educational experience as well. Student residences can be regarded as an extension of classrooms. If suitably designed and introduced, they can

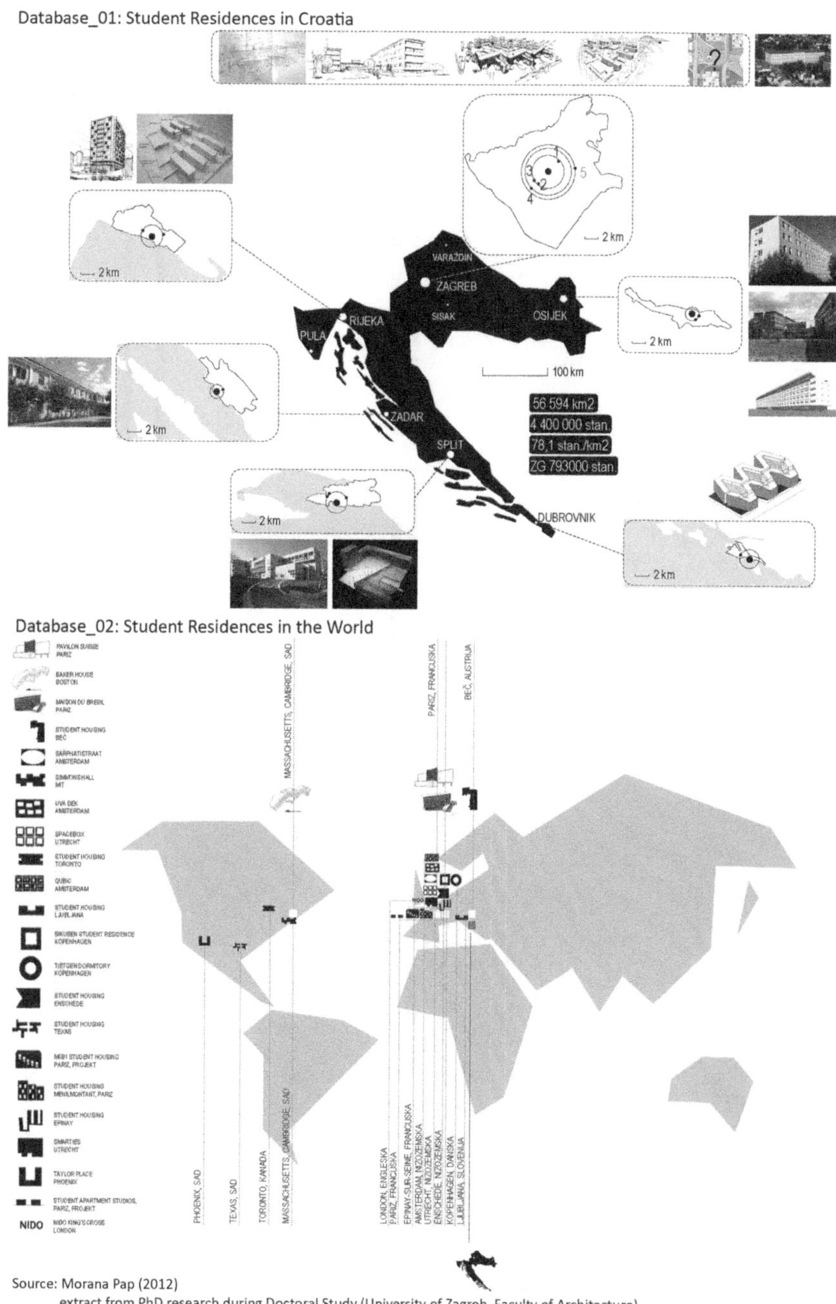

Figure 10.2 Database of student residences. (Source: Pap, M. 2012, from PhD research University of Zagreb, Faculty of Architecture.)

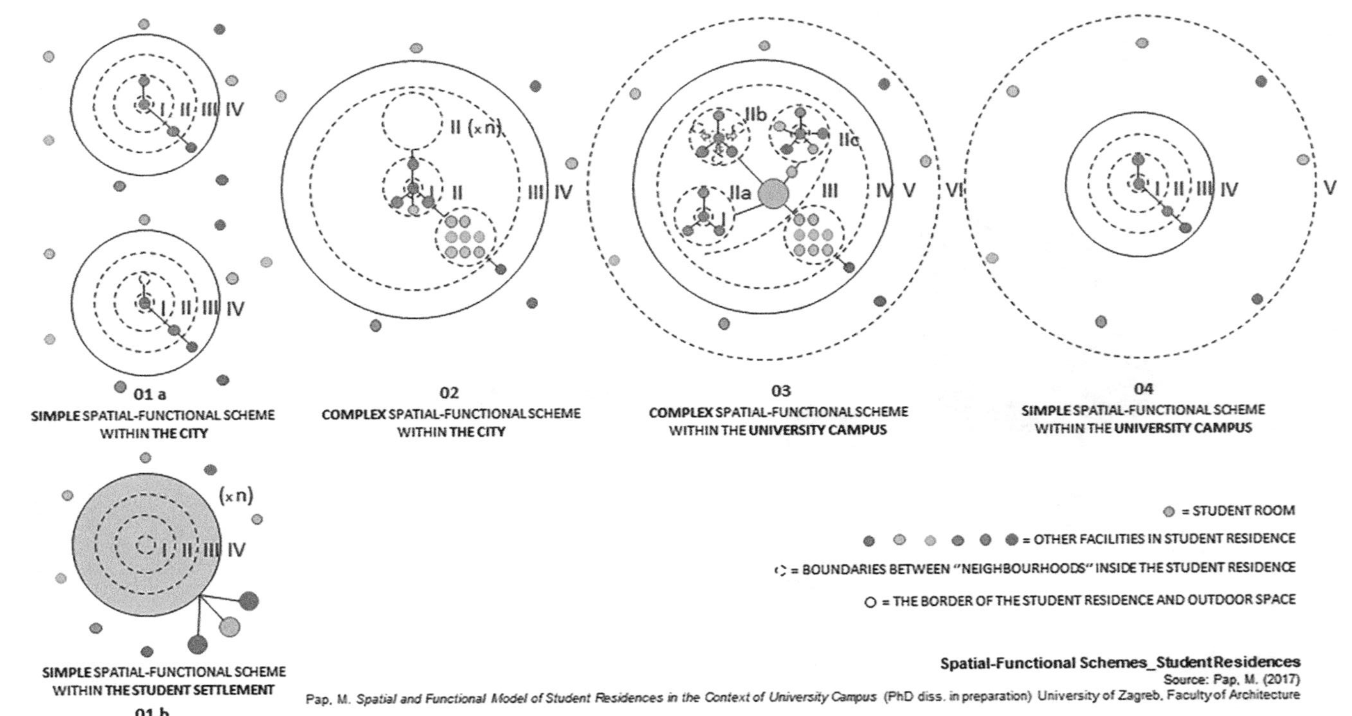

Figure 10.3 Spatial–functional schemes of student residences. (Source: Pap, M. 2019, from PhD dissertation University of Zagreb, Faculty of Architecture.)

become a teaching instrument about the future housing that students will occupy once they have left university.

Our aim was to develop a program for a student pavilion as an innovative green building model that could be an educational, research and social tool for trying out advanced energy, technology and living proposition. In this respect, the new university campus becomes a research and testing ground for future scenarios that will, with the involvement of business and public sector partners, take the role of a living laboratory for the new and emerging sustainable society.

10.5 *Case study of Campus Borongaj*

In 2006, the Government of Croatia decided to convert an army barracks in the wider centre of Zagreb to a new campus area. The University of Zagreb was invited to take part in its development. With facilities in the historic part of the city, the university, of more than 70,000 students, planned to reduce the present demand for new space and new programmes by developing the Borongaj campus. The campus would be a mixed-use innovation park in Zagreb. The latest opportunity for Zagreb to develop a dedicated campus area for the academic community comes in the moment when there is an international recognition of the importance of the university campus as a stage for new educational, research, social and economic processes that could influence the development of the society as a whole.

The new campus must be a stage for the four **i** (innovation, interdisciplinary collaboration, interaction and involvement). The living laboratory approach offers a conceptual framework for this. The university is viewed by sustainable development, and living laboratory advocates as a major player in the future changes. The consequence of this is that the traditional campus physical planning process is evolving to include many new considerations and stakeholders. In its 2009 vision for the Campus Borongaj, the University of Zagreb stated that the defining characteristics of the campus will be: sustainability (a campus that demonstrates the latest in green technologies and human practices), entrepreneurship (a campus renowned for creativity and invention) and focus on Croatia (a campus that serves Croatia locally, regionally and nationally). This would be achieved through interdisciplinary research and practices, institution–industry partnerships, permeable and inviting boundaries and the promotion of Croatian identity and influence. Implementing the highest standards of sustainable development on the campus was defined as a request early in the process and was seen as a common goal for all the institutions on the campus. The three defining characteristics of the campus: sustainability, entrepreneurship and focus on Croatia later received updated programmatic definitions. Sustainability was referenced to the framework of the 2015 UN SDGs, entrepreneurship was in 2012 linked to the new context of living laboratory practices and focus on Croatia was reinterpreted within the 2014 EU policy of Smart Specialization Strategy (S3).

10.5.1 Campus Borongaj as a new lifestyle experience

The Campus Borongaj planning and design process addressed, among other topics, the new reality of climate change, sustainable campus design, technology transfer, need for social innovation, urban quality indicators, impact on economy, information technologies, shift in educational paradigm and low university integration. The new campus would complement the existing urban campuses in Zagreb, but as a 90-ha brown field development, it should represent a game changer for the university. The planning process had to advance campus programming, architectural character and energy standards (Figure 10.4).

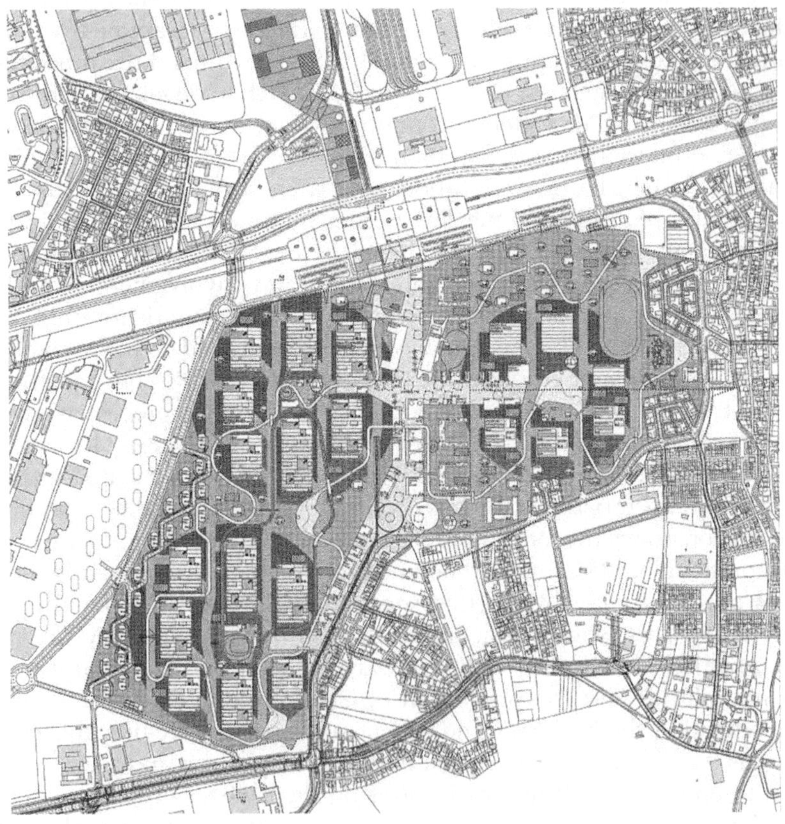

Figure 10.4 Barongaj campus masterplan by Njirić + Architects, 2011. (Source: University of Zagreb.)

Programming for the new Campus Borongaj has set the standards high: promoting innovation, bio-consciousness, creativity, interdisciplinary, knowledge transfer and a new educational experience. It had to be urban, inclusive, outreaching and cost-effective. For students, campus had to be inviting, sustainable and smart. For the university, it had to bring about its new identity of an institution promoting research and overcoming academic challenges, supporting creativity, integration and interdisciplinarity. For the city, it had to provide an urban quality to the surrounding neighbourhood and become an example of sustainable urbanization and community. For the Republic of Croatia, it would become a proponent of green technologies, and as a living laboratory, it would represent a reference point for central and south-eastern Europe for the EU.

The first challenge was to motivate the academic community who had the means to facilitate changes and to apply a conceptually progressive approach. Also, having in mind the socio-economic context, the new campus had to be, above all, flexible regarding space, time and cost. To achieve that, it was necessary to introduce various design strategies, such as time-based landscaping, hybrid spaces, modular development, seasonal urbanism or bioclimatic low-tech elements. The 2011 architectural master plan for Campus Borongaj by architect Hrvoje Njirić (Njirić, 2011) put forward the main planning question "Design: form and function follow climate. Can ecology provoke invention of forms and patterns on campus that will provide new social, cultural and political interpretation?" (Njirić, 2011). It respected the local bioclimatic conditions, took advantage of the on-site renewable energy sources and formed an educational built environment that would be inspirational for developing sustainable lifestyles. As a response to the large site, partial programme and segmental development through time, Njirić developed many simple but very interesting architectural strategies.

The energy standard builds on the ETH Zurich methodology and the goal of the 2,000-watt society (2,000 W × 8,760 h/a is 17,520 kWh/pa). It should be an example to the highest energy requirements with the consumption of the 98 kWh/m^{2}a energy according to the German energy certificate. It should be CO_2 zero (all energy is produced from renewable energy sources on-site) and possibly become CO_2 minus (with the use of CO_2 captured on-site). To achieve this, Campus Borongaj would be a closed circle of production and consumption of energy. It would use contemporary trigeneration technologies, biomass would be cultivated and gasified in partner projects, the excess energy would be used all year round for agricultural production in campus glass houses, algae would be grown on campus 24 h/day and the excess energy would be sold to the city grid. The energy proposal for Campus Borongaj, developed in 2013, represented an innovative approach considering the contemporary technological developments in Europe.

10.5.2 Student housing for Campus Borongaj

The University of Zagreb plans to build additional 4,000 beds in ten-student housing pavilions on Borongaj. The process of programming and designing student housing for 2030 is surrounded by new cultural concepts, practices, technologies, work ethics and policies which indicate an important paradigm shift.

When addressing programming, design and use for the new student housing on a sustainable campus demanded management and architects to "think out of the box". The new housing programme needs to extend from only dormitory (present practice) to include spaces that promote co-working, skill development (maker spaces, entrepreneurial training and creative expression) and social initiatives (urban gardening, second-hand exchange, alternative food choices and volunteering activities).

Looking at the examples of student residences, it is obvious that the conventions from the past are being replaced by a share culture of the new generation. The characteristics of the future/new student housing are heterogeneity and multi-functionality. The value parameter becomes the possibility of choice, the ideal is a quantitative offer of quality solutions. The future student housing must promote an increased experience of community life during study. The assimilation of technology in all aspects of student life is transforming the higher education system and will create new challenges for the maintenance of the environment in which the aim is to realize the integration of involvement, learning and housing. Planning of the future student housing is evolving from providing separate facilities related to academic, social and other segments of student life in the model that is mixing common areas to promote synergy of student activities. Gradation of space in the new typology system is inevitable: private, semi-private, semi-public, public spaces: boundaries are blurring.

The design of the buildings should be informed by the past housing traditions but should also boost the sustainable character of the campus through energy performance (nZEB, energy production and storage) and sustainable lifestyle choices. They should accommodate social innovation initiatives and should be a showcase for new efficient construction methods and materials. Our interest was also in the growing presence of the new IT technologies (IoT, Big Data, smart assistants, digital fabrication) and their implementation within student housing, as a future notion of a thinking building.

Within ten-student pavilions planned for Campus Borongaj, having different level of living laboratory criteria included, one pavilion will be progressive and experimental – a full living laboratory environment. While all pavilions conform to the nZEB standard, the criteria that define the further performance distribution between the pavilions are occupant's study level, their interests, functional organization, internal space organization, the requirements of urban location and the level of IT presence in the building.

The experimental student pavilion, part of the campus living laboratory activities, should be open to research, change, testing and monitoring. It should be flexible in its organization, adaptive to change of building elements and technologically equipped. The students could apply for this kind of residence, to be involved in research and implementation of sustainable and technological solutions as part of the living laboratory practice. The use of an advanced green building would also be a learning and social experience for its occupants. As an experimental facility, it would be connected to research and development activities (based on departments, research labs, centres or techno park) on campus. In our research, special interest was dedicated to user adoption of the new environment and technologies.

10.5.3 *Evaluating the new propositions for sustainable living*

It was important to check how inclined were the current students to embrace the changes towards new sustainable living propositions. It was important to detect if Croatian students had the affinity to adapt to proposed lifestyle changes.

According to a recent study on the student population in Croatia (Institute for Social Research Zagreb, 2014), present Croatian students have proved to be the elite within the young population, but more due to their socio-economic status and resources than to their sociocultural capacities. Findings of previous research also showed that students were more in harmony with the (post) modern values on a sociocultural and ideological level than other youth subgroups. Today's students are more inclined to see the younger generation as the bearer of innovative and creative potential than as a mediator of acquired values whose main exponents are older generations.

As a part of the research project Campus Living Lab, we organized two workshops. The first workshop was focused on the will to innovate and take part in the makers' culture, while the second probed into the possible lifestyle preferences. In the first one, a design thinking workshop "Making & Sharing", 12 students of diverse profile (gender, age, profession, background) were put in a maker space environment and were given simple creative and collaborative tasks. Workshop concluded that students tend to work together, do projects and work in general. They presented easy adaptation to a new environment and new tasks. Mixing of the disciplines worked well and created an unexpected result. They had significant interest for the use of new technologies and an overall capacity

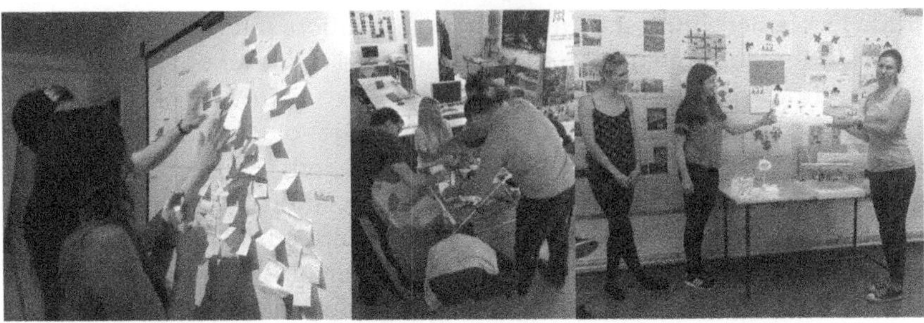

Figure 10.5 *Making & sharing* student workshop. (Source: CLL project 2014–2017.)

Figure 10.6 *Student living in innovative green campus* student workshop. (Source: CLL project 2014–2017.)

for makers working and learning concept. In terms of student life, along with living standard, they expressed the importance of social interactions and "erasing" borders (Figure 10.5).

The second workshop on student living in innovative green campus aimed to examine the capacity and affinity of the student population for the basic social, cultural, organizational and personal characteristics of contemporary campus lifestyle. Another diverse group of students were faced with contemporary student lifestyle trends and sustainability requirements (cycling, energy saving, gardening and vegetarian diet, recycling…), as well as different spatial organization models for housing (Figure 10.6). In conclusion, students are open to alternative forms of transportation, vegetarian diet, growing their own food and so on but will oppose if being forced and lose their freedom of choice. In terms of dormitory space plan organization, they all welcome introduction of sharing spaces in the dorm for working, socialization and leisure, but if it doesn't jeopardize privacy and serenity provided by their own, even if small, rooms.

In general, the workshops have shown that students are actively open for experimenting but with having freedom of choice; it is recommended that there is a diversity of lifestyle options as well as housing organization systems on a campus.

10.5.4 *Evaluating the design for the sustainable campus and student housing*

To evaluate Campus Borongaj sustainability performance and to evaluate the best locations for the student housing regarding environmental quality, we used the *Eco-systemic urbanism certification* (UEAB, 2012) criteria. As a result, Campus Borongaj plan has shown very high performance in terms of urban sustainability, having majority of the criteria fully acknowledged by the plan and a minor number of them only partially visible. However, some of the criteria were not fulfilled by the plan, the ones that depend on the surrounding infrastructural capacity: city waste management model, urban processes management with local government, as well as city management instruments in urban transformation processes. In addition to that, three locations on Campus Borongaj were chosen for the evaluation of their suitability for student housing. Criteria were acoustic and thermal comfort, visibility of urban greenery, proximity of transportation and services and general urban connectivity. All three locations have proved acceptable but with different levels of suitability, a good foundation for different lifestyle models of the dorms: better urban setting, more natural and so on.

We have, also, used the SDGs' evaluation methodology to test the campus and the methodology itself. Croatia, as the socio-economic context for Campus Borongaj, is positioned in the group of developed countries with high to very high UN Human Development Index (HDI=0.827) and reasonably low ecological footprint per person (3.92) compared to other European countries (Figure 10.7).

Presented as a reference and action-defining framework, 17 SDGs are subdivided into 168 sub-goals to also serve as a checklist tool for diverse range of activities. Each activity, positioned in the centre of the SDG circle, can present intensity of its reference to one or more SDGs and their sub-goals. For our research, the university campus and the new programme for student housing have been positioned within the SDG circle, to determine their capacities for contributing to SDG efforts. The result has shown that the expected extensive influence of the living laboratory approach at the universities and sustainable campus planning on SDGs is

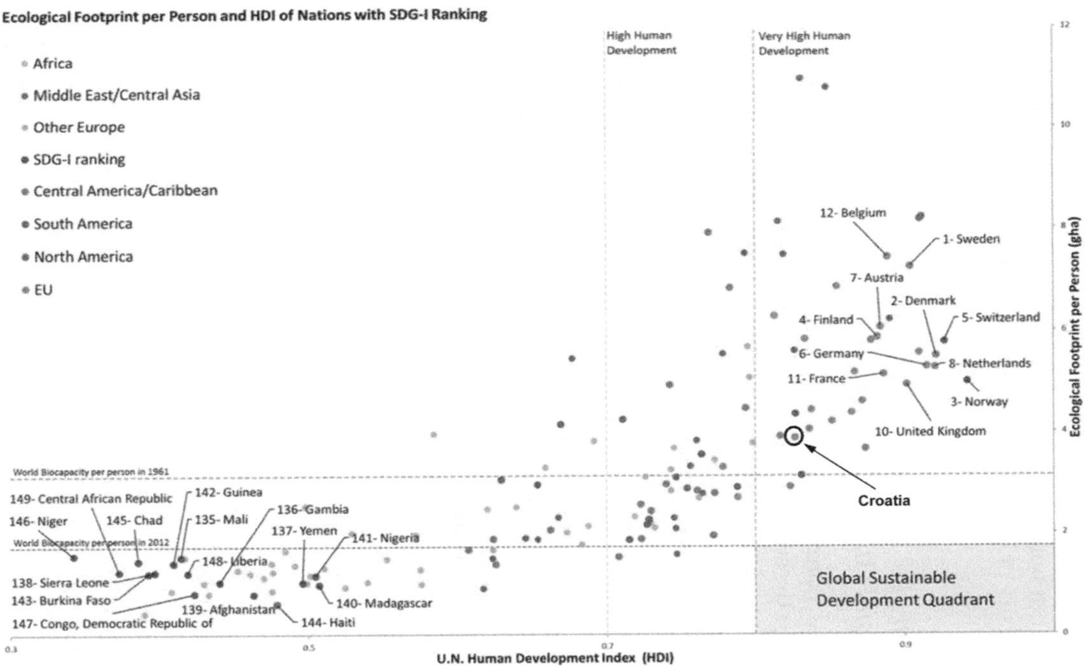

Figure 10.7 GDI/ecological footprint chart and the position of Croatia. (Source: https://doi.org/10.3389/fenrg.2017.00018.) Copyright: © 2017 Wackernagel, Hanscom and Lin. This is an open-access article distributed under the terms of the Creative Commons Attribution License (CC BY). The use, distribution or reproduction in other forums is permitted, provided the original author(s) or licensor are credited and that the original publication in this journal is cited, in accordance with accepted academic practice. No use, distribution or reproduction is permitted which does not comply with these terms.

closely followed by student housing influence, which is not only a sustainable construction but a tool for future sustainable lifestyles (Figure 10.8).

The SDG methodology used in this way has opened up some questions and dilemmas. When trying to associate an activity to some of the 168 specific sub-goals, we do not find the connection we expected when reasoning on the level of 17 SDGs. This leads to a conclusion that sub goal list could be more comprehensive. In addition to that, since it should be of higher importance in the developing world, when sustainability for the developed world is observed in a holistic manner, we may find some sustainability criteria of high importance absent or at least "well hidden" in the sub-goals' list. In our opinion, SDG is a strong platform that will lead the sustainable development in years to come, but at the same time there should also be some improvements made to it along the way.

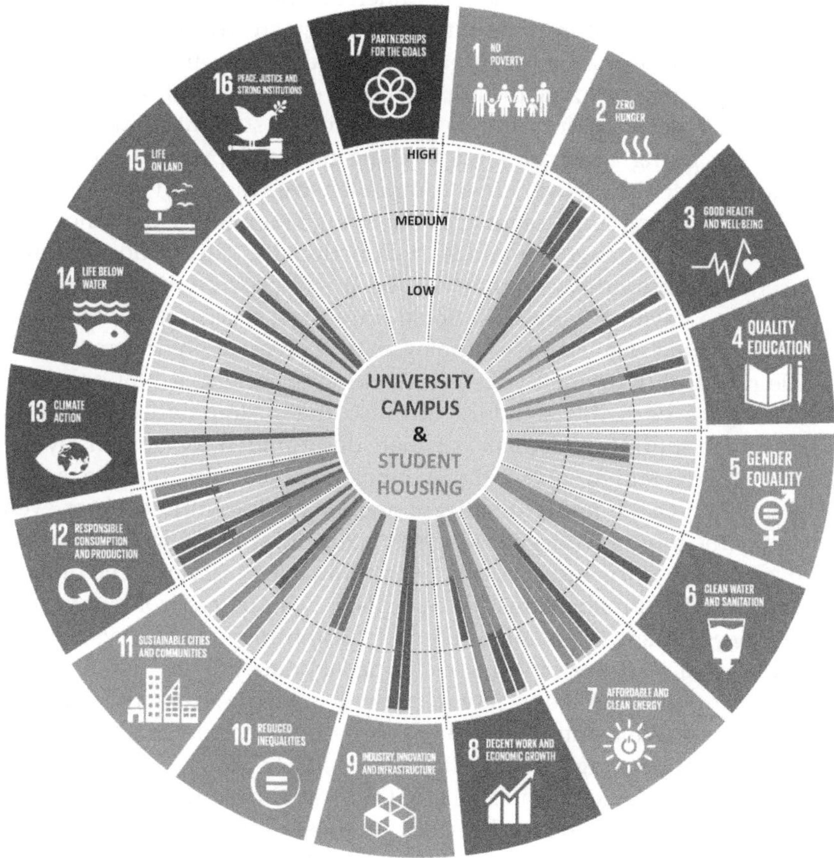

Figure 10.8 SDG wheel measuring the university and student housing potential. (Source: CLL project 2014–2017.)

Although University of Zagreb is a city university, dispersedly integrated within the urban fabric, introducing campus on Borongaj represents a unique opportunity to give students different and advanced learning and living experience. Furthermore, this research has proven that its ambitions, to become sustainable living laboratory game changer for the Croatian context, are possible.

10.6 Conclusion

The path towards sustainable development primarily starts from designing the system to support it, together with efforts in education and lifestyle changes. There is no doubt that universities here play a prominent role, 13.6 million EU students present an enormous potential for future paradigm changes on the subcontinent. This research has shown that today's students have all the prerequisites for the task: inherent values of sustainability, collaborative skills and affinity for interdisciplinary creative work. As future environmentally and socially conscious individuals, they are strong on freedom of choice, emphasizing the importance of diversity in lifestyle options.

So, what does the future look like, from the student point of view, considering living, studying and working on a campus? This research tried to provide the answers in a comprehensive way, with strong focus on student housing as the source of the living laboratory experience and a point where all changes should start. Functionally, student dorm design will serve a community of residents; and it will provide multifunctional and diverse layouts with additional informal spaces for interaction and unexpected use. Diverse typologies and creative environment will enable rich and dynamic living and working experiences. Not only will the buildings be nZEB standard but also the sustainability principles will become founding elements of their design; identity, spatial layout, location choice and so on. As for technological features, buildings will be automated (basic level), intelligent and smart (medium level) or even thinking (the most advanced level). In conclusion, student housing will be a "diverse laboratory and classroom of sustainable lifestyles". The university will further broaden this into a holistic campus environment; from open spaces design, alternative and ecological transportation systems, to education and work possibilities – a living laboratory.

Wide influence on the society, as presented, has its strong reflection on SDGs. Sustainable campus planning has its influence on 13 goals, while future student housing directly empowers 10 of 17 SDGs. We are advancing systems to change our lifestyles, at the same time, our lifestyle defines the systems that are offered. To speed up these circles in the continuous development process, Sustainable Campus Knowledge Base plays an important role. There, new student housing models and their experiences become examples for comparison and inspiration.

We may say that just as university campuses in their structure act as small cities, student dormitories, on the other hand, embody all basic principles of a sustainable campus. In other words, sustainability principles and practices embedded into student housing are the foundation for sustainable campuses, and furthermore, they represent a seed of a sustainable city or the society in general.

References

Arup Foresight. 2013. It's Alive. www.driversofchange.com/projects/its-alive/ (accessed September 1, 2015).

Baletić, B. and Holme Samsoe, M. 2013. How can physical campus planning support universities in their development and ultimately help cities change? In *Regenerative Sustainable Development of Universities and Cities: The Role of Living Laboratories*, ed. A. Koenig, pp. 209–235. Cheltenham; Northhampton: Edward Elgar Publishing.

Baletić, B., Lisac, R., and Vdović, R. 2015. Interactive knowledge base of the university campuses - Planning based on urban and cultural heritage. In *International Conference Proceedings, Cultural Heritage - Possibilities for Spatial and Economic Development*, ed. M. Obad Šćitaroci, pp. 598–603. University of Zagreb, Faculty of Architecture, Zagreb.

Baletić, B., Lisac, R., and Vdović, R. 2017. Campus living lab knowledgebase: A tool for designing the future. In *Handbook of Theory and Practice of Sustainable Development in Higher Education*, Vol 4, ed. W. Leal Filho, U. Azeiteiro, F. Alves, P. Molthan-Hill, pp. 459–474. Berlin: Springer International Publishing.

Campus Living Lab (CLL). 2017. CLL Project Report - Innovative Green Building Research in the Campus Living Laboratory, ed. B. Baletić. University of Zagreb, Faculty of Architecture.

European Commision. 2018. Nature-Based Solutions. https://ec.europa.eu/research/environment/index.cfm?pg=nbs.

Friedman, A. 2016. *Innovative Student Residences*. Mulgrave: The Images Publishing Group Pty Ltd.

Green Monday. 2016. Green Monday. http://greenmonday.org/about-us/.

Institute for Social Research Zagreb. 2014. *Sociological Portrait of Croatian Students*, ed. V. Ilišin. Zagreb: IDIZ.

International Living Future Institute. 2018a. Living Building Challenge. https://living-future.org/lbc/.

International Living Future Institute. 2018b. Living Community Challenge. https://living-future.org/lcc/.

ISCN – GULF. 2010. Guidelines to the ISCN-GULF Sustainable Campus Charter, Zurich. www.international-sustainable-campus-network.org/charter-and-guidelines (accessed February 18, 2016).

Lisac, R. 2012. Guidelines System for University Campuses Sustainable Planning. *PhD dissertation*, University of Zagreb, Faculty of Architecture.

Njirić, H. 2011. *Campus Borongaj - Park B. Competition Booklet, Winning Proposal from Njirić + Arhitekti*. Zagreb: University of Zagreb.

Pap, M. 2012a. Student Housing in Croatia. Study, University of Zagreb, Faculty of Architecture.

Pap, M. 2012b. Student Housing in the World. Study, University of Zagreb, Faculty of Architecture.

Pap, M. 2019. Spatial and Functional Model of Student Residences in the Context of University Campus. *PhD dissertation* in preparation, University of Zagreb, Faculty of Architecture.

SPREAD. 2012. Scenarios for Sustainable Lifestyles 2050: From Global Champions to Local Loops. www.sustainable-lifestyles.eu/fileadmin/images/content/D4.1_FourFutureScenarios.pdf.

The Economist Group. 2018. The Economist Intelligence Unit (EIU). www.eiu.com/home.aspx.

The Student Room Group. 2014. Future Student – What to Expect in the Next Ten Years. http://tsrmatters.com/future-student-what-to-expect-in-the-next-ten-years/.

UN. 2015. Transforming our world: the 2030 Agenda for Sustainable Development. www.un.org/en/development/desa/population/migration/generalassembly/docs/globalcompact/A_RES_70_1_E.pdf.

United Nations University. 2009. 2,000 Watt Society. https://ourworld.unu.edu/en/2000-watt-society.

Urban Ecology Agency of Barcelona (UEAB). 2012. Ecosystemic Urbanism Certification. http://publicacionesoficiales.boe.es/detail.php?id=019516112-0001 (accessed September 15, 2015).

Wackernagel, M., Hanscom, L., and Lin, D. 2017. Making the sustainable development goals consistent with sustainability. *Frontiers in Energy Research*. Vol. 5. p. 18. https://www.frontiersin.org/article/10.3389/fenrg.2017.00018. DOI: 10.3389/fenrg.2017.00018. ISSN: 2296-598X.

chapter eleven

Bike-Friendly Campus, new paths towards sustainable development

José Carlos Mota, Frederico Moura e Sá, Catarina Isidoro, and Bernardo Campos Pereira
University of Aveiro

Contents

11.1 Introduction

The occurrence of a year without Summer in 1816 may have been a relevant catalyst towards the invention of the bicycle and the main event responsible for its development. This meteorological anomaly arose due to the brutal eruption of the Tambora volcano in Indonesia, producing a significant temperature drop during the summer months, with snow and ice prevailing throughout much of North America and Europe (Stommel & Stommel, 1983). Those phenomena produced losses in agricultural and animal production, including a reduction in the number of horses, either as a direct consequence of the bad weather or indirectly from the consequent price increase of oats (Oppenheimer, 2015). Such constraints on city mobility – and anticipating the risks of horse dependency – have been claimed to have prompted Baron Karl von Drais' development of the first generation of bicycles, in Mannheim, Germany, in 1817 – the *Laufmaschine* – presented publicly the following year, in Paris.

Two centuries after von Drais' *Laufmaschine,* the planet is currently facing the challenges of another global climate change phenomenon – a permanent growth in climate risk as the greenhouse effect is increasing temperatures towards a possible point of no return – and once again, the bicycle seems to appear as an important tool for the development of new actions towards a more sustainable pattern of development. The bicycle's relevance in the global agenda is significant, due to its ability to answer serious problems related to excessive automobile dependency in daily work/school commutes. Significantly, car-use dependence represents in general over 56% of all commuter trips currently travelled in Europe (Fiorello, Martino, Zani, Christidis, & Navajas-Cawood, 2016) with a significant increase observed over the last three decades.

As identified in several research articles, contemporary car dependence aggravates trends associated with increasing: (1) obesity in the population (in Europe almost 20% of adults are considered obese (EUROSTAT, 2016)), (2) road injuries (in Europe over 1 million incidents and over 25,000 fatalities in 2016 (European Commission/Directorate General for Mobility and Transport, 2017)), (3) traffic congestion and loss of personal autonomy for non-motorized citizens, (4) public urban space deterioration (which requires interventions to ensure a greater humanization of public space), and (5) exponential CO_2 emissions – the transport sector represents 32% of the energy consumed in the EU28 and is responsible for 893.1 million tonnes of CO_2 greenhouse gas (GHG) emissions, 72% related with road transport (CE, 2000; Damert & Rudolph, 2018).

Thus, bicycle use can be the trigger to reverse the complex dynamics of growing motorization and the starting point for leveraging a combined effort for effective public policies, capable of contributing to the construction of a more sustainable development path, answering to new public policy orientations set forth as the United Nations (UN) Sustainable Development Goals (SDGs).

The goal of this article is to discuss how cities and particularly university campuses can provide relevant contributions in achieving the UN SDG by analysing Bike-Friendly Campus (BFC) initiatives promoted by the University of Aveiro (UA) and look at which outcomes can be relevant and useful in other contexts. A methodology based on literature review and case study analysis was used, with research structured in six parts: (1) the relation between cycling and the UN SDG, (2) the role of cities and BFC and their specific importance, (3) the contribution of BFCs in achieving the UN SDG, (4) an analysis of recent policies to promote cycling in

Portugal, (5) a discussion of the initiatives aimed at developing a BFC at the UA, and (6) the Conclusion.

11.2 Cycling and the UN SDG

The UN resolution entitled "Transforming Our World: 2030 Sustainable Development Agenda" comprising of 17 goals (SDGs), with 169 targets, approved by world leaders, came into effect on the September 25th, 2015 (UNRIC, 2016). The new goals established "seek to build on the Millennium Development Goals and complete what they did not achieve".[1] The Millennium Development Goals (MDGs) (2000–2015) focused on poverty, education and diseases (Pedersen, 2018), to compensate the concentration of wealth and assure the dignity of poor people (Gupta & Vegelin, 2016). Despite the important social achievements of the MDGs (Pedersen, 2018), it became explicit that regarding global environmental problems more was required: "Efforts to ensure global environmental sustainability have shown mixed results throughout the last 15 years ...Environmental sustainability is a core pillar of the post-2015 agenda and a prerequisite for lasting socioeconomic development and poverty eradication" (United Nations, 2015, p. 61).

According to Pedersen (2018, p. 22), to achieve global sustainable development, it is fundamental to respond to the new challenges imposed by "fast-growing industrial production, consumption and urbanization" and engage the private sector in the development of the sustainable development agenda since "the revenues of large companies exceed the GDP of many countries". The UN recognized the importance of designing a plan that includes "the links between socioeconomic and environmental sustainability and protects and reinforces the environmental pillar" (United Nations, 2015, p. 61). In face of this background, the 2030 Sustainable Development Agenda is «a plan of action for people, planet and prosperity»,[1] and the 17 SDGs (Figure 11.1) cover the three dimensions of sustainable development: economic growth, social inclusion and environmental protection (United Nations, 2015).

The UN recognizes that the multidimensionality and complexity of these challenges lead to a strong focus on the means of implementation, from "the mobilization of financial resources as well as capacity-building and the transfer of environmentally sound technologies".[1] Moreover, it will require cooperation between governments, international organizations, the private sector and civil society (ECF & WCA, 2015), acknowledging "the role of the diverse private sector, ranging from microenterprises to cooperatives and multinationals, and that of civil society organizations and philanthropic organizations in the implementation of the new Agenda".[1]

[1] https://sustainabledevelopment.un.org/post2015/transformingourworld.

Figure 11.1 United Nations Sustainable Development Goals. (www.undp.org/content/undp/en/home/sustainable-development-goals.html.)

Another relevant issue is the evaluation of the established goals. The assessment of progress must be carried out regularly by each country, involving governments, civil society, companies and interest groups. A set of global indicators is used and the results published in an annual report (UNRIC, 2016). The fact that the SDGs are a voluntary agreement may be a disadvantage: states can feel free to bypass commitments. On the other hand, these can work as an opportunity because without legal obligations, they may be willing to implement a more ambitious agenda (Pogge & Sengupta, 2016).

Despite the great optimism surrounding this agenda, several authors (Gupta & Vegelin, 2016; von Braun, Hornidge, & Borqemister, 2014; Pogge & Sengupta, 2016) draw attention to crucial issues that are essential to assure a successful implementation of the SDGs. Gupta & Vegelin (2016, p. 435) highlight the importance of the coherence "between objectives and the means", to "challenge the business-as-usual approach to growth", and to "set implementation goals to improve the decision-making framework". von Braun et al. (2014) recommend that despite their global dimension, SDG should give "a lot of attention to local realities", with an interlinked agenda supported by a strong evidence-based framework.

Regarding the inclusion of different sectors, evolution is observed since the previous MDGs, which did not consider the transport dimension (especially sustainable transport); this was a missed opportunity when considering the strong link between transport and economic development (Neun, 2016). In 2014, the United Nations Secretary-General Ban Ki-moon announced the creation of a High-Level Advisory Group on Sustainable Transport "to provide a focused set of recommendations on how the transport sector can advance sustainable development with poverty eradication at its core, promote economic growth, and bolster action against climate change. The outcome of this effort resulted in Mobilizing Sustainable Transport for Development, the first ever Global Sustainable Transport Outlook Report" (United Nations, 2016, p. 7). Ban Ki-moon's document forward claimed that "Sustainable transport is fundamental to progress in realizing the promise of the 2030 Agenda for Sustainable Development and in achieving the 17 Sustainable Development Goals" (United Nations, 2016).

Moreover, as Neun (2016) pointed out, through an important work developed by the European Cyclists' Federation/World Cycling Alliance (ECF/WCA), "transport impacts achieving the SDGs were identified to be safe, affordable, accessible, efficient, resilient, and minimizing carbon and other emissions and environmental impacts"; "content wise the SDGs' overarching End of Poverty was contributed by cycling economics significantly" and the "systematic analysis of the cycling benefits in the SDGs was based on the Active Mobility Agenda as an evaluation matrix" (Neun, 2016).

In a noteworthy event organized by the ECF Scientists for Cycling Network at UA the results of the WCA and ECF's "Cycling Delivers on the Global Goals" report (ECF, 2016) showed that cycling could contribute significantly to 12 of the 17 SDGs, and simultaneously assumed the organizations' commitment in unleashing the potential of cycling aimed at achieving those goals» (see Table 11.1). In fact, cycling was claimed as "already contributing to the necessary global change, by making transport more sustainable and that active mobility should be a human right at all scales – including the right to cycle". That is why "governments at all levels should provide secure access to public space, protect those who walk and cycle and ensure – through mobility – equal participation in society" (ECF & WCA, 2015, p. 1). As Manfred Neun mentioned, "the benefits cycling can bring to a community or to societies, recognized as sustainable benefits, are already measured" (Neun, 2016, p. 57):

The ECF/WCA's commitment to the UN of having "more people cycling more often" implies working worldwide to integrate cycling into strategies and policies at all levels, mobilizing citizens to cycle, promoting cycling and engaging with authorities in the development of bicycle-friendly policies and infrastructure (ECF & WCA, 2015, p. 6).

Regarding the UN SDG, it is significant, as mentioned above, that cycling policies and measures are only a piece of a global strategy to implement sustainable urban travel. Ultimately, the principal goal is to

Table 11.1 Relation between cycling and SDG

UN SDG	Cycling
GOAL 1: No poverty	X
GOAL 2: Zero hunger	X
GOAL 3: Good health and well-being	X
GOAL 4: Quality education	
GOAL 5: Gender equality	X
GOAL 6: Clean water and sanitation	
GOAL 7: Affordable and clean energy	X
GOAL 8: Decent work and economic growth	X
GOAL 9: Industry, innovation and infrastructure	X
GOAL 10: Reduced inequality	
GOAL 11: Sustainable cities and communities	X
GOAL 12: Responsible consumption and production	X
GOAL 13: Climate action	X
GOAL 14: Life below water	
GOAL 15: Life on land	
GOAL 16: Peace and justice strong institutions	
GOAL 17: Partnerships to achieve the goal	X

Source: Cycling delivers on the global goals (ECF & WCA, 2015).

improve the entire transport network's sustainability, thus, cycling measures must be integrated and coherent with other policies addressing land use, environment, physical health and even finance.

In fact, the effectiveness of cycling strategies depends on the coherence of mobility planning policies. It is crucial to combine automobility dissuasion measures (e.g. parking fees, urban tolls, traffic restrictions in neighbourhoods), traffic calming measures and other policies that reinforce the role of sustainable modes of transport, especially cycling and walking. This approach is widely published in scholarly literature (Pucher & Buehler, 2012; Oldenziel, Emanuel, de la Bruhèze, & Veraart, 2016) and currently highly validated by several benchmarking analyses.

Yet obstacles around cycling strategy implementation are common. The OECD (2004) points to financial constraints, institutional barriers, safety concerns, lack of technical knowledge and frail public awareness as major challenges in achieving effective policy-making for cycling.

11.3 Cities and BFC

The creation of bike-friendly places (cities, campuses, enterprises) emerges among a wide range of solutions as one of the most important. Basically, because bike-friendly locations establish a global vision for mobility in a specific place-based area first and only afterwards define a set of multisectoral measures that necessarily require coherence, integration and articulation.

BFCs are one of the most common and interesting cases of how a mobility vision can change a specific location. The BFC concept has emerged mainly in the highly motorized North American urban context, seeking to encourage environmentally friendly travel modes and consequently urban settings that are more appealing. The BFC is a relatively recent concept, organizing a coherent set of initiatives promoted by higher education institutions that aim at increasing bicycle use in daily commutes.

The goal is to increase the modal share of cycling in the mobility system by establishing a series of interventions focusing on hard measures such as infrastructural issues, complemented by soft measures aimed at engaging the entire academic community. In sum, BFCs create infrastructure to facilitate cycling on campus (especially cycleways and bicycle parking) and simultaneously develop a set of measures to raise awareness of cycling by students, teachers and staff, through education, and even by supporting bicycle purchase, lease, or rentals.

Significant evidence in literature is consensual that the transfer to active travel modes (cycling and walking) requires a multidisciplinary approach combining the introduction or improvement of adequate infrastructure, and campaigns focused on the benefits of active travel behaviour, particularly in and around university campuses (Bopp et al., 2017).

The success of BFC programs is associated with the nature and specific characteristics of the target audience. Firstly, many university students live and conduct their activities on campus or in the surrounding neighbourhoods, many of which are mixed use and more livable than sprawled metropolitan areas. As a result, university populations' travel behaviour and daily commuting patterns tend to be closer and more accepting of active travel modes (Wang, Khattak, & Son, 2012).

Academic communities are a relevant target group to achieve modal-split change because of their higher education levels and greater sensitivity to societal issues (e.g. climate change). Foremost, these communities include a significant proportion of new arrivals, without predetermined travel habits. Consequently, academic communities are more susceptible to accept new and more sustainable modes of transport such as cycling.

The benefits of a BFC are many. BFCs assure a significant presence of cycling and provide solutions in several different areas such as safety, sustainability, cost savings, mobility, health and well-being. The BFC also represents an opportunity to intervene and qualify existing public space with non-motorized travel programs. In this sense, BFC works like a Trojan horse; behind the specific goal of the BFC, a wider and multisectoral "hidden agenda" striving for the humanization of public space and the transformation/regeneration of urban settlements is at play. "Riding a bicycle is a popular, environmentally friendly, and cost-effective alternative to automobiles as a method of transportation as well as an activity for recreation and fitness" (Lavetti & McComb, 2014, p. 1), and if framed by an effective BFC program, it may produce very comfortable and safe spaces, that can improve the entire urban area and its functions.

There are several examples of initiatives developed by certified universities. Essentially, there are: (1) automobility dissuading measures (e.g. removing or increasing the cost of parking spaces, media campaigns that apply to the environmental and social conscience of university members); (2) and attraction policies (e.g. building bike lanes and trails linking universities and faculties, schools, parks and intermodal transport hubs, the development of bike-share programs, implementation of traffic calming measures, free bicycle-use classes, bike-to-work days and road-safety campaigns for all road users).

Based on the success of the American programs and assuming that cycling is the "ideal mode of transportation for university campuses – it's quiet, it's clean, it's inexpensive, it's sustainable and it's space efficient" (Gilpin, 2012, p. 8), the question that remains for realities such as the Portuguese context (where bicycle-use is relatively low) is: how to start the transformation of a university campus for promoting bicycle use?

Gilpin (2012, p. 8) considers that the best way to "get the wheels spinning" is to follow eight steps that can be freely adapted as follows: (1) Develop a mobility plan particularly focused on cycling (a blueprint revealing what the campus will look like); (2) work with local authorities to improve bicycle facilities (ensuring bicycle-friendly infrastructure for daily commutes and "local cycling errands" is crucial); (3) innovate, the best or the most adequate solutions may still come from out of the toolbox; (4) provide adequate, safe and secure bicycle parking facilities; (5) evaluate the implementation of a campus bicycle-sharing system (BSS; it may become a "valuable resource and demonstrates a strong commitment to bicycling at the university"); (6) develop bicycle programs such as classes, bicycle shops, talks, bicycle rentals, lease or purchase programs, or other incentives to cycle; (7) "set up evaluation/monitoring programs" to measure the global impact of each action and effort; and (8) "apply for a bike-friendly recognition", to promote increasing engagement in the university community.

Recent studies reveal that at several university campuses, "recreation programs and facilities were supportive of healthful lifestyles for obesity prevention, but policies and the built environment were not" (Horacek et al., 2014, p. 2). In this sense, adequate improvements of the campus environment are highly recommended.

One of the major learning outcomes of the American Bicycle-Friendly University (BFU) experience is that beyond the fun of cycling, university members increase bicycle use if they feel it has a positive effect on the reduction of their daily travel and parking costs (Miller & Handy, 2012).

Other results also underpin the importance of an agenda that is somehow démodé: land-use patterns matter! Yet students living on or near campus reveal a significantly higher probability of walking and cycling to campus and are less likely to drive an automobile (Wang et al., 2012). Therefore, the travel patterns of university members are highly related to campus location within the city. One of the major long-term challenges is to assure that campus locations remain central within the city's geography, thus taking advantage of greater accessibility to activities and to a walking and bicycle-friendly environment.

11.4 *Contribution of the BFC program to the UN SDG*

The BFC program is an opportunity to "improve campus connectivity and accessibility, to increase safety for all transport modes; to reduce the carbon footprint, pollution and congestion; to lower health care and parking costs, to make happier and healthier students and staff; to enhance campus quality of life; to attract students and employees who want better transportation options and finally to become a model for the rest of the country",[2] then a BFC could probably also be the perfect space to test the implementation of the UN SDG, answering some of the criticisms or doubts regarding its viability.

Looking at ECF and WCA (2015), where Cycling Delivers on the Global Goals are analysed, we can identify four critical dimensions, responding for 10 of the 17 goals, where BFC can make a difference (Table 11.2):

11.4.1 *A sustainable (and healthy) campus*

Experiences related with BFCs in a wider sustainability framework show how they can respond to the demands of UN SDG #3, #7, 13, focused on good health and well-being, affordable and clean energy and climate action.

The sustainable campus focused on mobility answers to decarbonize travel, labelled as "greening" campus strategies, by reducing car use, CO_2 emissions and gas, transforming the transport system towards greater energy efficiency, the base of sustainability and of the climate climate change action agenda (EU policies). But it is also important to act within a wider context, improving environmental behaviour and impact on campus (improving energy efficiency, water and green space management)

[2] https://bikeleague.org/.

Table 11.2 BFC and the UN SDG

Bike-Friendly Campus (BFC)	UN SDG
A sustainable (and healthy) campus	GOAL 3: Good health and well-being GOAL 7: Affordable and clean energy GOAL 13: Climate action
The campus as part of an integrated cycling network and "free" bike-share scheme	GOAL 11: Sustainable cities and communities GOAL 5: Gender equality GOAL 1: No poverty
The campus as an open lab and incubator of new forms of sustainable economic growth	GOAL 8: Decent work and economic growth GOAL 9: Industry, innovation and infrastructure GOAL 12: Responsible consumption and production
The campus as a common partnership for change	GOAL 17: Partnerships to achieve the goal

Source: José Carlos Mota, Frederico Moura e Sá, Catarina Isidoro, and Bernardo Campos Pereira.

and issues related to campus/building planning, construction and management (Hooi, Hassan, & Mat, 2012; Ribeiro et al., 2017; Lipschutz, De Wit, & Lehmann, 2017: p. 14).

The healthy campus strategy is another important university policy to be developed. Concerned with several food and risk behaviours, especially from students (Booth & Anderson, 2017; Brandão, 2010), universities have been stimulating healthy behaviours and active lifestyles, where cycling has appeared as an important instrument.

11.4.2 *The campus as part of an integrated cycling network and «free» BSS*

University campuses are relevant because of their size (in terms of area and of population), location (central or peripheral; one or several campuses) and relation with the cities where they are located; therefore, it is crucial to link and equip these facilities with adequate cycling infrastructure (Pitsiava-Latinopoulou, Basbas, & Gavanas, 2013).

The traditional approach, especially the American model, is focused on improving the quality of service regarding cycling infrastructure – cycleways and bicycle parking – not only on campus but also in the city where it is located. The principles of the cycleway network focus on safety, speed and comfort, and intermodality with other transport modes (McClintock, 1992; Balsas, 2002).

Recent approaches have shown the importance of offering more than infrastructure, namely: bicycle maintenance services, campaigns to promote cycling, and even BSS to promote low-carbon strategies and sustainable multimodal urban mobility to university members. These efforts show the relevant role of universities in close collaboration with local authorities, leading innovative projects that can highlight and be helpful in achieving more sustainable cities and communities (UN SDG #11).

Gender equality is fundamental for accomplishing all the goals (United Nations, 2015). Women tend to cycle less than men, and as Gabriele Prati (2017, p. 373) emphasizes, "despite the different barriers and motivators for cycling among women, the findings suggest that women's under-representation in positions of power and the traditional task division between the genders for childcare and household responsibilities may inhibit women's participation in transport cycling". Campuses have a high percentage of women studying, and they are also places where gender equality is politically and academically debated. BFCs can provide the inclusive dimension contributing to UN SDG #5.

Different studies indicate that "the constraints faced by poor people in their journeys to work or school depend upon distance, cost and the ease and availability of different forms of transport. The poor often try to minimize their cash expenditures by walking or cycling" (Starkey & Hine, 2014, p. 41). Although most universities and governments have policies to help low-income families and students, transport cost is not always taken into consideration or it's related to a public transport system that doesn't always answer customers' needs. In this context, an academic BSS could meet several needs and contribute partially to UN SDG #1.

11.4.3 *The campus as an open lab and incubator for new forms of sustainable economic growth*

University campuses are becoming particularly active in their third mission, in which collaboration with enterprises and communities through knowledge and technology transfer is central. This happens because universities are places of excellence for research and innovation, with experts, information and technology, and there is a growing need from society for new solutions to respond to the enormous challenges facing the planet (Rodrigues, 2001; Romero & Mülfarth, 2017).

Cycling has been part of that new demand, and a growing number of universities are creating labs or platforms to experiment new planning concepts for more sustainable and resilient cities and to test innovative solutions to upgrade transport systems and improve intermodality (Balsas, 2002; Lipschutz et al., 2017). In this context, campuses can be the incubators for new businesses, products and services associated with cycling, promoting self-employment and/or incorporating students in a rapidly expanding activity sector, with relevant impacts upon UN SDGs #8 and #9.

The fact that many members of academia live close to campus or have good transport connections available increases potential for consuming local products and services. Furthermore, studies reveal cycling's important role in local economies (Neun & Haubold, 2016), with potential contributions to achieve UN SDG #12.

11.4.4 The campus as a common partnership for change

The SDGs cover several domains of public action and involve several actors, not always sharing common understandings and actions. To achieve an alignment between all agents involved, new partnerships and innovative methods of stakeholder involvement are essential, as are the bases for establishing consensus building.

Universities appear as important agents within this process, with a vast field of knowledge in several dimensions and with a mediation role that can be crucial in achieving policy and implementation change. According to Filho (2015, pp. 4–5), "due to their positioning and strategic nature, there is solid evidence that higher education institutions can and should make a strong contribution to sustainable development as a whole".

Regarding cycling and the UN SDG #17, universities have the possibility to «increase the availability of high-quality, timely and reliable data on cycling; to support the global development and dissemination of successful, environmentally sound cycling technologies and the development and implementation of cycling policies in developing countries» (ECF & WCA, 2015, p. 2). The existing knowledge and expertise available from universities should "(…) help society identify and implement the social and technical solutions to the environmental challenges they have helped identify" (Filho, 2015, p. 5).

11.5 Cycling in Portugal

Despite the recent attention dedicated to bicycle use, especially in the media, data from the 2011 national census reveals daily bicycle commutes accounting for less than 1% of the total daily commutes in Portugal's mobility system, representing approximately 31,000 regular users nationally, a figure well below the European average of 8.7% (Euro-Barometer on Transport Policy, 2016).[3]

[3] Preliminary results in a recent mobility survey of Portugal's two largest metropolitan areas – Lisbon and Porto – reveal a relevant increase in cycling in comparison to the Census 2011 results, with 41,300 bicycle users just in these two cities (IMob, 2017). Yet these results are still quite low in comparison to the European average (Euro-Barometer on Transport Policy, 2016) and ECF objectives for 2020 (ECF, 2012).

In Portugal, there are over 4 million cars, and automobility represents over 60% of daily trips; in 20 years, the number of automobiles has doubled, and car trips have increased threefold (INE, 1991/2001/2011). In 1950, 40% of vehicle traffic counted on national roads were bicycles (Ministério das Obras Públicas, 1965, pp. 22, 27, 29), and some city centres may have had even higher bicycle use. However, a significant reduction in cycling mode share from 40% to 17% occurred between 1950 and 1965, suggesting the impact of a series of social and cultural factors, reflected in and resulting from road infrastructure policies implemented during this period and persisting, with cycling decreasing to below 1% since then (Pereira, 2016).

The linked infrastructure policies supported by national and European programs have contributed to aggravate a dispersed territory based on widespread car use. At the same time, the weight of car ownership and the automotive industry in the national economy has increased significantly. Taxes directly associated to automobility, for car use (IUC), car purchase (ISV) and fuel tax (ISP), generate 15%–20% of all indirect taxes in Portugal.

The increase of motorization has contributed to inactive lifestyles, less physical activity, which associated with a poor diet led to two-thirds of Portugal's citizens revealing signs of obesity or being overweight,[4] 40% of the population (over 4 million people) live with «hypertension (HBP), … (and) cerebral vascular accidents (CVA) are the leading cause of death and disability in adults in Portugal».[5] Another serious problem is associated to road incidents. According to recent data, Portugal has the highest number of road deaths per capita in Western Europe. In 2017, 602 people died from road incidents in Portugal (ANSR, 2017), equivalent to 58.4 people per 1 million inhabitants, from a total of 34,416 traffic accidents with victims (ANSR, 2017, p. 6). Worrisomely, Portugal also registered a 6.9% increase in road deaths between 2016 and 2017 (ANSR, 2017), possibly from increased automobile use.

Nonetheless, in recent years, some important developments have been made related to cycling, mostly public investment in cycleway networks and mainly focused on leisure, being promoted by municipal governments. Other relevant projects are expected in the near future; according to regional policies financed with European funds, over 300 million euros will be spent under the Portugal 2017–2020 framework (Ministério do Ambiente, 2018)

BSSs have also increased cycling significantly. Aveiro, Cascais, Torres Vedras, Vilamoura and Lisbon are the most prominent BSS in Portugal,

[4] Estratégia Nacional para a Promoção da Atividade Física, da Saúde e do Bem-Estar para dez anos, Direcção-Geral da Saúde, maio 2016.

[5] Portrait of Health 2013.

with a significant impact in reinforcing bicycle use as a relevant mobility mode in urban areas.

At a national level, the government approved the official Institute of Mobility and Transport's (IMT) plan for bicycle promotion and other soft modes 2013–2020 (*ciclAndo*) in 2012, but no effective action has been taken so far. Sustainable mobility is identified in the 2014–2020 Partnership Agreement as one of the strategic themes (promoting low-carbon and ecological mobility) and a national program named UBIKE was launched, co-financed by the PO SEUR – Operational Program for Sustainability and Efficient Use of Resources, within the Portugal 2020 and the EU – Cohesion Fund.

The UBIKE program aims at promoting low-carbon strategies and stimulates sustainable urban mobility, with a focus on cycling in higher education public institutions (IES [Instituições de Ensino Superior]), delivering 3,300 public bicycles to academic community members at 15 universities in Portugal. By the end of 2018, UBIKE bicycles were estimated to have travelled at least 2,412,141 km, saving the equivalent of 166.34 tons of gas energy and reducing CO_2 emissions by 505 tons.

Although bicycle users represent a very low mode share within the overall national mobility system, bicycles assume a key position in Portugal's economy. Portugal's bicycle industry is one of Europe's top 3 in bicycle accessory and component production, and one of the top 5 European bicycle manufacturers. A significant part of this industrial production is in the Aveiro region, with an industry worth €200 million in exports (Rodrigues, 2011). There is also an emerging set of bicycle-related trade and services activities (small and microenterprises and organizations) distributed throughout the country (whose GDP value is unknown).

11.6 Aveiro BFC

In recent years, the UA has played a key role regarding cycling and bicycle use in its scientific and academic activities, internal organization and engagement with the city and the surrounding region.

In 2014, UA's Rectory established a Mission Group for Sustainable Development[6] aimed at creating and developing an Environmental Management System (EMS) applied to teaching, research, cooperation with society, support services and interface, which has been certified according to ISO14001:2015, aiming above all "to protect the environment by preventing/mitigating adverse environmental impacts, mitigating the adverse effects of environmental conditions at UA and improving its environmental performance".

[6] https://www.ua.pt/campusmaissustentavel.

Considering that Aveiro region has the highest cycling mode share in Portugal (2.8% of daily commutes, eight times higher than the national average, Census 2011), that it has a unique institutional ecosystem and that the promotion of sustainable mobility is integrated with the regional bicycle cluster (Strategic Study Region Aveiro 2020), UA created the Technology Platform for Bicycle and Soft Mobility[7] (PTBMS) as a structure integrated within these regional efforts. The PTBMS is an informal structure composed of researchers from different areas and research units. The Platform's mission is to support favourable conditions for cycling and other active/soft mobility modes, striving to improve the environment, economy, territory and the lives of citizens and communities.

To respond to these challenges, UA has been developing a set of initiatives that fit the above-mentioned philosophy, namely of a friendly campus for those who want to walk or cycle, and also to create a friendly campus for those who want to develop new products and services linked to cycling and mobility. The principles outlined above are clearly aligned with the UN SDG goals, as summed in the Table 11.3.

To ensure that the concept was not limited to Aveiro region, the national "Bicycle Commitment" challenge was launched by UA, with the motto "More Bicycles, Better Cities, and More Sustainable Economy and Society" aimed at challenging companies, municipalities, public administration and third sector organizations to promote activities aimed at increasing regular bicycle use by employees, customers, visitors, associates and/or the general public. To encourage and guide participation, nine challenges were identified that served as a guideline for the design of activities. This network was developed in 2016 with the support of several partners (UVP/FPC, FPCUB, ABIMOTA, MUBi, Ciclaveiro and the ECF).

Furthermore, as pointed out in the UN SDG, the importance of promoting sustainable economic growth (#8) and encouraging open laboratories (#11) was in fact the origin of PTBMS. In this context, UA's collaboration with ABIMOTA and Soft Mobility Programs I and II (2012–2015) was fundamental in establishing a closer relationship between the bicycle industry and the university, subsequently creating the Portugal Bike Value program.[8] Along the same line of action, but aimed at a different target group, UA collaborated in the *Empreende Já* (endeavour now) program launch,[9] aimed at young unemployed graduates and empowering them within a work and business setting, in an event with over 25 participants contacting with cycling-related businesses.

Within the UA campus, several academic and scientific projects have been carried out to create products and services related to bicycle use

[7] http://www.ua.pt/ptbicicleta/.
[8] http://portugalbikevalue.pt/.
[9] https://eja.juventude.gov.pt.

Table 11.3 UN SDG and UA BFC

0UN SDG	Bike-Friendly Campus/Actions
#13-A sustainable campus	UA sustainable campus • Commitment to cycling
#08-The campus as an incubator of new forms of sustainable economic growth #11-The campus as an open lab	Business and innovation-friendly campus • PT bicycle • Bicycle economic value (ABIMOTA, Portugal bike value) • Research (technologies, spatial planning) • Public debates on mobility • UA CYCLING LAB and cooperation projects of the departments (Schist bicycles)
#01-The campus offering cheap means of transport – bicycles #09-The campus as part of an integrated cycleway network	Bicycle user-friendly campus • UAUBIKE BIKE • Bicycle space – repairs • Network of city bike lanes and campus bicycle parking infrastructure
#17-Governance	A people's campus • Partnerships with local stakeholders Ciclaveiro, Cycle Centre, PSP – Polícia de Segurança Pública (Police), Commercial Association, INOVARIA (Associação de Empresas para uma Rede de Inovação em Aveiro) agency and municipalities *Campus and Cities Bike-Friendly* network • 2016 Scientist for Cycling Conference UN SDG at UA • Bike-Friendly Campus and Cities Forum

Source: José Carlos Mota, Frederico Moura e Sá, Catarina Isidoro, and Bernardo Campos Pereira.

and cyclists (from design to materials, health to tourism, ICT to energy, education to territory) using the UA campus as a space for experimentation.

Looking at the UN SDGs related to low-cost transport modes (#1) and integrating the cycling network (#9), the UA applied to Portugal's national UBIKE program (UAUBIKE in Aveiro), through which it will make 142 conventional and 97 electrically assisted bicycles (pedelecs) available to the UA academic community through a free long-term lending regime, aimed at creating regular bicycle use a common mode of mobility, as well as promoting energy efficiency and reducing energy consumption. Besides supplying 239 bicycles, the UAUBIKE project includes the creation of adequate infrastructural conditions, installing 200 new *Sheffield* bicycle parking racks, a significant increase from the 300 existing bicycle parking racks.

Considering the main objective, bicycles will be given preferentially to those who use individual automobile transport to travel to the university. All bicycles are equipped with GPS technology, since tracking kilometres travelled is essential to measure the impact of each UAUBIKE bicycle and the program's overall performance. Another important aspect of implementing UAUBIKE is bicycle maintenance, which is essential to ensure proper bicycle operation and safety. Thus, in addition to regular maintenance, which can be performed by the users themselves, or in specialized workshops, a thorough examination is realized before returning the bike, thus ensuring that each UAUBIKE bicycle is in good condition. After one semester of use, the bicycle must be returned to UA, for transfer to the next user. The goal is to achieve the largest number of academic community members cycling for mobility purposes and, thus, contribute to changing mobility habits on campus.

Project implementation methodology involves the active participation of the academic community, allocating bicycles to students, teachers and university employees. The UAUBIKE program also includes several activities to promote bicycle use. These activities are targeted at all bicycle users, non-users and potential users – aimed at engaging the academic community and placing cycling and active/soft mobility on the UA agenda. The first action to publicize the project was the creation of a wide-ranging publicity campaign, including the UAUBIKE website with project information (rules, applications, contacts, initiative schedule, etc.) but also useful project information: how to use the bicycle, campus map with cycling infrastructure, the national road code, local bicycle support organizations, means for users or anyone interested to follow and contact the project team using social networks (Facebook and Instagram) and the UAUBIKE newsletter.

In addition to these dissemination tools, activities and events that follow the different phases of project implementation were also created. Training sessions were organized to teach cycling in urban areas, specifically targeted at future UAUBIKES users. These sessions included cycling classes, road code training, and other useful information such as teaching proper secure bicycle parking. Designed for anyone interested in cycling and active/soft mobility, UAUBIKE Lectures are regular conversations scheduled at lunchtime, promoted to stimulate and raise awareness of different subjects related to cycling (bicycle use, advantages, obstacles, etc.) and innovation.

An important step to transform the UA campus into a BFC was the creation of the *"Espaço da Bicicleta"* (The Bicycle Space) a centrally located on-campus space for knowledge and information exchange, encouragement and assistance aimed at bicycle users and potential users. *"Espaço da Bicicleta"* is a co-working space serving as the UAUBIKE front office, where users obtain assistance and solve UAUBIKE program-related issues, also with a repair shop equipped to handle most bicycle repairs from AAUAv Bike Nucleus (NBICLA) volunteers, providing free or low-cost bicycle mechanical services to anyone interested.

The UN SDGs are at the heart of the creation of the UA Bicycle Strategy. The UN SDGs inspired the 2016 Scientist for Cycling Conference, organized in collaboration with the ECF, structuring a series of lectures[10] at UA's Department of Environment and Planning, from February 2017 to October 2018, and finally, it will frame a BFC and Cities Forum aimed at encouraging a reflection upon the role of cities and university campuses in promoting bicycle-friendly territories.

In short, results show that the path that has been taken so far, characterized by the involvement of the academic community, multisectoral and multilevel institutions and even citizen groups, allowed to affirm the promotion of cycling as a central agenda for the UA – politically, today the UA assumes the bicycle as one of its main identity brands. This result is very interesting especially because at the moment, the UA campus is still deeply motorized. In other words, the promotion of cycling (and the implementation of the BFC program) is a strategy against the current pattern and at the same time a tendency that seems to have reached a point of no return. Another relevant fact is that the university, by its influence and symbolic power (even more evident in medium-sized cities such as Aveiro), has also influenced the mobility strategies of the surrounding areas and territories. Consequently, one of the great results of the BFC at the UA is its contribution to the promotion of cycling as a more general, recommended, necessary and rationally irrefutable practice (as it is now also widely supported by several international incentives and strategies, with emphasis on SDG).

11.7 Conclusions

The UN SDGs provide a relevant set of goals to help cities and society build new paths towards sustainable development. Recent research points to cycling's significant role towards 12 of the 17 SDGs, as outlined above. Nevertheless, despite some efforts in terms of public policies, urban mobility is still mostly motor based, with the largest portion of mode share from energy-intensive automobility, still a heavily carbonized and oil-dependent mode.

Within this context, the BFC concept, characterized by a coherent set of initiatives conducted by universities seeking to increase on-campus cycling and, in the cities where they are located, suggest university campuses as optimal locations to test the implementation of UN SDGs, answering some of the criticism and doubts regarding the viability of

[10] https://uaonline.ua.pt/upload/med/joua_m_4612.pdf.

implementing such measures and achieving effective mode change towards sustainable urban travel modes.

To achieve the UN SDG goals, four main principles were identified: (1) a sustainable (and healthy) campus, (2) the campus as part of an integrated cycling network and a "free" bicycle-share system, (3) the campus as an open lab and incubator for new forms of sustainable economic growth, and (4) the campus as a common partnership for change. The literature review and several examples highlight some of the possible actions to be taken.

Cycling in Portugal has achieved important landmarks over the last decade. Taking into consideration its low modal figures (only 0.5% of commuter trips, according to the 2011 National Census (IMT, 2014)), an important public and private effort has been accomplished. Strong investment in cycleways and bike-share schemes in cities is underway, with the largest city, Lisbon, at the lead, and programs in universities (UBIKE) are creating a new favourable framework to decarbonize mobility and ensure more sustainable development.

UA has been leading efforts to create a bike-friendly environment for regular bicycle users but also to achieve greater economic and social value around cycling, following the creation of the Technological Cycling Platform (PTBMS), answering most of the SDG challenges. Three main sets of actions have been achieved: a network of cycleways and bicycle parking in articulation with the municipality; a bicycle-share system with 239 bicycles and pedelecs; and a relevant number of projects involving industry and researchers (PTBMS), advocacy groups and municipalities (at a national and local level).

Regional and local actors have assumed cycling as a key issue in terms of public policies and investments. By 2019, most of UA's and Aveiro Municipality's investments will be implemented, transforming concepts into reality. Portugal Bike Value, the Portuguese bicycle industry's economic cluster, was launched in a close collaboration between academia and industry, an ongoing partnership that is developing even further.

From a balance of the last 4 years, some significant learnings are clarified by UA's bicycle strategy: The importance of vision and leadership in a collaborative partnership, specially conducted by UA's Rectorship in connection with the Region of Aveiro; a significant number of actions developed in a short period of time, aligned around the four dimensions, with local, national and international impact (ECF); international and national networks, involving ABIMOTA (bicycle industry), local advocacy groups in the city and academia (Ciclaveiro, NBICLA) and ECF (Scientist for Cycling).

Although these initiatives bring some evidence of the possible important outcomes, they are still fragile and require more time and perseverance. Nevertheless, these illustrate principles, methodologies and actions than can inspire others pursuing the implementation of the UN SDG.

Nomenclature

ABIMOTA	National Association of Two Wheels, Industries, Hardware, Furniture and Related products
BFC	Bike-Friendly Campus
BFU	American Bicycle-Friendly University
BSS	Bike-Share systems
CO_2	Carbon Dioxide
CVA	Cerebral Vascular Accidents
ECF	European Cyclists' Federation
EMS	Environmental Management System
FPCUB	Portuguese Federation of Cycletourism and Bicycle Users
GDP	Gross Domestic Product
GHG	Greenhouse Gas
HBP	High Blood Pressure
ICT	Information and Communications Technology
IES	Higher Education Public Institutions
IMT	Institute of Mobility and Transports
ISP	Tax on Oil Products
ISV	Vehicle Tax
IUC	Car Circulation Unified Tax
MDG	Millennium Development Goals
MUBi	Association for Urban Bicycle Mobility
NBICLA	NBICLA University of Aveiro Students' Bicycle Association
OECD	Organisation for Economic Co-operation and Development
PO SEUR	Operational Program for Sustainability and Efficient Use of Resources
PTBMS	Platform for Bicycle and Soft Mobility
SDG	Sustainable Development Goals
UA	University of Aveiro
UN	United Nations
UVP/FPC	Portuguese Cycling Federation
WCA	World Cycling Alliance

References

ANSR. (2017). Relatório Anual Sinistralidade Rodoviária 2017 Vítimas a 30 Dias.

Balsas, C. (2002). Sustainable transportation planning on college campuses. *Transport Policy*, 10, 35–49.

Bopp, M., Sims, D., Matthews, S. A., Rovniak, L. S., Poole, E., & Colgan, J. (2017). Development, implementation, and evaluation of active lions: A campaign to promote active travel to a university campus. *American Journal of Health Promotion*. doi: 10.1177/0890117117694287.

Brandão, M. P. M. (2010). Estudo epidemiológico sobre a saúde de estudantes universitários. Dissertação de Doutoramento. Universidade de Aveiro.

CE. (2000). *Cidades para Bicicletas, Cidades de Futuro*. Luxemburgo: Serviço das Publicações Oficiais das Comunidades Europeias. Bruxelas

Damert, M., & Rudolph, F. (2018). Policy options for a decarbonisation of passenger cars in the EU: Recommendations based on a literature review. Wuppertal papers (Vol. 193).

ECF. (2016). Cycling delivers to the UN Global Goals. *Scientists for Cycling Colloquium*, Aveiro, November 2016.

ECF & WCA. (2015). Cycling delivers to the Global Goals – Shifting towards a better economy, society, and planet for all. Brochure published in 2015 by ECF and WCA: https://ecf.com/sites/ecf.com/files/The%20Global%20Goals_internet.pdf.

European Commission. (2016). Eurobarometer on Transport Policy.

European Commission/Directorate General for Mobility and Transport. (2017). Road safety evolution in EU.

EUROSTAT. (2016). European Health Interview Survey. Almost 1 adult in 6 in the EU is considered obese. EUROSTAT: Newsrelease, 1–5.

Filho, W. L. (2015). *Transformative Approaches to Sustainable Development at Universities*. doi: 10.1007/978-3-319-08837-2.

Fiorello, D., Martino, A., Zani, L., Christidis, P., & Navajas-Cawood, E. (2016). Mobility data across the EU 28 member states: Results from an extensive CAWI survey. *Transportation Research Procedia*, 14, 1104–1113.

Gupta, J., & Vegelin, C. (2016). Sustainable development goals and inclusive development. *International Environmental Agreements: Politics, Law and Economics*, 16(3), 433–448. doi: 10.1007/s10784-016-9323–z.

Hooi, K. K., Hassan, F., & Mat, M. C. (2012). An exploratory study of readiness and development of green university framework in Malaysia. *Procedia-Social and Behavioral Sciences*, 50, 525–536.

Horacek, T. M., White, A. A., Byrd-Bredbenner, C., Reznar, M. M., Olfert, M. D., Morrell, J. S., … Thompson-Snyder, C. A. (2014). PACES: A physical activity campus environmental supports audit on university campuses. *American Journal of Health Promotion*. doi: 10.4278/ajhp.121212-QUAN–604.

IMT. (2014). Mobilidade em cidades médias. Retrieved from www.imt-ip.pt/sites/IMTT/Portugues/Observatorio/Relatorios/MobilidadeCidadesMedias/Documents/IMT_Mobilidade_em_Cidades_Medias_vrevista_atualizada.pdf.

INE. (2017). IMob – Inquérito à Mobilidade nas Áreas Metropolitanas do Porto e de Lisboa.

Lavetti, E., & McComb, S. (2014). Examining bicycle safety on a college campus: Observations and rationale for unsafe cycling. In *Proceedings of the Human Factors and Ergonomics Society*. doi: 10.1177/1541931214581286.

Lipschutz, R. D., De Wit, D., & Lehmann, M. (2017). Sustainable cities, sustainable universities: Re-engineering the campus of today for the world of tomorrow. In W. Leal Filho, et al. (eds.), *Handbook of Theory and Practice of Sustainable Development in Higher Education*, World Sustainability Series. New York: Springer. doi: 10.1007/978-3-319-47889-0_1.

McClintock, H. (ed.), (1992). *The Bicycle and City Traffic, Principles and Practice*. London: Belhaven Press.

Miller, J., & Handy, S. (2012). Factors that influence university employees to commute by bicycle. *Transportation Research Record: Journal of the Transportation Research Board*, 2314, 112–119. doi: 10.3141/2314-15.

Ministério do Ambiente. (2018). Portugal Ciclável 2030.

Ministério das Obras Públicas. (1965). *Estatística do tráfego nas estradas nacionais.* Lisboa: Junta Autonoma das Estradas.

Neun, M. (2016). Cycling and the UN Sustainable Development Goals – Contributions based on the Active Mobility Agenda. *Scientists for Cycling Colloquium*, Aveiro, November 2016.

Neun, M., & Haubold, H. (2016). *The EU Cycling Economy - Arguments for an Integrated EU Cycling Policy.* Brussels: European Cyclists' Federation.

Oldenziel, R., Emanuel, M., de la Bruhèze, A. A., & Veraart, F. (2016). *Cycling Cities: The European Experience.* Eindhoven: Foundation for the History of Technology. Retrieved from https://books.google.es/books?id=_tKiAQAACAAJ&dq=cycling+cities+the+european+experience&hl=pt-PT&sa=X&ved=0ahUKEwiOovTfsqLYAhXB1RQKHUy4AjAQ6AEIKDAA.

Oppenheimer, C. (2015). Eruption politics. *Nature Geoscience*, 8, 244.

Pedersen, C. S. (2018). The UN Sustainable Development Goals (SDGs) are a great gift to business! *Procedia CIRP*, 69(May), 21–24. doi: 10.1016/j.procir.2018.01.003.

Pereira, B. C. (2016). Bicycle mobility amongst women and children in the Lisbon - Cascais corridor. In University of Aveiro (ed.), *Scientists for Cycling (s4c) Colloquium 2016 in Europe "Cycling Delivers to the UN Global Goals".* Aveiro. Retrieved from www.slideshare.net/ClusterBicicleta/22-bernardo-campos-pereira.

Pitsiava-Latinopoulou, M., Basbas, S., & Gavanas, N. (2013). Implementation of alternative transport networks in university campuses. *International Journal of Sustainability in Higher Education*, 14(3), 310–323, Permanent link to this document: doi: 10.1108/IJSHE-12-2011-0084.

Pogge, T., & Sengupta, M. (2016). Assessing the sustainable development goals from a human rights perspective. *Journal of International and Comparative Social Policy*, 32(2), 83–97. doi: 10.1080/21699763.2016.1198268.

Prati, G. (2017). Gender equality and women's participation in transport cycling. *Journal of Transport Geography.* doi: 10.1016/J.JTRANGEO.2017.11.003.

Pucher, J., & Buehler, R. (2012). *City Cycling* (R. B. John Pucher, ed.). Cambridge, MA: The MIT Press. doi: 10.1080/01441647.2013.782592.

Ribeiro, J. M. P., Casagrande, J. L., Sehnem, S., Berchin, I. I., da Silva, C. G., da Silveira, A. C. M., Zimmer, G. A. A., Faraco, R. Á., & de Andrade Guerra, J. B. S. O. (2017). Promotion of sustainable development at universities: The adoption of green campus strategies at the University of Southern Santa Catarina, Brazil. In W. Leal Filho, et al. (eds.), *Handbook of Theory and Practice of Sustainable Development in Higher Education*, World Sustainability Series. New York: Springer. doi: 10.1007/978-3-319-47868-5_29.

Rodrigues, C. (2001). Universidades, Sistemas de Inovação e Coesão Regional, Universidade de Aveiro.

Rodrigues, P. (2011). Perspetivas positivas para a indústria nacional de bicicletas, Revista Portugal Global, dezembro, pp. 16–20, Lisboa.

Romero, M. de A., & Mülfarth, R. C. K. (2017). University of São Paulo: Sustainability masterplan for policies, plans, goals and actions. In W. Leal Filho, et al. (eds.), *Handbook of Theory and Practice of Sustainable Development in Higher Education*, World Sustainability Series. New York: Springer. doi: 10.1007/978-3-319-47877-7_34.

Starkey, P., & Hine, J. (2014). Poverty and sustainable transport: How transport affects poor people with policy implications for poverty reduction. *A Literature Review.*

Stommel, H., & Stommel, E. (1983). *Volcano Weather: The Story of 1816, the Year without a Summer [Indonesia]*. Newport, Rhode Island: Seven Seas Press.

United Nations (2015). The millennium development goals report. www.un.org/development/desa/publications/mdg-report-2015.html.

United Nations (2016). Mobilizing sustainable transport for development - Analysis and policy recommendations from the United Nations Secretary-General's High-Level Advisory Group on Sustainable Transport. https://sustainabledevelopment.un.org/index.php?page=view&type=400&nr=2375&menu=1515

United Nations Region Information Center (UNRIC) (2016). 17 Objetivos para Transformar o Nosso Mundo. Guia sobre Desenvolvimento Sustentável, 1–38.

von Braun, J., Hornidge, A.-K., & Borgemeister, C. (2014). Sustainable Development Goals need priorities and a stronger science base. *ZEF Annual Report 2013/2014* - University of Bonn.

Wang, X., Khattak, A., & Son, S. (2012). What can be learned from analyzing university student travel demand? *Transportation Research Record: Journal of the Transportation Research Board*. doi: 10.3141/2322–14.

chapter twelve

The implementation of the Sustainable Development Goals at the national level

The case of the SDG Alliance Portugal

S. Leal
Polytechnic Institute of Santarém

U. M. Azeiteiro
University of Aveiro

F. Seabra
ISCAL Lisbon Accounting and Business School, Lisbon Polytechnic Institute & IJP Portucalense Institute for Legal Research

Contents

12.1 Introduction

The 2030 Agenda for Sustainable Development set 17 Sustainable Development Goals (SDG) that should be nationally implemented. This implementation is based on the effective translation of sustainable development policies into concrete actions at the national level (and regional frameworks) (United Nations 2015). Portugal's multilevel and multidimensional sustainability challenges (forestry, clean and efficient energy, and climate action, among others) impose a new agenda and require the redefinition of strategies (and missions) that implicate the local, regional and national engagement in meeting the SDGs, namely by promoting local and regional intervention (always with high societal relevance and sustainability purposes) to cope with the targets established in 'Transforming our world: the 2030 Agenda for Sustainable Development' (United Nations 2015).

In Portugal, the United Nations Global Compact Network Portugal (UN-GCNP), with the support of the United Nations Global Compact (UNGP), found the SDG Alliance Portugal (http://globalcompact.pt/alianca-ods), a national multi-stakeholder platform, whose mission is to raise awareness, inform, implement, monitor, and assess the contribution of the business sector to SDG at the national level. It is as important to define goals and targets, as it is to create performance indicators to help in controlling and evaluating the SDGs (United Nations 2015).

The research question that supports this chapter is: "How is the SDG Alliance Portugal engaging and increasing the contributions of the business sector and other civil society partners to turn into practice the SDGs actively?" Following this, the aims of the study are: (1) to describe the objectives, strategy, and practical initiatives of the SDG Alliance Portugal to contribute to the 2030 Agenda for Sustainable Development, (2) to describe the collaborative network and multi-stakeholder approach implemented to achieve the SDGs and targets. This research question and these goals will be addressed following the method of a single-case study (Yin 2003).

12.2 2030 Agenda for Sustainable Development

12.2.1 Sustainable Development Goals

The SDGs are a universal set of goals (17), targets (169), and indicators that were adopted on September 25th, 2015, to "end poverty in all its forms" by 2030 "and balance the three dimensions of sustainable development: the economic, social, and environmental" (United Nations 2015, p. 1). The SDGs follow and expand the eight Millennium Development Goals (MDGs), which were launched in 2001, considered targets for developing countries to achieve, with financial support from wealthy states, and expired in 2015 (Sachs 2012). The SDGs aim to be universal – that is, applicable to all countries and not only to developing countries – and to serve as guideposts for a global transition to the sustainable development world.

The "Sustainable Development Goals Report" of 2017 (United Nations 2017a) showed that all countries in the world have an enormous journey to accomplish what was promised at 2030 Agenda for Sustainable Development: "no one will be left behind" (United Nations 2015, p. 1).

Moreover, the last United Nations' report asserts that "while people overall are living better lives than they were a decade ago, progress to ensure that no one is left behind has not been rapid enough to meet the targets of the 2030 Agenda. Indeed, the rate of global progress is not keeping pace with the ambitions of the Agenda, necessitating immediate and accelerated action by countries and stakeholders at all levels." (United Nations 2018, p. 4). That is, the progress observed in the last years is insufficient to meet the Agenda's goals and targets by 2030 (United Nations 2018).

As highlighted by the United Nations SDGs Reports, although some indicators show a positive evolution, most of them confirm how urgent it is to introduce changes in our world (United Nations 2017a, 2018). In 2013, 11% of the persons in the world were still living in extreme poverty; in 2016, about 815 million people were undernourished; in 2017, about 51 million children under age 5 suffered from wasting (low weight for height); in 2016, about 5,6 million children under age 5 died worldwide (most from preventable causes); and in 2016, the participation rate in early childhood and primary education was only 70%. Fifty-eight percent of children and adolescents worldwide are not achieving minimum proficiency in reading and mathematics (estimated in 2017). Between 2000 and 2016, women spent roughly three times as many hours in unpaid domestic and care work as men. One in five ever-partnered women and girls were subjected to physical and/or sexual violence by an intimate partner. In 2015, more than 2 billion people were affected by water stress. In 2015, six in ten people lacked access to safely managed sanitation facilities. In 2016, 41% of the world's population were still cooking with polluting fuel and stove combinations; 89% of the countries with data available, the hourly wages of men are, on average, higher than those of women, with a median pay gap of 12.5%. In 2016, 4.2 million people died from ambient air pollution. Damage to housing due to natural disasters showed a statistically significant rise between 1990 and 2013. The per capita "material footprint" of developing countries grew from 5 Mt in 2000 to 9 Mt in 2017. The year 2017 was one of the three warmest on record and was 1.1°C above the pre-industrial period. In 2013, 31% of marine fish stocks were overfished. The Red List Index shows an alarming trend in biodiversity decline for mammals, birds, amphibians, corals, and cycads; and in 2014, 71% of victims of trafficking were women and girls.

The "SDGs offer a unique opportunity to reinvigorate the international research agenda" (Leal Filho et al. 2018, p. 136). Implementation implies a precise definition of research needs, establishing priorities and collaboration, promoting innovation and coordination at various levels, and integrating existing institutional structures, namely synergies between academia and society in the promotion of a sustainability research culture (Suni et al. 2016).

The goals and targets aim to stimulate proactive actions to be put into practice until 2030, at several levels (global, regional, national, and local), embracing Governments, the private sector, civil society, United Nations, and all stakeholders (United Nations 2015). The goals cover five areas (the so-called 5 P's) of critical importance for humanity and the planet (United Nations 2015): people, planet, prosperity, peace, and partnership. The complete list of the 17 goals is available in Table 12.1. To help to put into practice, each goal was set a list of targets, 169 in total.

Table 12.1 List of the SDGs

Goal number	Goal short designation	Goal meaning	Number of targets	Number of SDG indicators
SDG 1	No poverty	End poverty in all its forms everywhere	7	14
SDG 2	Zero hunger	End hunger, achieve food security and improved nutrition and promote sustainable agriculture	8	13
SDG 3	Good health and well-being	Ensure healthy lives and promote well-being for all at all ages	13	27
SDG 4	Quality education	Ensure inclusive and equitable quality education and promote lifelong learning opportunities for all	10	11
SDG 5	Gender equality	Achieve gender equality and empower all women and girls	9	14
SDG 6	Clean water and sanitation	Ensure availability and sustainable management of water and sanitation for all	8	11
SDG 7	Affordable and clean energy	Ensure access to affordable, reliable, sustainable and modern energy for all	5	6
SDG 8	Decent work and economic growth	Promote sustained, inclusive and sustainable economic growth, full and productive employment and decent work for all	12	17
SDG 9	Industry, innovation, and infrastructure	Build resilient infrastructure, promote inclusive and sustainable industrialization and foster innovation	8	12

(*Continued*)

Table 12.1 (Continued) List of the SDGs

Goal number	Goal short designation	Goal meaning	Number of targets	Number of SDG indicators
SDG 10	Reduced inequalities	Reduce inequality within and among countries	10	11
SDG 11	Sustainable cities and communities	Make cities and human settlements inclusive, safe, resilient and sustainable	10	15
SDG 12	Responsible consumption and production	Ensure sustainable consumption and production patterns	11	13
SDG 13	Climate action	Take urgent action to combat climate change and its impacts	5	8
SDG 14	Life below water	Conserve and sustainably use the oceans, seas and marine resources for sustainable development	10	10
SDG 15	Life on land	Protect, restore and promote sustainable use of terrestrial ecosystems, sustainably manage forests, combat desertification, and halt and reverse land degradation and halt biodiversity loss	12	14
SDG 16	Peace, justice, and strong institutions	Promote peaceful and inclusive societies or sustainable development, provide access to justice for all and build effective, accountable and inclusive institutions at all levels	12	23
SDG 17	Partnerships for the goals	Strengthen the means of implementation and revitalize the global partnership for sustainable development	19	25
Total			*169*	*244*[a]

Source: United Nations (2015, 2017b).

[a] Although the total number of indicators listed in the global indicator framework of SDG indicators is 244, the total number of the indicators is 232 (nine indicators repeat under two or three different targets).

12.2.2 *Stakeholders engagement as a critical step and the implementation of the SDGs*

Since 1992, when the first United Nations Conference on Environment and Development occurred, "it was recognized that achieving sustainable development would require the active participation of all sectors of society and all types of people" (United Nations - Department of Economic and Social Affairs 2018, np). No government or country could introduce massive changes without the commitment of all stakeholders (e.g., NGOs, local authorities, workers and trade unions, business and industry, local communities, volunteer groups, and foundations). The involvement of stakeholders is even more relevant if we recall that the objectives of the SDGs are multidimensional. For instance, "improved access to safe drinking water (Goal 6) can promote health (Goal 3) and food security (Goal 2), while increasing the use of land for agriculture to help end hunger can actually undermine efforts to curb loss of biodiversity (Goal 15)" (OECD 2015, p. 4).

Engagement of stakeholders is at the same time a challenge and a condition for success (Jäger 2009, Leal Filho et al. 2018), and implementing sustainability at the national level is a fundamental step (e.g., Sardain et al. 2016) with high impact for governance (Husted and Sousa-Filho 2017, Patterson et al. 2017). The sustainability problems are complex, dynamic, and complicated to solve. They are best approached through collaborative and network practices (Husted and Sousa-Filho 2017). Nevertheless, to achieve a solution is a challenging exercise due to the interdependencies, uncertainties, circularities, and, most of the time, conflicting stakeholders implicated (Lazarus 2009). Countries with a more stakeholder-oriented culture (Dhaliwal et al. 2012) are more prepared to embrace all stakeholders, from business to civil society, in the developing of change strategies to promote the SDGs.

Implementation is a critical step in ensuring success, and the debate on how to implement 'the SDGs' is fundamental (Spangenberg 2017). An indicator framework and associated monitoring systems is a (challenging) key prerequisite to achieve the goals. To monitor progress towards the sustainable development it is necessary to have the suitable indicators (Sustainable Development Solutions Network 2015, p. 2). The 2017 and 2018 United Nations' reports show "that the rate of progress in many areas is far slower than needed to meet the targets by 2030" (United Nations 2017a, p. 4), demonstrating how vital it is to have indicators to access where we are and where we want to go. The 2030 Agenda requires the mobilization of all means of implementation and a strong follow-up and review mechanism to ensure progress and accountability (European Commission 2016). What is not measured, can unlikely be managed and assessed. In the Resolution 71/313 adopted by the General Assembly on 6 July 2017, the United Nations (2017b) presented a global indicator framework for the

SDGs and targets of the 2030 Agenda for SD, embracing 232 indicators (see in Table 12.1 the number of indicators per goal).

However, at the local and national levels, most of the indicators do not have available collected data, and the local and regional statistical institutes are struggling to get data to produce such indicators. Operationalization, implementation strategies, resources, monitoring, and evaluation of progress are critical steps, together with funding for implementing the SDGs (Lebada 2016) (see also The United Nations Conference on Trade and Development's Investment Policy Framework for Sustainable Development).

12.2.3 *Portugal and the 2030 Agenda for Sustainable Development*

In Portugal, the general coordination of the 2030 Agenda for Sustainable Development is done by the Ministry of Foreign Affairs, together with the Ministry of Planning and Infrastructures. This work is done with the collaboration of the remaining ministries according to their attributions relatively to the SDGs. For each SDG, a minister was allocated and he or she is responsible for its implementation, monitoring, and review (Ribeiro 2017b). For operational questions, a network of focal points from different government departments has been established (Ministry of Foreign Affairs 2017).

The Portuguese implementation of the 2030 Agenda started with a country's baseline analysis "with the collection of data and information in relation to all 17 SDGs and, as a result, leading to a mapping of national policies contributing to the implementation of the 2030 Agenda" (Ministry of Foreign Affairs 2017, p. 9). Additionally, there were several initiatives, workshops, seminars, and online inquiries with the participation of local authorities, non-governmental organizations of the Portuguese civil society and citizens that aim to collect information, to share visions and involve all stakeholders for the 2030 Agenda implementation, at the national level.

At the Portugal Statistics (https://www.ine.pt/), the institution for production and diffusion of official statistics in Portugal, a thematic file was created dedicated to the SDGs where data for 75 indicators is already available (Portugal Statistics 2017).

Furthermore, a multi-stakeholder platform has been created, coordinated by the UN-GCNP: the SDG Alliance Portugal. This multi-stakeholder platform will develop partnerships, projects, programmers, and actions within the framework of the 2030 Agenda.

In the 5th High-Level Political Forum of the United Nations, held in New York, on 18 July 2017, the Portuguese Government was one of the 43 countries that presented the Voluntary National Review in 2017 (Ministry of Foreign Affairs 2017, United Nations - Department of Economic and Social Affairs 2017). This document is an implementation support tool, and it gives baseline information on the main national and regional policies, plans, and strategies that contribute to the achievement of the 17 SDGs (Ribeiro 2017a). Although all SDGs are relevant, the Portuguese Government defined as being strategic to the implementation of the: SDG 4 (quality education), SDG 5 (gender equality), SDG 9 (industry, innovation, and infrastructure), SDG 10 (reduced inequalities), SDG 13 (climate action), and SDG 14 (life below water) (Ministry of Foreign Affairs 2017). The Portuguese National Reform Program embraces several medium-term strategies related to it.

12.3 Method

A descriptive single-case study method (Yin 2003) was selected as a research strategy, which aims "to describe an intervention and the real-life context in which it occurred" (Yin 2003, p. 15). In this case, the approach of SDG Alliance Portugal will be described for the implementation of the SDGs, at the national level. The SGD Alliance Portugal is a distinctive and representative case (Yin 2003), in Portugal, of a multi-stakeholder platform set up to embrace different stakeholders to put into practice the SDG goals and targets.

A case study means that multiple sources of information are used for data collection (Yin, 2003). For data collection, an in-depth interview was carried out. The interview was made to the Executive President of the Alliance, a key informant in this study. The questions were designed as open-ended to provide flexibility in the discussion (Table 12.2), and the interview was tape-recorded. The interview was made in Lisbon, in February of 2018, and it lasted 1 h. First, the interview was transcribed by

Table 12.2 List of the open-ended questions applied at the interview

Q1: What are the main characteristics of the SDG Alliance Portugal?
Q2: Is the organizational model of the SDG Alliance Portugal inspired by identical models or is it something innovative?
Q3: Are there in the Alliance entities with diverse juridical nature and organizational characteristics? Is it challenging to match such different visions and expectations?
Q4: The Alliance is rooted in collaborative work. Will the Portuguese organizations have enough experience to comply with the challenge addressed by the Alliance?
Q5: The Alliance has a network of ambassadors to pursuit of their goals. What was the criteria followed to select the current ambassadors?
Q6: Concerning the pursuit of the objectives for sustainability, it is expectable that there are "different speeds" and a different pace of work between entities like private companies, public administration, NGOs, and universities. Are these "different speeds" a further obstacle for the Alliance?
Q7: At this stage, is it already possible to point out some relevant projects to the pursuit of the Alliance's objectives?

two of the authors; then, the transcriptions were compared, and a working document was sent to the interviewee for validation. The documents published on the website of the SDG Alliance Portugal (http://globalcompact.pt/alianca-ods) were used for triangulation of the interview data.

12.4 SDG Alliance Portugal

The SDG Alliance Portugal is the response of UN-GCNP to the 2030 Agenda for the SD (SDG Alliance Portugal 2018) in Portugal. The Alliance was created on 20 January 2016 as a first step to strengthen the means of implementation and revitalize the partnerships for sustainable development, as advocated by SDG 17.

The SDG Alliance Portugal is a multi-stakeholder platform that aims to "raise awareness, inform, implement, monitor, and evaluate the contribution of the private sector and other civil society partners to the SDGs at [the] national level" (Ministry of Foreign Affairs 2017, p. 10). The specific goals of the Alliance embrace, among others (SDG Alliance Portugal 2018):

a. To provide instruments developed by the UNGP and other UN Agencies;
b. To create and maintain channels for the exchange of information;
c. To support the involvement of stakeholders;
d. To promote and facilitate bilateral contacts according to specific interests;
e. To create and maintain specialized commissions by SDGs;
f. To promote the cooperation with other entities involved in SGDs;
g. To organize events;
h. To support academic, business, public administration, and civil society initiatives;
i. To recognize entities for their contribution to SDGs;
j. To cooperate with similar entities in other countries;
k. To promote and encourage entities to participate in UNGC and UNGC-Network Portugal; and
l. To promote the objectives of the UN, its programs, initiatives, and agencies along with the business sector.

As the President of the Alliance affirmed, at the 5th High-Level Political Forum of the United Nations, the Alliance "organizes and manages a 'partnership of partnerships' where employers confederations, workers unions, consumer associations, chambers of commerce, academia, local authorities, NGOs, and Social Economy, join efforts for specific partnerships to put in practice the goals and targets of the SDGs. The governance of such a large partnership was entrusted to a Stakeholder Representative Group" (Silva 2017, p. 1).

In the first 3 years, 57 organizations joined the Alliance, namely associations, city councils, state entities, NGOs, foundations, employers' confederations, higher education institutions, and some other entities (e.g., commissions, cooperatives, and labour unions). As asserted by the interviewee (first question), the Alliance is a space "almost without limits for any entity of Portuguese society", that integrates associations and entities of identical nature. As defended by their president, the Alliance "is an entity of associations and not one entity of entities" (Silva 2018). The rule of the Alliance is to accept associations, NGOs, confederations of associations, or large entities with the representation of a high number of people or organizations (e.g., labour unions and employers' confederations), as well as companies with a large impact on the economy or large non-governmental organizations.

Relatively to the working method (second question), the Alliance adopted the method of the mirror committees also used and tested in the workgroup that developed the International Standard ISO 26000 on social responsibility (International Organization for Standardization 2006), where Portuguese associations were represented (namely the Portuguese Association of the Business Ethics). The mirror committees are committees established for sharing information and exchanging views in a particular area, which "mirror" the structure of the international working groups, regarding stakeholder representation (International Organization for Standardization 2006). As the interviewee recalls, these mirror committees allow entities, that in public typically have opposite opinions, to converge and work together for the same goals (Silva 2018). The Alliance wants to "make people understand that together they do more than they do alone" and motivate people to "converge on what we have to give, instead of diverging in what we have to receive" (second question; Silva 2018). Although the entities integrated into the Alliance have different missions, organizational characteristics, and juridical natures, "all of them understand that is the time to cooperate" (third question; Silva 2018).

In the business sector, Portugal "does not have a collaborative culture" (fourth question; Silva 2018). In general, the Portuguese have the idea that "the secret is the soul of business" and, in general, the "vocation to collaborate is low. One of the key messages that the Alliance is keen on promoting is the need to collaborate, to join efforts, and this first step is progressing well" (Silva 2018).

To put into practice the 2030 Agenda for the SD in Portugal, the Alliance invited several individuals to be ambassadors. Three years after the beginning of the Alliance, 75 ambassadors had already expressed their commitment to the 2030 Agenda for SD (SDG Alliance Portugal 2018) (this is an open process, which means that the number of ambassadors is dynamic with a high probability of increasing in the future). However, the ambassadors are not representing their organizations but acting as individuals (fifth question). At the end of 2018, 20% of the ambassadors acted on behalf of the SDG 4 (quality education); 10% on behalf of SDG 5 (gender equality); 10% on behalf of SDG 9 (industry, innovation, and infrastructure); 8% on behalf of SDG 11 (sustainable cities and communities); 7% on behalf of SDG 16 (peace, justice, and strong institutions); and 10% embraced more than one SDG. On the other hand, eight SDGs had only one ambassador representing it (SDG Alliance Portugal 2018).

To opt for the figure of ambassadors seemed interesting because not only is it a frequent figure in the United Nations but individuals also feel free to give their testimony, with high emotional commitment and engagement (Silva 2018). Ambassadors tend to have more credibility and authenticity when working as individuals than as organizational representatives because when they act in the context of their organizational role, they are probably limited to actions that are aligned with the mission of their organizations (Silva 2018).

Even though the significant progress is achieved in the first years of activity, "the developing of the net of ambassadors is an ongoing progress. It is necessary to broaden the regional coverage of this network" (Silva 2018). The different entities could also have different approaches and speeds relative to the SDG (sixth question), but that is natural and expectable in the point of view of the interviewee. For instance, "the environmental associations would expect that changes in the care of environment be implemented faster and have greater scope while the employee entities would defend the viability of the labour market" (Silva 2018).

Relatively to the main projects of the Alliance (seventh question), it is first pointed out that the Alliance should, preferably, act in the domains that the entities are not able to work on. As the interviewee highlighted, "the biggest challenges for the Alliance are those in which entities cannot naturally produce a response. The Alliance must respect the principle of subsidiarity and only act where their members do not act spontaneously".

Nevertheless, there are goals where just one organization is not able to embrace all the targets. It is the case of SDG 16 (pace, justice, and strong institutions). In this context, one project is highlighted by the interviewee, called Guide of Good Governance to small and medium-sized enterprises, supported by the Portuguese Association for Business Ethics with the backing of the UN-GCNP (Associação Portuguesa de Ética Empresarial 2017). In the quality education domain (SDG 4), a second project is also highlighted, which aims to train teachers on SDGs. Additionally, the SDG Alliance Portugal maintains the efforts to develop the partnerships implied to SDG 17 (partnerships for the goals).

12.5 Conclusion

This chapter presented one innovative approach, which contributes to the implementation of the SDGs in Portugal: the SDG Alliance Portugal. The

Executive President of the SDG Alliance Portugal was interviewed as a key informant.

The above-referenced Alliance is a multi-stakeholder platform, which aims to give a practical contribution to the 2030 Agenda, namely, to support the involvement of stakeholders and to promote the cooperation between entities involved in SDGs.

With activity of only 3 years, 57 organizations and 75 ambassadors expressed their formal support to the Alliance. The Alliance was able to congregate various efforts and motivations, from civil society to large entities and also contributes to stimulate the associative movement. Relatively to the working method, the Alliance organized itself with mirror committees. The mirror committees allow entities to work together for the same goals, even when they have divergent business approaches.

Portugal is only slightly oriented to stakeholders (Dhaliwal et al. 2012) which can justify the opinion of the President of the Alliance that the Portuguese have a low collaborative culture. The Alliance also helps to change this mindset and to introduce more collaboration among stakeholders. The mindset change of entities and their members is necessary to embrace the complex and dynamic problems of sustainability. Synergetic approaches are needed to develop innovative solutions to it.

Besides the stakeholders' and ambassadors' network and respective projects, the Alliance has two other projects, one related to the quality education domain intended to train teachers on SDGs and the other related with strong institutions, which aim to promote the good governance in small and medium-sized enterprises. There are also a different set of other projects going on but led by the ambassadors or the entities associated with the Alliance.

This study, as all studies, has some limitations. Firstly, the results are specific to the reality studied and could not be generalized to the population. Secondly, only a key informant was used to collect the information, and only one interview was performed. Lastly, the reality described is still new (only 3 years), and limited examples are available to describe it. Notwithstanding the difficulties, an example of a Portuguese multi-stakeholder platform was presented to address the SDG, which can be seen as an excellent example to replicate in other regions or countries.

References

Associação Portuguesa de Ética Empresarial. 2017. Guia de Boa Governação Para PME: Orientações e requisitos. In Lisboa: Associação Portuguesa de Ética Empresarial. http://cite.gov.pt/pt/destaques/complementosDestqs2/APEE_guia_boa_governacao_PME.pdf.

Dhaliwal, Dan S., Suresh Radhakrishnan, Albert Tsang, and Yong George Yang. 2012. Nonfinancial disclosure and analyst forecast accuracy: International evidence on corporate social responsibility disclosure. *The Accounting Review* 87(3): 723–759. doi: 10.2308/accr-10218.

European Commission. 2016. Next steps for a sustainable European future: European action for sustainability. In *Communication from the Commission to the European Parliament, the Council, the European Economic and Social Committee and the Committee of the Regions (COM(2016) 739 final)*. Strasbourg: European Commission. https://ec.europa.eu/europeaid/sites/devco/files/communication-next-steps-sustainable-europe-20161122_en.pdf.

Husted, Bryan W., and José Milton de Sousa-Filho. 2017. The impact of sustainability governance, country stakeholder orientation, and country risk on environmental, social, and governance performance. *Journal of Cleaner Production* 155: 93–102. doi: 10.1016/j.jclepro.2016.10.025.

International Organization for Standardization. 2006. Participating in the future International Standard ISO 26000 on social responsibility. www.iso.org/files/live/sites/isoorg/files/archive/pdf/en/iso26000_2006-en.pdf.

Jäger, J. 2009. Sustainability science in Europe, background paper prepared for DG research. http://seri.at/wp-content/uploads/2009/11/Sustainability-Science-in-Europe.pdf

Lazarus, Richard J. 2009. Super wicked problems and climate change: Restraining the present to liberate the future. *Cornell Law Review* 94: 1153–1234.

Leal Filho, Walter, Ulisses Azeiteiro, Fátima Alves, Paul Pace, Mark Mifsud, Luciana Brandli, Sandra S. Caeiro, and Antje Disterheft. 2018. Reinvigorating the sustainable development research agenda: The role of the sustainable development goals (SDG). *International Journal of Sustainable Development & World Ecology* 25(2): 131–142. doi: 10.1080/13504509.2017.1342103.

Lebada, Ana Maria. 2016. UN launches SDGs financing platform. International Institute for Sustainable Development. http://sdg.iisd.org/news/un-launches-sdgs-financing-platform/.

Ministry of Foreign Affairs. 2017. National report on the implementation of the 2030 Agenda for Sustainable Development, on the occasion of the voluntary national review at the United Nations high-level political forum on sustainable development. Portugal. https://sustainabledevelopment.un.org/content/documents/15766Portugal2017_EN_REV_FINAL_29_06_2017.pdf.

OECD. 2015. The Sustainable Development Goals: An overview of relevant OECD analysis, tools and approaches. www.oecd.org/dac/The%20Sustainable%20Development%20Goals%20An%20overview%20of%20relevant%20OECD%20analysis.pdf.

Patterson, James, Karsten Schulz, Joost Vervoort, Sandra van der Hel, Oscar Widerberg, Carolina Adler, Margot Hurlbert, Karen Anderton, Mahendra Sethi, and Aliyu Barau. 2017. Exploring the governance and politics of transformations towards sustainability. *Environmental Innovation and Societal Transitions* 24: 1–16. doi: 10.1016/j.eist.2016.09.001.

Portugal Statistics. 2017. Sustainable Development Goals Indicators -2030 Agenda -2017. www.ine.pt/ngt_server/attachfileu.jsp?look_parentBoui=292257844&att_display=n&att_download=y.

Ribeiro, Teresa. 2017a. Presentation of the national voluntary review to the UN high level political forum on sustainable development. *5th High-Level Political Forum*, New York, July 18th.

Ribeiro, Teresa. 2017b. Speech of the state secretary for foreign affairs and cooperation. *5th High-level Political Forum for Sustainable Development*, New York, July 18th, 2017.

Sachs, Jeffrey D. 2012. From millennium development goals to sustainable development goals. *Lancet* 379(9832): 2206–2211.

Sardain, Anthony, Cécile Tang, and Catherine Potvin. 2016. Towards a dashboard of sustainability indicators for Panama: A participatory approach. *Ecological Indicators* 70: 545–556 doi: 10.1016/j.ecolind.2016.06.038.

SDG Alliance Portugal. 2018. SDG Alliance Portugal website. http://globalcompact.pt/alianca-ods.

Silva, Mário Parra. 2017. *Presentation at 5th High-Level Political Forum of the United Nations*. Last Modified July 18th. http://globalcompact.pt/images/pdf/UNDISCURSO.pdf.

Silva, Mário Parra. 2018. Interview to the Executive President of the SDG Alliance Portugal.

Spangenberg, Joachim H. 2017. Hot air or comprehensive progress? A critical assessment of the SDGs. *Sustainable Development* 25(4): 311–321. doi: 10.1002/sd.1657.

Suni, Tanja, Sirkku Juhola, Kaisa Korhonen-Kurki, Jukka Käyhkö, Katriina Soini, and Markku Kulmala. 2016. National Future Earth platforms as boundary organizations contributing to solutions-oriented global change research. *Current Opinion in Environmental Sustainability* 23: 63–68. doi: 10.1016/j.cosust.2016.11.011.

Sustainable Development Solutions Network. 2015. Indicators and a monitoring framework for the Sustainable Development Goals: Launching a data revolution. http://unsdsn.org/wp-content/uploads/2015/05/150612-FINAL-SDSN-Indicator-Report1.pdf.

United Nations - Department of Economic and Social Affairs. 2017. 2017 voluntary national reviews: Synthesis report. https://sustainabledevelopment.un.org/content/documents/17109Synthesis_Report_VNRs_2017.pdf.

United Nations - Department of Economic and Social Affairs. 2018. Major groups and other stakeholders. https://sustainabledevelopment.un.org/mgos.

United Nations. 2015. Transforming our world: the 2030 Agenda for Sustainable Development. Resolution adopted by the General Assembly on 25 September 2015, A/RES/70/1. http://www.un.org/ga/search/view_doc.asp?symbol=A/RES/70/1&Lang=E.

United Nations. 2017a. The Sustainable Development Goals Report 2017. New York. https://unstats.un.org/sdgs/files/report/2017/TheSustainableDevelopmentGoalsReport2017.pdf.

United Nations. 2017b. Work of the Statistical Commission pertaining to the 2030 Agenda for Sustainable Development. Resolution adopted by the General Assembly on 6 July 2017, A/RES/71/313. https://undocs.org/A/RES/71/313.

United Nations. 2018. The Sustainable Development Goals Report 2018. New York. https://unstats.un.org/sdgs/files/report/2018/TheSustainableDevelopmentGoalsReport2018-EN.pdf.

Yin, Robert K. 2003. *Case Study Research: Design and Methods*. 3rd ed. Thousand Oaks, CA: Sage Publications.

chapter thirteen

Contribution of advanced training for real problem solutions within Sustainable Development Goals

The case of an e-learning PhD

Carla Padrel de Oliveira
Universidade Aberta
CQE/IST/Universidade de Lisboa

Sandra Caeiro
Universidade Aberta
CENSE/Universidade NOVA de Lisboa

Jorge Trindade
Universidade Aberta
IGOT/Universidade de Lisboa

Contents

13.1 Introduction

Education for sustainable development (SD) is a long-term process, and as Lozano and Huisingh (2011) highlighted, the time dimension plays a key role in real, long-term changes towards sustainability.

Higher Education Institutions (HEIs) keep progressing towards the integration of sustainability, having as main drivers' pressure of internal and external stakeholders, global challenges and treaties, the credibility of institutions and the environmental crises that demand a greater commitment (Blanco-Portela et al., 2017, Dlouhá et al., 2018). An increasing number of HEIs have indeed been engaged in integrating SD in their practices, including, community outreach, operations, assessment and reporting, university collaboration, institutional framework, campus experiences and of course, education and research (Lozano et al., 2013).

E-learning in higher education can be of great relevance in effective lifelong learning towards education for SD in a population of students who are simultaneously full-time employees, allowing them to pursue their studies, in a flexible, collaborative and interactive way. So, it can contribute to and have a role in the transition to sustainable societal patterns (Bacelar-Nicolau et al., 2012, Amador et al., 2015, Azeiteiro et al., 2015).

Universidade Aberta (UAb), the Portuguese Distance Learning University established three decades ago, became a reference European institution in the area of innovative e-learning and online learning through the recognition of its Virtual Pedagogical Model (Pereira et al., 2008). UAb has developed programs and courses, both formal and non-formal, in environmental/sustainability sciences, which aims to actively promote education for SD. Within the formal context, UAb offers a bachelor's degree in Environmental Sciences; a master's degree in Participation and Environmental Citizenship and a PhD degree in Social Sustainability and Development.

The focus of the present work is the PhD program that aims to qualify professionals in advanced studies in SD, with emphasis on environmental and social aspects. The students enroled in this program are from different and diverse nationalities and backgrounds and engaged in a professional life, mainly with full-time jobs. Hence, their research interests and projects cover local, regional and national to worldwide sustainability research topics. As pointed out in the UNESCO publications (UNESCO, 2014, 2017), sustainability research topics need to be interdisciplinary, with critical and reflective thinking, collaborative but also need to give solutions to the real and actual society challenges, namely the ones defined by the 2030 Agenda. This is the approach that we have implemented in this PhD program, focusing on theoretical and fundamental issues but promoting research projects based on a more practical and applied problem-solution designs. It is our belief that the use of advanced and innovative e-learning associated with advanced training programs is the seamless combination namely for those that cannot attend face-to-face programs. Furthermore, this setting is also suitable for applied research through the development of networks of different researchers in different countries and areas of knowledge.

In this work, we intend to evaluate the contribution of this e-learning PhD program on Social Sustainability and Development for worldwide real problem solutions, in particular related with the achievement of the Sustainable Development Goals (SDG). The research was conducted for those research plans submitted between 2014 and 2016. A content analysis was applied to those students' research plans.

A brief characterization of the student's population is given, and following the content analysis performed, the results are presented and discussed based on the relations between the students' research plans, students' professional profile and the contribution for SDG.

13.2 Advanced training and research and Education for SDG

In the context of the Decade for Educations for Sustainable Development 2005–2014 (ESD) (UNESCO, 2005), issues related to sustainability started to be looked upon as priorities, particularly in terms of poverty reduction, equality, health promotion, environmental protection, human rights, peace, responsible production and consumption, among others. This comprehensive and inclusive approach to UNESCO sustainability topics facilitates the application of systemic and interdisciplinary methodologies, promotes an understanding of a complex reality and develops an innovative socio-educational background for responsible citizenship that develops in several axes (Imbernon, 2002): democratic citizenship (culture of peace and the search for peaceful solutions to conflicts); justice and social equity (effective equality of opportunities, opposing inequality between genders, races, religions, etc.); social citizenship (fighting poverty and social exclusion); and environmental citizenship (responsibility and respect for the environment).

After the ESD Decade and since the SDGs have been adopted, the focus of the academic debates, as well as policy and civil society, is on how to implement 'the SDGs' (Spangenberg, 2017). SDGs are transversal, interdisciplinary and at the same time vague (Spangenberg, 2017), therefore, difficult to integrate in teaching and other university activities. Theoretical discussions on education for SD and competence-based teaching are related with the international policy processes of the SDG implementation where 'the knowledge and skills needed to promote sustainable development' are expressed in particular in the SDG 4.7 (Dlouha and Pospísilov, 2018).

ESD is holistic and transformational that addresses learning content and outcomes, pedagogy and the learning environment. Thus, ESD should create interactive, learner-centred teaching and learning settings. What ESD requires is a shift from teaching to learning. It asks for an action-oriented, transformative pedagogy, which supports self-directed learning, participation and collaboration, problem orientation, interdisciplinarity and transdisciplinarity and the linking of formal and informal learning. Only such pedagogical approaches make possible the development of the key competencies needed for promoting SD (UNESCO, 2017). Indeed, competences are considered to be one of the cornerstones of systemic educational transformation (Dlouha and Pospísilov, 2018).

Competencies in sustainability are related to acquire knowledge, skills and attitudes that enable successful task performance and problem solving with respect to real-world sustainability problems, challenges and opportunities (Barth et al., 2007).

Higher education for SD aims at enabling people to not only acquire and generate knowledge but also to reflect on further effects and the complexity of behaviour and decisions in a future-oriented and global perspective of responsibility. Higher ESD must participate in the discussion about sustainable ways of living and working. University as a learning and life world where a culture of teaching should be superseded by a culture of learning that combines the learning processes in academic formal and informal settings and that includes competencies developed in extracurricular settings (Barth et al., 2007). The incorporation of SD into curricula requires systems thinking and interdisciplinary approaches and calls for pedagogical innovations that provide interactive, experiential, transformative and real-world learning (Lozano et al., 2017, Annan-Diaba and Molinari, 2017).

According to Lozano et al. (2017), case studies, project and/or problem-based learning, community service learning, interlinked teams and participatory action research are examples of pedagogic approaches with a good coverage with the sustainability competences.

Addressing sustainability problems from local to global scales and across geographical and cultural contexts has become an important task for sustainability research and education. This implies finding research and educational formats to study sustainability in different locations, making use of digital technologies for collaboration and providing appropriate institutional support for implementation (Caniglia et al., 2017). Sustainability research should involve interdisciplinary, multidisciplinary and transdisciplinary research (linked to areas such as people, planet, justice, prosperity and dignity) to jointly find solutions and design strategies that can contribute to creating good lives for the community today and in the future (Leal Filho et al., 2018). Sustainability research should then have a transformational and solution-oriented research agenda (Miller et al., 2014). By reinvigorating SD research, a considerable step forward may be taken to integrate it into existing political structures and thus help deliver the goals of the SDGs (Leal Filho et al., 2018).

The university, as well as a training space, is also a place for experimentation with new educational proposals and a platform for the dissemination of changes in perceptions, attitudes and behaviours towards a new and more SD. It is also assumed that the university must scaffold students' learning experiences by providing opportunities to work across disciplines and also through combined efforts to facilitate students' perspective shifts, their capacity for innovation and their ability to unify knowledge from many disciplines and sectors. PhD programs and specially those aimed at working/professional population can play an integrator role to establish an active dialogue with other forms of knowledge, bringing together professional communities, corporate perspectives and civil society with different disciplinary perspectives to solve the myriad of complex problems facing humanity (McGregor and Volckmann, 2013). Owing to its time and space flexibility, e-learning programs such as the one we are offering at UAb can reach a variety of professionals from different countries and cultures and can actively contribute for the accomplishment of SDGs.

The question is to evaluate how e-learning and advance training about SD is currently contributing to SDGs targets achievement.

13.3 Methodology

The sample for this study comprises 28 students of the PhD program on Social Sustainability and Development of UAb. These are the total number of research plans submitted in the academic years of 2014/2015 and 2015/2016 and that were then analysed. A characterization of the students of the PhD program will be given in the next topic.

Three coders using predefined categories conducted a content analysis independently. Following the issue of data making, as defined by Krippendorff (2013), an effort on unitizing, sampling, coding and reducing data was carried out. For this analysis, five categories were defined and discussed among coders according to the main subject topics present in literature (see Table 13.1) (Stake, 1995; Merriam, 2009; Yin, 2014; Brooks and King, 2017; Goodwin, 2017), including: (1) case study approach, (2) applied study approach, (3) students' professional activities, (4) sustainability approach present on the PhD study and (5) the study's SDG compatibility.

Each category was scaled from 0 to 2: 0 stands for not related to the appraised category and 2 stands for totally related within the appraised category. The procedure proposed by Hernández Sampieri et al. (2006) has been followed in order to determine the level of agreement between the coders as given by (eq. 1).

$$\text{CoderVal} = \frac{C_{1,2} + C_{1,3} + C_{2,3}}{3} \tag{13.1}$$

Data analysis was carried out based on mode analysis of the coder's answers. This means a clear choice for repetition on the understanding

Table 13.1 Systematization of the subject topics in the categories

Categories	Subject topics/examples	References
Case study approach	• Focus on what is studied (the case) rather than how it is studied (the method); • Study the uniqueness or complexity of a single case; • An in-depth description and analysis of a bounded system; • Focus on a particular issue; • The product of an investigation should be descriptive and heuristic in nature; • Study of individual units as persons or communities and their relation to the environment.	Stake (1995) Merriam (2009) Yin (2014)
Applied study approach	• Seeks to solve practical problems; • Addresses real life/world issues; • Often problems have unique characteristics; • Leads to an immediate and practical application of the findings.	Brooks and King (2017) Goodwin (2017)
Students professional activities	• Straightforward analysis of the students curriculum vitae	
Sustainability approach	• Triple bottom line approach • Environmental, social and economic pillars • SD as an objective • Sustainability as a process	Elkington (1994) Elkington (2004)
SDG compatibility	• Straight forward analysis of the United Nations SDGs defined after A/RES/70/1 resolution in 2015	UN (2016)

of the coders and not for the mean or median point of understanding that could correspond to an intermediate value with no correlation to out scale of classification and to a false precision.

Finally, a clustering approach was applied to find the more common keywords of each thesis plan, using content analysis as a basis for information acquisition. NVIVO software was used for this analysis.

13.4 The PhD program on Social Sustainability and Development

It is within the framework of multidimensional nature of the concepts of SD and sustainability that requires new study approaches that the PhD program on Social Sustainability and Development arises. In this

PhD program on Social Sustainability and Development, it is intended to (1) deepen the knowledge, through the use of a current and interdisciplinary theoretical body; (2) develop analytical skills, evaluation and critical reflection on concrete situations; (3) develop capacities and competences to carry out research autonomously in the area of studies and (4) develop skills and competences within the study area to contribute to the advancement of knowledge and to social and cultural progress.

The pedagogical offer of the UAb follows the principles of the Bologna reform, and the PhD programs are structured in two parts: a curricular segment, with the duration of two semesters, and research, conducive to the thesis, with the duration of four semesters, making a total of six semesters for the full-time student.

Throughout its 8 years of existence, this advanced studies program has had a demand of about 160 students, many of whom are part-time, given their nature as working students. Figure 13.1 shows the distribution per gender and per age. About two-thirds of the students are males and the majority – 67% – are between 30 and 49 years of age (Figure 13.1).

Figure 13.2 shows the distribution of students per country of residence, and becomes evident the importance of program being offered online. Even if the majority of students – 46% – live in Portugal (distributed throughout the territory of the continent and Madeira and Azores islands), 42% of the students are distributed between Mozambique and Brazil.

For the current work, we have considered those students that have presented and discussed their thesis project and working plan in the academic years of 2014/2015 and 2015/2016. In this case, we have 28 students, 75% male and geographically distributed as shown in Figure 13.3.

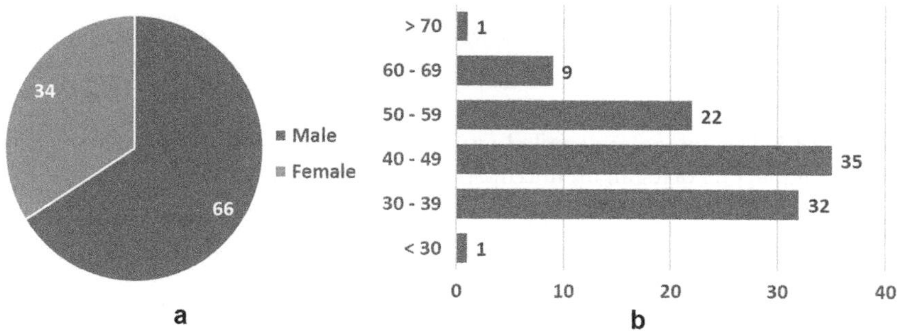

Figure 13.1 Percentage of students according to gender (a) and age (b).

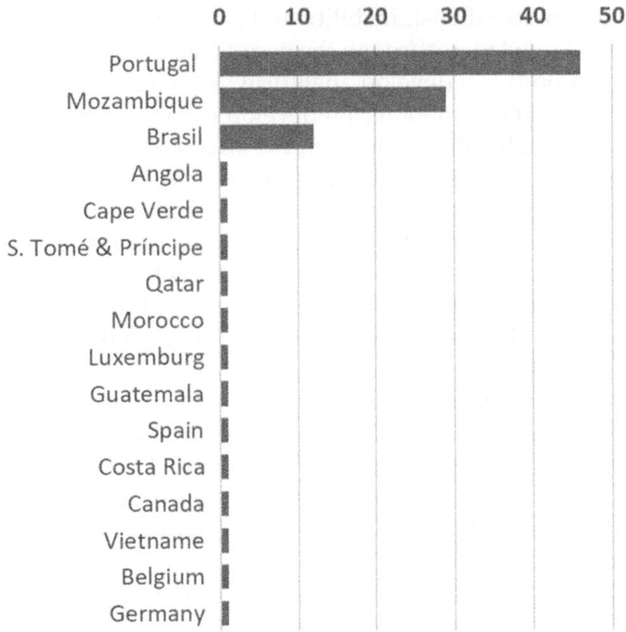

Figure 13.2 Percentage of students according to the country of residence.

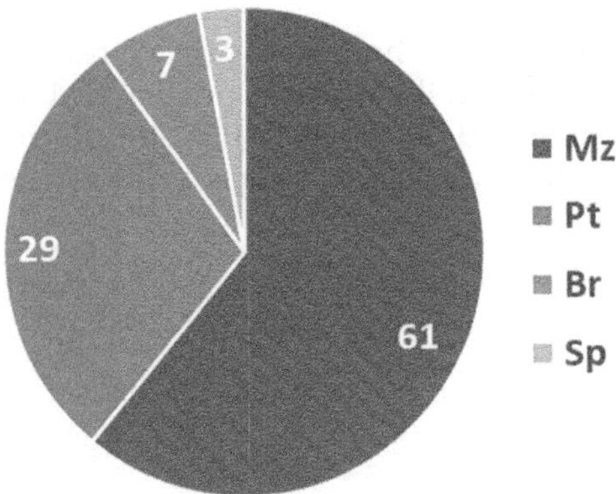

Figure 13.3 Percentage of sampled students according to their nationality. Mz, Mozambique; Pt, Portugal; Br, Brazil; Sp, Spain.

13.5 Results

The coder validation showed that only a reduced percentage was obtained for total disagreement between all the judges (a maximum of 17.9 in the SDG classification – see Figure 13.4).

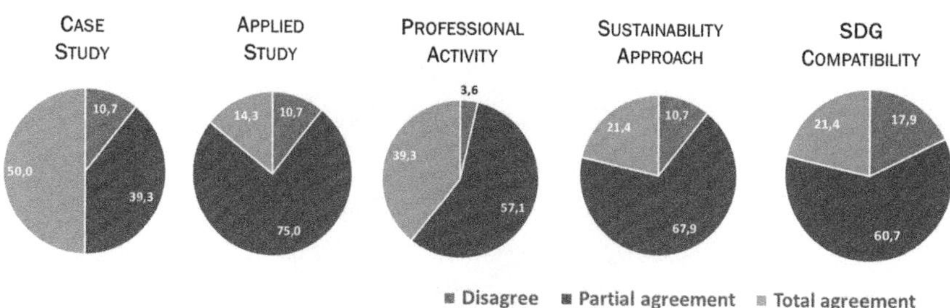

Figure 13.4 Content analysis coder validation.

Table 13.2 Category classification results

	Not related (%)	Partially related (%)	Totally related (%)
Case study	10.8	21.7	**67.5**
Applied study	14.5	**50.6**	34.9
Professional activity	31.3	**36.1**	32.5
Sustainability approach	28.6	**48.8**	22.6
SDG compatibility	8.3	**58.3**	33.3

Higher value of each category are in bold.

The majority of the students' research plans are based on case studies related with local problem solutions. A total agreement between the coders was obtained in 50% of the cases, and a total disagreement occurs in only 11%, representing 3 cases in 28 (Figure 13.4). This disagreement can be explained by student's misclassification of their studies as case study (Table 13.2). Nevertheless, case study research within ESD is useful to secure the capacity of the collaborative learning process to benefit both the actors involved in the process and those external parties who may be interested in gaining knowledge from it (Karatzoglou, 2013) and contribute to acquisition of competencies for sustainability (Lozano et al., 2017).

Also, almost all the students' research is related to applied studies to the country or region where the students live at the moment (most of the students come from Mozambique – see Figure 13.3). Sixty-nine percentage of the research plans are partially or totally related with their professional activity. Most common professional activity of the sample students is teaching, although also coming from private and public sector (Figure 13.5). UAb has an agreement with Catholic University of Mozambique for capacity building of their permanent staff, which justifies the high percentage of teachers among the students. Nevertheless, in this sample, there are students coming, for example, from Agriculture

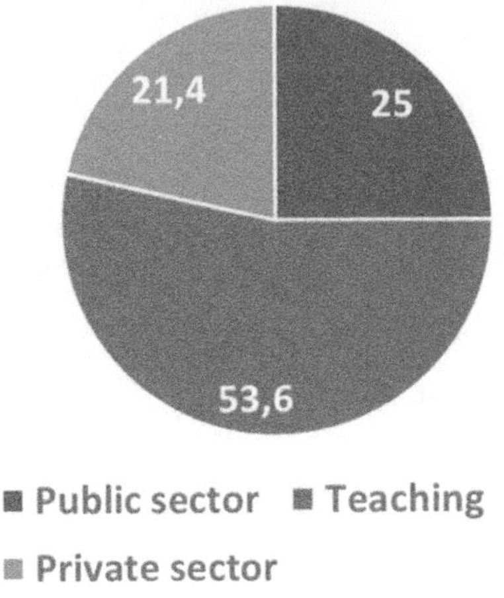

Figure 13.5 Students professional activity.

Ministry that are working on Sustainable Agriculture or working in Federal Policy investigating illegal forest practices.

Most of the students only focus their research on up to two dimensions of sustainability (being the social component always present), what can justify a 48.8% of partially related sustainability approach and 28.6% of not related (Table 13.2 and Figure 13.6). Although students are being taught during the first year of the PhD of the importance of a holistic and integrated approach of sustainability, some still do not follow it on their research. This integration in an interdisciplinary way and, in practical terms, is not always easy. Although the benefits of interdisciplinarity are known, and interdisciplinarity education has been challenging, in general, the inclusion of social, economic and environmental aspects depends on the students, sometimes prompted by their teachers (Annan-Diab and Molinari, 2017). In addition, transdisciplinarity – complementary to

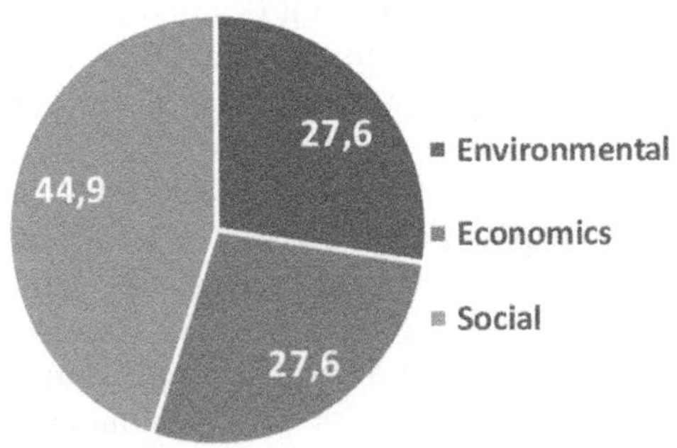

Figure 13.6 Sustainability approach present on the PhD study.

traditional disciplinary knowledge – is also considered a key competence for universities that intend to play a role in SD namely through introducing sustainability in curriculums. Transdisciplinary approaches can thus not only be associated with a type of reasoning that is more fluid and ad hoc than problem solving in most sciences but can also be used to help to overcome mono-disciplinarity (Tejedor et al., 2018).

Most of the PhD researches are related with the SDG (92% are totally or partially related – Table 13.2). Nevertheless, this category was the one where a higher percentage of disagreement between coders was obtained. The different concepts and diverse understanding due to the large application of the SDG and even vague context (as highlighted by Spangenberg, 2017) can explain different interpretations from the coders. The more targeted SDGs are related with economic growth and related social issues (SDG 8), sustainable cities (SDG 11) and terrestrial ecosystems (SDG 15) (see Figure 13.7). PhD thesis related with economic growth typically addresses issues about social responsibility in local companies or tourism development. Theses related with terrestrial ecosystems are those linked with agricultural problems, forest management or environmental impacts of mining. Floods and land-use management are the themes usually related with SDG 11. Those investigations are usually related to problems with impact in the origin countries. Brazil, Portugal and Mozambique are the most frequent countries where the focus on solving SDG issues needs to be addressed according to United Nations 2030 Agenda (UN, 2016).

Those more targeted SDGs are in line with the most common keywords used on student's research plans (see Figure 13.8). Clustering these most common terms, keywords like social, development and management are key concepts present in almost all SDGs. On a lower level, but with a very significant importance, are keywords whose concepts are clearly related to the sustainable use of ecosystems, human settlements and sustained economic growth as mentioned on SDGs 15, 11 and 8. In addition, keywords like citizenship, communities, public, policies and responsibilities are also very common on the student's research. As highlighted by Dlouha and

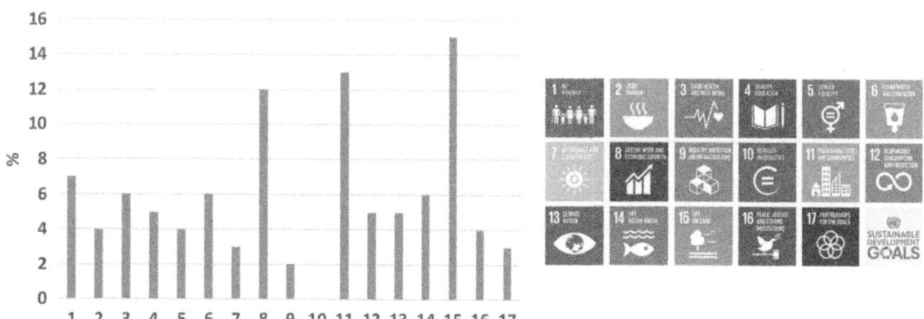

Figure 13.7 PhD study compatibility with SDG.

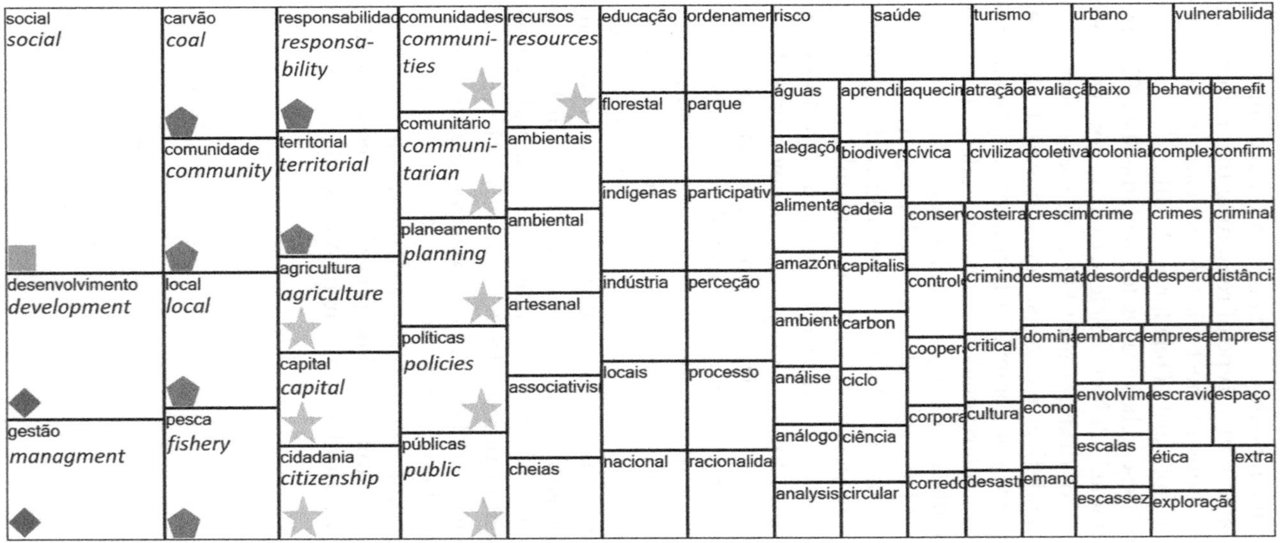

Figure 13.8 Keywords more used in the PhD students' research plans. Different symbols correspond to different clusters (only marked for the more frequent keywords).

Pospísilov (2018), the specific contribution of sustainability science to the desired social change assisted by the dialogue of different actors and processes of knowledge co-generation with diverse input can lead to multiple values and perspectives. Also, according to the same authors, these multilateral information flows not only increase understanding of complex social systems but also support sociopolitical decision-making in practice.

At the stage of developing their research, students sometimes feel isolated, or are discouraged with their research when facing so many obstacles particular at the data collection, and simultaneously have to work at their own jobs and also devote time to family affairs. For that reason, the PhD coordination team developed strategies towards students' motivation and dropout mitigation, besides an individual monitoring of the research work. Those strategies are (1) development of a virtual collaborative learning environment for thesis research process, (2) development of specific hands-on laboratories (e.g. about statistical analysis), (3) face-to-face annual seminars where students present and debate their research and (4) short web seminars about specific tools and data acquisition to support students' research. This collaborative environment that e-learning allows, brings new dimensions to traditional education, when it comes to adult learning and increases the motivation of students to learn about sustainability issues. Moreover, it may increase their readiness to learn if the students are allowed to move into new social roles along the training and research (Wilson et al., 2011, Azeiteiro et al., 2015).

For a better link and application with education for SDG, UAb could also develop more partnerships and networks with the organizations where the students work or develop their research. Networking may be an important mechanism for systemic change in higher education, and the nature and role of social capital embedded in higher education networks, stimulating the policy environment, consciously contributing to shared (sustainability) goals (Dlouhá et al., 2018). Making effective use of digital technologies, capitalizing on cultural and national differences and making the best of available resources at universities are important drivers to advance research and education for sustainability (Caniglia et al., 2017). These actions can contribute to universities curricular innovation and research to achieve the SDGs.

13.6 Conclusions

On this research, an evaluation was conducted in a sample of students of an e-learning PhD in Social Sustainability and Development as advanced training for worldwide problem solutions, in particular related with the achievement of the SDGs. The majority of the students' research plans are based on case studies, related with local problem solutions and focus on up to two dimensions of sustainability.

An online education based on a clear pedagogical model supports a collaborative PhD learning environment and a community of practice between students. This program clearly shows that working people with professional experience in the research field can develop their thesis in their workplaces, making use of their knowledge and expertise and at the same sharing it with others as well as learning about other realities and contribute to different approaches.

The work carried out also demonstrated that advanced training about SD taught online can be an important driver for SDG achievement, in particular, in developing countries. Nevertheless, students should be well aware that if they want to contribute to real problem solutions within SDGs, their research should be developed according to a holistic and systematic sustainability approach and based on intradisciplinary and interdisciplinary perspectives and transdisciplinary methods. Transdisciplinary research is issue- or problem-focused and follows responsive or iterative methodologies, requiring innovation, creativity and flexibility and often employing participatory research design strategies. This is certainly an added challenge for any PhD program if we also consider the students with very different backgrounds and weak knowledge of research methodologies.

Further developments should include more students as well a questionnaire survey to validate and complement the research plans analysis. It would also be important to follow up these students' activities after finishing their PhD to assess whether their proposed research is being applied for local problem solving.

References

Amador, F., Martinho, A. P., Bacelar-Nicolau, P., Caeiro, S., Oliveira, C. P. (2015). Education for sustainable development in higher education: Evaluating coherence between theory and praxis. *Assessment & Evaluation in Higher Education* 40(6), 867–882.

Annan-Diab, F., Molinari, F. (2017). Interdisciplinarity: Practical approach to advancing education for sustainability and for the Sustainable Development Goals. *The International Journal of Management Education* 15, 73–83.

Azeiteiro, U. M., Bacelar-Nicolau, P., Caetano, F. J. P., Caeiro, S. (2015). Education for sustainable development through e-learning in higher education: The Portuguese experience. *Journal of Cleaner Production* 106, 308–319.

Bacelar-Nicolau, P., Martinho, A. P., Amador, F., Caeiro, S., Azeiteiro, U. M. (2012). Online learning for sustainability: The student perception in an environmental science post-graduation. In: Gonçalves, F., Pereira, R., Leal Filho, W., Azeiteiro, U.M. (Eds.), *Contributions to the UN Decade of Education for Sustainable Development*, series Umweltbildung, Umweltkommunikation und Nachhaltigkeit e Environmental Education, Communication and Sustainability, vol. 33. Peter Lang, pp. 281–294.

Barth, M., Godemann, J., Rieckmann, M., Stoltenberg, U. (2007). Developing key competences for sustainable development in higher education. *International Journal in Sustainability in Higher Education* 8(4), 416–430.

Blanco-Portela, N., Benayas, J., Pertierra, L. R., Lozano, R. (2017). Towards the integration of sustainability in higher education institutions: A review of drivers of and barriers to organisational change and their comparison against those found of companies. *Journal of Cleaner Production* 166, 563–578.

Brooks, J., King, N. (2017). *Applied Qualitative Research in Psychology*. London: Palgrave.

Caniglia, G., Luederitz, C., Groß, M., Muh, M., John, B., Keeler, L. W., Wehrden, H., Laubichler, M., Arnim Wiek, A., Lang, D. (2017). Transnational collaboration for sustainability in higher education: Lessons from a systematic review. *Journal of Cleaner Production* 168, 764–779.

Dlouhá, J., Henderson, L., Kapitulcinová, D., Mader, C. (2018). Sustainability-oriented higher education networks: Characteristics and achievements in the context of the UN DESD. *Journal of Cleaner Production* 172, 4263–4276.

Dlouha, J., Pospísilov, M. (2018). Education for Sustainable Development Goals in public debate: The importance of participatory research in reflecting and supporting the consultation process in developing a vision for Czech education. *Journal of Cleaner Production* 172, 4314–4327.

Elkington, J. (1994). Towards the sustainable corporation: Win-Win-Win Business Strategies for sustainable development. *California Management Review* 36, 90–100.

Elkington, J. (2004). Enter the triple bottom line. In A. Henriques, J. Richardson (Eds.). *The Triple Bottom Line: Does it All Add Up*. London: EarthScan.

Goodwin, C. J. (2017). *Research in Psychology Methods and Design*. New York: John Wiley & Sons.

Hernández Sampieri, R., Férnandez Collado, C., Baptista Lucio, P. (2006). *Metodologia De Pesquisa [Research Methodology]*. São Paulo: McGraw-Hill.

Imbernon, F. (Coord.) (2002). *La investigación educativa como herramienta de formación del profesorado: Reflexión y experiencias de investigación educativa*. Barcelona: Graó.

Karatzoglou, B. (2013). An in-depth literature review of the evolving roles and contributions of universities to education for sustainable development. *Journal of Cleaner Production* 49, 44–53.

Krippendorff, K. (2013). *Content Analysis: An Introduction to Its Methodology*. Thousand Oaks, CA: Sage Publications.

Leal Filho, W., Azeiteiro, U., Alves, F., Pace, P., Mifsud, M., Brandli, L, Caeiro, S., Disterhelft, A. (2018). Reinvigorating the sustainable development research agenda: The role of sustainable development goals. *International Journal of Sustainable Development & World Ecology* 25(2), 131–142.

Lozano, R., Huisingh, D. (2011). Inter-linking issues and dimensions in sustainability reporting. *Journal of Cleaner Production* 19, 99–107.

Lozano, R., Lukman, R., Lozano, F., Huisingh, D., Lambrechts, W. (2013). Declarations for sustainability in higher education: Becoming better leaders, through addressing the university system. *Journal of Cleaner Production* 16(17), 10–19.

Lozano, R., Merrill, M. Y., Sammalisto, K., Ceulemans, K., Francisco, J., Lozano, F. J. (2017). Connecting competences and pedagogical approaches for sustainable development in higher education: A literature review and framework proposal. *Sustainability* 9, 1889. doi: 10.3390/su9101889.

McGregor, S. L. T., Volckmann, R. (2013). Transversity: Transdisciplinarity in higher education. In G. Hampson & M. RichTolsma (Eds.). *Leading Transformative Higher Education* (pp. 58–81). Olomouc: Palacky University Press.

Merriam, S. B. (2009). *Qualitative Research: A Guide to Design and Implementation* (2nd ed.). San Francisco, CA: Jossey-Bass.

Miller, T., Wiek, A., Sarewitz, D., Robinson, J., Olsson, L., Kriebel, D., Loorbach, D. (2014). The future of sustainability science: A solutions-oriented research agenda. *Sustainability Science* 9(2): 239–246.

Pereira, A., Mendes, A.Q., Morgado, L., Amante, L., Bidarra, J. (2008). *Universidade Aberta's Pedagogical Model for Distance Education* © (p. 109). Lisbon: Universidade Aberta.

Spangenberg, J. H. (2017). Hot air or comprehensive progress? A critical assessment of the SDGs. *Sustainable Development* 25, 311–321.

Stake, R. E. (1995). *The Art of Case Study Research.* Thousand Oaks, CA: Sage Publications.

Tejedor, G., Segalàs, J., Rosas-Casals, M. (2018). Transdisciplinarity in higher education for sustainability: How discourses are approached in engineering education. *Journal of Cleaner Production* 175, 29–37.

UN (2016). Transforming our world: the 2030 Agenda for Sustainable Development. A/RES/70/1. United Nations.

UNESCO (2005). International implementation scheme. United Nations decade of education for sustainable development (2005–2014). UNESCO Education Sector, ED/DESD/2005/PI/01, October.

UNESCO (2014). UNESCO roadmap for implementing the Global Action Programme on Education for Sustainable Development.

UNESCO (2017). *Education for Sustainable Development Goals. Learning Objetives. Education 2030.* Paris: United Nations Educational, Scientific and Cultural Organization. ISBN 978-92-3-100209-0.

Wilson, G., Abbott, D., De Kraker, J., Perez, S., Terwisscha Van Scheltinga, C., Willems, P. (2011). The lived experience of climate change: Creating open educational resources and virtual mobility for an innovative, integrative and competence-based track at Masters level. *International Journal of Technological Enhanced Learning* 3(2), 111–123.

Yin, R. K. (2014). *Case Study Research: Design and Methods.* Los Angeles, CA: Sage Publications.

chapter fourteen

The influence of food choices determinants on the achievement of the sustainable food consumption goal

An emerging subject in food consumer sciences in an e-learning environment

Ana Pinto de Moura and Luísa Aires
Universidade Aberta

Contents

14.1 The relevance of sustainable food consumption on the sustainable development

In 2000, the UN Millennium Declaration was signed by 189 countries and translated into eight Millennium Development Goals (MDGs) and then, into wide-ranging practical, measurable and time-bound targets, for the period of 2000–2015. They have played an important role for development and poverty eradication especially for developing countries/low-income countries, mobilizing globally country policies and programmes and society (UN 2000). To a large degree, these goals (Table 14.1) addressed the basic needs for human survival and other deprivations that should be considered as basic human rights (Easterly 2009).

Although some of these goals were achieved (UN 2015a), it is generally agreed that fighting against poverty and hunger should continue beyond 2015 (UN 2015b). As reflected upon Oliveira et al. (2016, p. 118) work: "furthermore, rising population levels combined with shifting dietary patterns in emerging economies put increasing pressure on global food supply: more food is necessary to feed more people. The United Nations (UN) predicts that the world population will reach 9.6 billion by 2050 (UN 2012), and this growth will require at least a 70% increase in food production but also an increase in the amount of crops used for biofuels (FAO 2009) or a more efficient use of natural resources and food production (European Commission 2014)".

Table 14.1 Elements covered by the MDGs and the SDGs

Elements	MDGs	SDGs
Dignity	**MDG 1** Eradicate extreme poverty and hunger **MDG 3** Promote gender equality and empower women	**SDG 1** End poverty in all its forms everywhere **SDG 2** End hunger, achieve food security and improved nutrition and promote sustainable agriculture **SDG 10** Reduce inequality within and among countries
People	**MDG 2** Achieve universal primary education **MDG 4** Reduce child mortality **MDG 5** Improve maternal health **MDG 6** Combat HIV/AIDS, malaria and other diseases	**SDG 3** Ensure healthy lives and promote wellbeing for all at all ages **SDG 4** Ensure inclusive and equitable quality education and promote lifelong learning opportunities for all **SDG 5** Achieve gender equality and empower all women and girls
Prosperity		**SDG 6** Ensure availability and sustainable management of water and sanitation for all **SDG 7** Ensure access to affordable, reliable, sustainable and modern energy for all **SDG 8** Promote sustained, inclusive and sustainable economic growth, full and productive employment and decent work for all

(Continued)

Table 14.1 (Continued) Elements covered by the MDGs and the SDGs

Elements	MDGs	SDGs
		SDG 9 Build resilient infrastructure, promote inclusive and sustainable industrialization and foster innovation
		SDG 11 Make cities and human settlements inclusive, safe, resilient and sustainable
		SDG 12 Ensure sustainable consumption and production patterns
Planet	**MDG 7** Ensure environmental sustainability	**SDG 13** Take urgent action to combat climate change and its impacts
		SDG 14 Conserve and sustainably use the oceans, seas and marine resources for sustainable
		SDG 15 Protect, restore and promote sustainable use of terrestrial ecosystems, sustainably manage forests, combat desertification and halt and reverse land degradation and halt biodiversity loss
Justice		**SDG 16** Promote peaceful and inclusive societies for sustainable development, provide access to justice for all and build effective, accountable and inclusive institutions at all levels
Partnerships	MDG 8	**SDG 17** Strengthen the means of implementation and revitalize the Global Partnership for Sustainable Development

Adapted from UN (2000); UN (2014b).

In the same way, it is generally accepted that many targets expressed in MDG 7 related to the environmental protection goal ("Ensure environmental sustainability") have not accomplished, including climate change as the result of human-caused greenhouse gases emissions (GHG) and global warming; loss of biodiversity caused by unsustainable demands for forests (logging for timber or wood fuel) and the conversion of forests into farms and pastures; depletion of key fossil resources (including oil, gas, coal) and groundwater; acidification of the oceans, caused mainly by the increased concentration of atmospheric carbon dioxide, with impacts on the marine food chain; rising sea levels due to global warming, stressed ocean fisheries and coastal degradation and massive environmental pollution as a result of heavy runoff of nitrogen-based and phosphorus-based fertilizers (UN 2014a).

All these trends lead to the proposal of a new agenda for sustainable development for the period 2016–2030 by the UN, being titled: "Transforming our world: the 2030 Agenda for Sustainable Development". This agenda sets out 17 Sustainable Development Goals (SDGs), with a total of 169 targets and 303 indicators, ranging from 5 to 12 targets per goal (UN 2015c). It has as a key principle of sustainable development the integration of environmental, social and economic concerns throughout the decision-making process (Holden and Linnerud 2007; Moldan et al. 2012; Sachs 2012). Moreover, this new agenda aimed to be targeted to all people (enterprises, financial institutions, governments, international institutions, local authorities, civil society, scientific and academic community) and applicable to all countries, both developing and developed countries, while taking into account different national realities, capacities and levels of development and respecting national policies and priorities. Its main finality is to serve as a guide whenever the international community decides on dealing with global changes over the next years in areas of critical importance for humanity and the planet. This guide can be framed into six dimensions: dignity, people, prosperity, planet, justice and partnership (Table 14.1).

Moreover, these goals cover a much broader range of issues: goals 1–6 were built on the core agenda of the MDGs, while goals 7–17 were a new set of goals (UNSD 2014). In this context, Le Blanc (2015) criticized the fact that sustainable consumption and production dimension have not been fully integrated into previous agendas, considering that it is a key ingredient of sustainable development paths and provide critical connections among other goals through multiple targets. According to Osborn et al. (2015), the sustainable consumption and production goal represents one of the largest transformational challenges that developed countries are facing as they need to transform their own economies to implement more sustainable patterns of consumption, production and lifestyle in order to reduce the pressure that demand has on limited or finite resources and the load they impose on the world through waste production, pollution and GHG emissions. This is particularly relevant considering that in current societies, the most familiar way to achieve prosperity is, in economic terms, by recommending a continual rise in national and global economic output, with corresponding increasing levels of people's incomes leading to an increase in consumption (Jackson 2009).

Taking into account SDG 12: "Ensure sustainable consumption and production patterns" and its relation to SDG 4: "Ensure inclusive and equitable quality education and promote lifelong learning opportunities for all", the purpose of this chapter is to describe the most important food consumption patterns that contribute to a sustainable diet particularly in developed countries, evaluating the main challenges related to them. The focus on "food system" can be explained by the fact that as contemporary food system's production is becoming ever more globalized and industrialized, it makes a significant contribution to climate changing GHG emissions, from agricultural production through processing, distribution, retailing and, likewise, food consumption (Garnett 2013). In this context, "food consumption" receives particular attention because consumers make the final choice of the goods and services they consume, and their lifestyles determine how they influence healthy and sustainable practices (Fischer and Garnett 2016). This analysis may contribute to the development of food consumption sciences courses by incorporating the sustainable food consumption perspective in their curricula (Moura and Aires 2018). The necessity of applying online learning is considered, taking into account the fact that e-learning is an important approach to strengthen an inclusive and equitable quality education, becoming widely accepted in formal and non-formal education (Azeiteiro et al. 2015). This task may contribute to the accomplishment of the SDG target 12.8: "By 2030 ensure that people everywhere have the relevant information and awareness for sustainable development and lifestyles in harmony with nature" (UN 2014b).

14.2 *Food choice determinants: the need for a sustainable diet*

As described by Moura and Cunha (2005, p. 206): "food choice is a complex behaviour influenced by many interrelating factors. These factors may be categorized as those related to food, the individual making the choice and the external economic and social environment within which the choice is made (Shepherd 1999). According to Kittler and Sucher (2004), choice of what to eat is typically made according to what is obtainable (food domain), what is acceptable (environmental domain) and what is preferred (individual domain). Moreover, some chemical and physical properties of food are perceived by the individual in terms of sensory attributes (taste, texture, appearance and smell), and the preference towards each of those attributes influences the choice of the food product. Within the limitations of dietary domains, personal preference is most often concerned with the more immediate aspects, such as sensory attributes, convenience, well-being, self-expression, variety or monetary constraints, i.e., the cost of the product against the income (Drewnowski 2002). In sum, individual food choice determinants range in scope from sensory preferences, psychological (mood, stress and guilt) to practical reasons (convenience, price/income, variety) and personal concerns (well-being, self-expression), all being interrelated as consumers are faced with several factors every time decisions about food are made".

Studies conducted in Western societies have shown that health is operating as an important individual food choice criterion (Cunha et al. 2018) because consumers are interested in feeling well (Carillo et al. 2013). In fact, more and more consumers believe that foods contribute directly to their health, and eating healthy products may prevent nutrition-related diseases and improve physical and mental well-being (Ares et al. 2015), even though health is not the only factor affecting food choice. Sensory attributes, and particularly taste, are reported as one of the most important factors that influence the individual food choice, especially for Western societies (Cunha et al. 2018). These results come in agreement with the fact that concerns about reductions in the "taste quality" of the diet are the most often mentioned obstacles to adopting a healthy diet (Moura and Cunha 2005). Moreover, changes in lifestyle, occupational patterns and urbanization have also contributed to changes in culinary practices and dietary habits, having increased the need for safe convenience foods. However, convenient solutions are perceived by consumers that buy these products as unhealthy and as highly caloric (Darian and Cohen 1995). In other words, consumers have become more health conscious in their food choices but have less time to prepare healthy meals. In addition, as identified by Moura and Cunha (2005, p. 207), food consumption patterns for low-income families are "characterized by a low consumption of fruits and vegetables and a high consumption of cereals (Krebs-Smith and Kantor 2001). Two main reasons could explain this unhealthy eating pattern: the low price of higher energy dense foods (often containing refined grains, added sugars and vegetable fats) and a taste preference for high-fat energy-dense foods (Drewnowski 2003, 2004; Drewnowski and Specter 2004). Generally speaking, diets based on added sugar, oil, margarine and refined grains are more affordable than the recommended diets based on lean meat, fish, fresh vegetables or fruit (Darmon et al. 2004; Drewnowski 2003, 2004; Drewnowski and Specter 2004). This is economically logical because cereals, added sugars and fats, which are dry and tend to have a stable shelf life are easier to produce, process, transport and store more than perishable meats, dairy products or fresh produce, with high water content". These drastic changes have led to an overall simplification of diets and reliance on a limited number of energy-rich foods, characterized by a high consumption of foods of animal origin, particularly red and processed meat and also refined grains, dairy products, processed and artificially sweetened foods, and salt, with minimal intake of fruits, vegetables, fish, legumes and whole grains (Cordain et al. 2005). As reflected upon the work by Moura and Aires (2018): "all these drivers promote the process of diet Westernization amongst other cultures/countries, namely in developing economies (Drewnowski and Popkin 1997), and even in populations which have rich and deeply rooted culinary traditions (Morinaka et al. 2013; Varela-Moreiras et al. 2010). Nevertheless, there is a great consensus among the scientific community that changing the Western diet will have a positive outcome for both people's health and environment (Friel et al. 2009)". In fact, this dietary pattern increases the risk of many chronic non-communicable diseases (NCDs): cardiovascular diseases, diabetes, cancers, chronic respiratory diseases, obesity and osteoporosis. By contrast, diets rich in minimally processed grains, legumes, fibre, vegetables, fruits and foods of plant origin protect against chronic diseases (Popkin and Du 2003).

Additionally, as contemporary food system production is becoming ever more globalized and industrialized, it also gives rise to major environmental impacts due to its reliance on high fossil-fuel energy use, chemicals, and energy inputs, long-distance transport, low-cost human work and cultural loss (Lairon 2012). In fact, the global food system contributes to some 20%–30% of anthropogenic GHG emissions (Vermeulen et al. 2012). Although the whole food chain (from farming through to transport, cooking and waste disposal) contributes to these problems, it is at the agricultural stage where the greatest impacts occur mainly from deforestation, agricultural emissions from livestock and soil and nutrient management, the latter causing land and soil degradation, a decline in biodiversity with substantial ongoing losses of populations, species and habitats (UNEP 2012) and accounting for the major source of water pollution and freshwater consumption (Smith and Bustamante 2014; Vermeulen et al. 2012).

Acknowledging that food, health and environment are interchangeable dimensions, the Food and Agriculture Organization of the United Nations (FAO) encourages the need to move towards dietary patterns that are both healthy and respectful of environmental limits and defines sustainable diets as: "diets with low environmental impacts which contribute to food and nutrition security and to a healthier life for present and future generations. Sustainable diets are protective and respectful of biodiversity and ecosystems, culturally acceptable, accessible, economically fair and affordable, nutritionally adequate, safe and healthy; while optimizing natural and human resources" (FAO 2012, p. 7). Such diet is centred on a diverse range of minimally processed tubers, whole grains, legumes and fruits and vegetables, with animal products (meat, dairy products and fish) being eaten sparingly. Since these foods are also rich in essential micronutrients (the case of fish), it will be important that reduced animal product intakes are compensated for with increases in the quantity and diversity of whole grains, legumes, fruits and vegetables, to ensure adequate nutritional intakes (Fischer and Garnett, 2016; Garnett 2014a,b). As identified by Moura and Aires (2018): "different studies suggested that respecting the dietary recommendations for a healthy diet would reduce the overall environmental impacts in developed countries particularly in terms of GHG emission and land use. The change would imply a reduction of meat consumption and would lead towards a more plant-based diet (Davis et al. 2010; Westhoek et al. 2014)". One possible strategy to address both environmental and health concerns is to consider the role of the consumer emphasizing the need for a dietary change (Fischer and Garnett 2016), namely by reducing (1) overconsumption, (2) meat consumption in favour of plant-based alternatives and (3) food waste, as revised by Moura and Aires (2018).

14.3 *Determinants of food choice module: exploring collaborative and individual learning in an online learning environment*

The increasing of online courses in higher education requires an online pedagogy for teachers and students. As referred by Moura and Aires (2018, p. 166): "with many advances in digital technologies, there have been tremendous impacts on the format and on the approach to teaching and learning, most notably in terms of online education programmes (Hay et al. 2004; Johnson et al. 2017). Considering the Moura et al. (2010, p. 552) description: "online education provides students an alternative method of study facing individuals' busy lifestyles, allowing students to be able to proceed at their own pace and identify their own personal course timeline (Shanley et al. 2004)". Following social constructivism, which emphasizes the social dimension of individual's learning (Vygotsky 1978), some researchers suggest that collaborative learning is crucial to enrich and increase online student learning experiences (Pallof and Prat 2007). e-learning offers a great number of opportunities of interaction and decision-making with peers, based on flexibility of format and easy access to knowledge, as well as engagement within the learning process (Aires et al. 2014; Dias et al. 2018). As discussed by Moura and Aires (2012, p. 200): "online students should have a greater inclination to transfer knowledge to a new domain, greater sense of community and communication and greater ownership of knowledge and independence (Garrison 2006; Dawson 2006; Hansen 2008; Peltier et al. 2007)". In fact, the online environment is particularly appropriate for collaborative learning approaches, since it emphasizes the learning potential of students working together and they take equal responsibility for the contribution to the teamwork progress (Ficapal-Cusía and Boada-Grau 2015). This interaction during the learning process promotes deep learning, and so high academic achievements, because through argumentation, students become aware of their knowledge gaps and lack of understanding, thus resulting in the asking of questions to verify the other's understanding or in asking for explanations

(Webb 1995). In such dialogical process, students share understandings, negotiate knowledge and co-regulate learning. Nevertheless, some studies on collaborative learning are critical with respect to its impact (Akpinar 2014). Some students like to work together, and some students prefer to work individually and self-regulate their learning (Akpinar 2014). For some students, collaborative learning requires extra planning and some are afraid because their marks were dependent on how well the team works together (Eijl et al. 2005). This means that learning approaches are full of diversity. Contextual, social, cultural, cognitive and metacognitive perspectives lead to different epistemologies and practices in online learning. If we focus on self and co-regulatory learning, we find a rather more critical issue (Shea et al. 2013) as it interferes either with the planning and development of learning processes or with the structural and organizational design of the online teaching and learning environments.

The Determinants of Food Choice module is part of the MSc in Food Consumption Sciences curriculum, offered by Universidade Aberta (UAb), the Portuguese public distance learning higher education institution, and is developed in the context of e-learning mode. The MSc course is formally organized, according to the European Credit Transfer and Accumulation System (ECTS), and with the virtual pedagogical model of UAb (Pereira et al. 2007). All UAb's courses are delivered through the open-source platform Moodle (http://elearning.uab.pt/), organized through closed learning communities available to the students enroled in the module and to the module teaching staff.

The MSc modules present a Learning Agreement that conveys the academic requirements and meets the students' needs and interests by scheduling and different activities and description of their workload (Pereira et al. 2007). In this context, a forum is used to allow for asynchronous exchanges: teacher guidelines, questions posed by students and discussion (Moura and Aires 2018).

The Determinants of Food Choice module intents to cover the complexity of individual food choice, covering the following topics (Moura and Aires 2012):

Topic1: Consumer's attitudes towards food. This topic considers the concept of healthy food defined by consumers and by experts and the advantages related to a healthy diet.

Topic 2: The individual food choice criteria. The individual food choice model was presented (Drewnowski 2002) and the main food choice determinants were identified: (1) sensory attributes, (2) healthy/well-being, (3) convenience and (4) monetary considerations. Each consumer food choice determinant was subsequently considered along the module.

Topic 3: Sensory attributes. The topic analyses the different drivers associated to the consumption of organic products: sensory, health, safety and environmental motives (Magnusson et al. 2003; Lee and Yun 2015;

Goetzke et al. 2014; Vega-Zamora et al. 2013). Concern about taste, health and safety can be regarded as egoistic (benefits the individual or his/her family), while consideration for the environment and animal welfare is more altruistic (benefits society rather than the individual) (Cunha and Moura 2004). This topic is particularly relevant as organic product sales continue to rise (Willer and Lernoud 2017), and many retailers have introduced lines of organic private label product, and food manufactures continue to introduce large number of new category products (Dimitri and Oberholzer 2009).

Topic 4: Healthy/well-being. In this topic, physical and mental well-being was analysed as part of one's food choice consideration (Ares et al. 2016), taking into consideration that physiological characteristics, including age, gender, body image and state of health often affect a person's food habits. Regarding this topic, one may consider the main food choice drivers of functional foods as categories of healthy foods perceived by consumers (Urala et al. 2003).

Topic 5: Convenience. The multidimensional phenomenon of convenience in food and the main factors that promote the move to convenience, particularly by members of urbanized societies, were analysed, considering principally the solution of eating out in restaurants (Cunha et al. 2011). In this context, the teacher asked students to analyse an example of a foodservice with a highly innovative dimension.

Topic 6: Monetary considerations. In this topic, the teacher asks students to analyse food cost as a barrier to dietary change, which is particularly relevant to low-income families as one of the main factors that could explain the unhealthy eating patterns of these low-income families. As an activity, students were asked to express their own opinion considering this approach, even though these food products are within the global economy, where margins are reduced and further cut, food prices still act as an obstacle to eat healthily and, therefore, to have a balanced diet.

The purpose of this study was to explore the preferred learning method of each student regarding the Determinants of Food Choice module and how well students responded to the task proposed by the teacher to achieve its outcome.

14.3.1 Methods

The work reported on this chapter corresponds to an exploratory study. In 2016/2017, 12 students (42% men), average age of 37.5 ± 11.3 years, from Continental Portugal and Islands, as well as from Portuguese-speaking African countries (Angola and Cape Verde) were enroled in this module with a diverse academic background, coming from disciplines such as engineering, environmental sciences, life sciences (nursing) or social sciences (Table 14.2).

Table 14.2 Characteristics of the students

Student code	Gender	Age (years)	Academic degrees (Higher Diploma)	Residential area
P1	Female	29	Food Engineering	Cape Verde
P2	Male	44	Environmental Sciences	North region of Portugal
P3	Male	25	Chemical and Biological Engineering	Cape Verde
P4	Female	45	Agroindustry Engineering	Southern region of Portugal/Angola
P5	Male	24	Environmental and Management Engineering	Azores (Island, Portugal)
P6	Male	62	Management	Southern region of Portugal
P7	Female	52	Environmental Sciences	Southern region of Portugal
P8	Female	40	Management and Spatial Planning	Central region of Portugal
P9	Male	35	Social Sciences	Azores (Island, Portugal)
P10	Female	26	Pharmacy	North region of Portugal
P11	Female	39	Applied Languages	Southern region of Portugal
P12	Female	29	Nutritional Sciences	Azores (Island, Portugal)

This study has focused on Topic 3 that analyses the influence of sensory attributes on organic food choice faced with other factors (namely health and safety), the task proposed to students was to assess how food companies and retailers communicate these products' attributes to consumers, to the tune of higher costs. For this, students were stimulated to visit different supermarkets chains and e-commerce platform solutions and exposed their experiences.

Support materials included the teacher's slide presentations and original teacher's documents, scientific publications (papers, research papers), website links and a digital video created by the teacher and produced by the UAb digital services related specifically to: (1) sensory attributes (particularly taste), (2) sustainable food consumption and (3) organic foods.

Moreover, for a period of 16 days, part of the students worked in a team, working collaboratively, and part of them worked individually on an assignment. The teamwork forums were separated forums composed by two members and supported work in smaller groups, where members commented the information transmitted by their colleagues and understood the subject matter by helping to produce an assessment that represents a common solution to the problem posed by the teacher. One may notice that it was the teacher who chose which students would benefit from working collaboratively or individually, according to the teacher's previous information about each one of the students' learning process and learning outcome. Students that were more autonomous worked individually. At the end, the representative of each group and individuals submitted their work to the platform, and students discussed the different outputs for 3 days. Students were aware about the evaluation criteria and weighting that would be used to evaluate their level of performance in tasks: assessment and enlarged forum discussion had the same weight, for the final grade of the topic.

In the same way, at the end of the module, students were invited to answer a website questionnaire with the following open questions, answering as "collectively" or as "individually", according to the specific way of studying given by the teacher (not for a mark):

1. Did this activity meet your expectations regarding the subject? Why?
2. How did you organize your study time during this activity?
3. On a five-point anchored scale (where 1 = "not useful" and 5 = "very useful"), how do you evaluate the results from the learning activities you have partaken in?
4. If you had developed the activity in the other learning modality ("individually", "collectively"), would the results had been the same? Why?
5. What were the advantages and disadvantages of having taken part in this individual or group activity?

14.3.2 *Results*

Eleven students that attended the course gave their answers to the questionnaire by email: six answered as "collectively" (students: S1, S2, S4, S6, S7, S12) and five as "individually" (students: S5, S8, S9, S10, S11). Next, we checked to what degree students that were collaborating or learning individually differed in their questionnaire answers. The students' quotes used in this study were translated into English by the research team. Quotations are used to illustrate the main findings.

Expectations. Both collectively and individually working students confirm that the acquisition of knowledge and skills is positive in the subject matter. They referred that following this task, they better understood the main reasons that lead consumers to buy organic foods and how retailers and manufactures use organic food attributes to catch the consumers, resulting in professional valorization. At the end, students stated that they now much better understand the reactions of others and their own actions when acting as consumers.

> The fact that I actively sought out information regarding the theme allowed me to understand that [organic foods] have ceased to be a market segment, only for vegans, vegetarians or people with some kind of specific diet, but such that brands take their chances in reaching new customers and make them loyal to the brand, presenting them with the advantages that organic agriculture can present.
>
> *S10, Individually*

Learning organization. Globally, during the assignment preparation, both student groups shared similar learning strategies. First, they read and studied the resources provided by the teacher, considering the learning objectives proposed for the task. Some students that accomplished the task individually, reported that they searched additional information related to the topic in online libraries (students: S8 and S9). Then, students wrote that they searched additional information in food companies' websites and professional/technical magazines about manufacturers and retailers of organic products. In this research, students were getting closer to what is requested in the activity by analysing the strategies of manufacturers and retailers of organic products. Subsequently, students reported that they visited stores to collect additional information and see in loco the organic brands' approach. Curiously, there were more students that worked collectively, which referred that they visited different supermarkets (67%) compared to students that worked individually (40% of them). Students who worked together additionally reported that they collaborated with their fellow in teamwork forums to discuss the work and built the assignment.

> I organised the study of this activity in the following way:
>
> 1st reading and analysing the documents proposed in the learning resources.
>
> 2nd find in the following supermarkets: Continente, Intermarché, Lidl e Pingo Doce from Elvas, Hipercor, Corte Inglês and Carrefour; the way of promotion of organic products.
>
> 3rd find catalogues and advertisements about organic products on the internet.
>
> 4th exchange of information and accomplishment of the work with the teammate.
>
> *S6, Collectively*

Learning outcome. Students positively evaluated their learning outcome for this activity. The mean rating for students that worked collaboratively was 4.3, and for students that worked individually was 4.8, of a total of 5 points. They mentioned a new understanding of the current organic market situation and how food companies valorized these products to consumers. They also mentioned that this subject was particularly interesting considering the growth of organic market in the last years, its impact on a daily diet and the fact that it also gave relevant sources for future research (the MSc Dissertation).

> I evaluate the result of my learning as profitable, I managed to understand how the organic food market is and how it is promoted, even if it is still "shy".
>
> *S4, Collectively*

> It is a very current topic, of general interest, and very directed at our daily diet.
>
> *S8, Individually*

Changing learning modality. The majority of students answered that their learning outcomes would have been different if they had changed their learning modality. By analysing their opinions, one can notice that they were very satisfied with their learning modality because changing learning modality would give worse results. This is true for all non-collaborating students and for three of the collaborating students.

> I doubt that I would be impaired, because I would have to wait on the results of the others, which has been a problem, sometimes persistent. It is much more effective to do work individually and to have discussions on these individual works, in my opinion.
>
> *S5, Individually*

The three students (S1, S2, S4) that expressed that changing their learning modality would change their learning outcome worked collectively, and they were disappointed because the team did not function as an effective team:

> Yes, because the activity ended up being done individually.
>
> *S1, Collectively*

Advantages and disadvantages. Students from both groups mentioned advantages and disadvantages for both learning modalities. Students who worked individually experience important advantages namely, planning of their own time schedule and having less pressure to conclude the assignment. Although these students recognized the lack of getting new insights that could be important to reaching the collective output, they found that collaborative learning requires extra time and energy because each team member has differing time managements and contribute differently to the task.

> The advantage is that I am able to adjust the time taken to study and the time for the delivery of the work. I work, I have children, I'm pregnant, and I am able to better manage the units of work when I work alone and I'm not dependent of others, or when I don't have to insist for them to send me their part of the work. The disadvantage is that because we are all from different areas and countries, each person has a different viewpoint and approach to this theme.
>
> *S11, individually*

Students that collaborated with their fellows referred as advantages of collaboration extra learning because they discussed different points of view that otherwise they would not have noticed and everyone's opinions are respected. The discussion supported dialogue between students so that comments, information, data and links were shared. Moreover, some students also said that because tasks were allocated and distributed among team members, this increased work productivity. They also reported that working collectively approximates team members and intensifies peer complicity. As disadvantages of collaboration, some students mentioned problems of time management, planning and extra anxiety because "they are always dependent on one another" (S2), especially when their messages were not replied to by the other team colleague during the teamwork preparation at team discussion forum. These students (S1, S2, S4), as referred previously, had a negative collaboration experience due to the lack of interaction among members.

> Group work is always more enriching than individual work, the exchange of knowledge are an asset in the learning process. When there isn't any interaction between the members of the group, it doesn't allow for a good contribution for the completion of the work, hindering learning.
>
> *S4, Collectively*

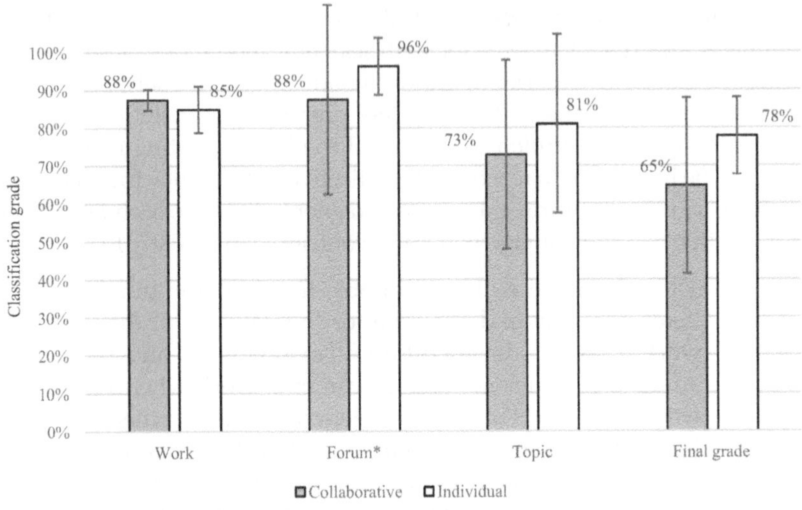

Figure 14.1 Mean and standard deviation for the classification of students according to the learning method, calculated for the assignment, forum class participation, topic mark and final module mark (*one individual student and two collaborate students did not participate in the forum).

Assessment results. A comparison of the marks of collaborating students and students working individually revealed that the mean mark of the collaborators was 88% (±3%) and 85% (±6%) for the individual students. This is particularly relevant considering that students who had been chosen by the teacher to work individually had highly performers at the open class forum, at the end of the topic and at the end of module (Figure 14.1).

14.4 Discussion

From the Determinants of Food Choice module, both collaborating and individually working students positively evaluated the learning method used in this activity that further engaged them in active learning, which was strongly related to their professional and personal life (Moura and Aires 2012). This could be explained by the fact that students apply the online module knowledge to the real world (Hansen 2008). In the end, each student in his/her daily life is a potential food consumer that has experienced food consumption decisions (Moura and Aires 2012). As consumers, our students have a major role in making food chains more sustainable, by the choices they make when buying food. They can support a critical reflection of what they learnt, increasing their critical thinking capabilities and reward more sustainable food production, like buying organic products, and punish less sustainable alternatives (Grunert 2011).

Despite the awareness of collaborative learning advantages, students that worked individually preferred maintaining this learning modality. Working individually gave them the flexibility of planning their time schedule and their time investment. Additionally, there were some collaborating students, those where the team did not function well, that complained about negative aspects of the collaborative learning, namely the fact that it requires more time and it is more stressful to accomplish the task (Ficapal-Cusí and Boada-Grau 2015). Nevertheless, working collaboratively provides good learning results, and this may be an argument that engages students to work together. This could be explained by the fact that exchanging experiences and knowledge on the different elements enriched the final output and added value to the learning process. Thus, one approach to promote collaborative learning is to give the students the possibility to decide on the choice of learning modality and on the choice of their small team members (Eijl et al. 2005). This may allow students to work together in a more responsible way, by respecting the time established to complete the assignment, the opinions of each group member and to be responsible for accomplishing the tasks that are assigned (Moura and Aires 2012).

Finally, one may stress that placing students in peer tasks or in independent tasks does not explain the complexities of online learning outcomes, as we are facing a multidimensional phenomenon that is strongly connected to institutional, social and individual variables. However, the study of learning regulatory processes may lead to better teaching and learning practices, to a better understanding of learning success or lack thereof during online courses (Sobocinski, Malmberg and Järvelä 2017). This is particularly relevant, considering that the cooperative interaction, the communication skills and team-based problem-solving skills are necessary to achieve the sustainable food consumption goal, regarding the complexity of food systems and the diversity of actors (Schaffner et al. 2003).

Acknowledgements

Authors thank José B. Cunha from Oporto British School for revising English usages and grammar throughout the manuscript.

References

Aires, L., P. Dias, J. Azevedo, M.A. Rebollo, and R. García-Perez. 2014. Education, digital inclusion and sustainable online communities. In *E-Learning and Education for Sustainability*, eds. U.M. Azeiteiro, W. Leal Filho, and S. Caeiro, pp. 263–273. Frankfurt: Peter Lang. doi: 10.3726/978-3-653-02460-9.

Ares, G., L. de Saldamando, A. Giménez, A. Claret, L.M. Cunha, L. Guerrero, A.P. de Moura, D.C.R. Oliveira, R. Symoneaux, and R. Deliza. 2015. Consumers' associations with wellbeing in a food-related context: A cross-cultural study. *Food Quality and Preference* 40: 304–315. doi: 10.1016/j.foodqual.2014.06.001.

Akpinar, Y. 2014. Different modes of digital learning object use in school settings: Do we design for individual or collaborative learning? *International Journal of Education and Development using Information and Communication Technology* (IJEDICT) 10(3): 87–95.

Azeiteiro, U.M., P. Bacelar-Nicolau, F.J.P. Caetano, and S. Caeiro. 2015. Education for sustainable development through e-learning in higher education: Experiences from Portugal. *Journal of Cleaner Production* 106: 308–319. doi: 10.1016/j.jclepro.2014.11.056.

Carrillo, E., V. Prado-Gascó, S. Fiszman, and P. Varela. 2013. Why buying functional foods? Understanding spending behaviour through structural equation modelling. *Food Research International* 50: 361–368.

Cunha, L.M., D. Cabral, A.P. Moura, and M.D.V. de Almeida. 2018. Application of the food choice questionnaire across cultures: Systematic review of cross-cultural and single country studies. *Food Quality and Preference* 64: 21–36.

Cunha, L.M., and A.P. Moura. 2004. Conflicting demands of agricultural production and environmental conservation: Consumers' perception of the quality and safety of food. In *Ecological Agriculture and Rural Development in Central and Eastern European Countries*, ed. W.L. Filho, pp. 137–157. Amsterdam: Nato Science Series, IOS Press. ISBN: 1-58603-439-1.

Cunha, L.M., A.P. Moura, and A. Frias. 2011. Valorisation of menu labelling at fast food restaurants: Exploring consumer perceptions. *Brazilian Journal of Food Technology* 7: 55–64. doi: 10.4260/BJFT201114E000107.

Cordain, L., Eaton, S.B., and Sebastian, A. 2005. Origins and evolution of the Western diet: Health implications for the 21st century. *American Journal of Clinical Nutrition* 81(2): 341–354.

Darian, J.C., and J. Cohen. 1995. Segmenting by consumer time shortage. *Journal of Consumer Marketing* 12(1): 32–44.

Darmon, N., E. Ferguson, and A. Briend. 2004. Do economic constraints encourage the selection of energy dense diets? *Appetite* 41: 315–322.

Davis, J., U. Sonesson, D.U. Baumgartner, and T. Nemecek. 2010. Environmental impact of four meals with different protein sources: Case studies in Spain and Sweden. *Food Research International* 43: 1874–1884. doi: 10.1016/j.foodres.2009.08.017.

Dawson, S. 2006. Online forum discussion interactions as an indicator of student community. *Australasian Journal of Education Technology* 22(4): 495–510.

Dias, P., L. Aires, and D. Moreira. 2018. e-learning diversification in higher education: Conceptions of participation. In *Climate Literacy and Innovations in Climate Change Education: Distance Learning for Sustainable Development*, eds. U. M. Azeiteiro, W. Leal-Filho, and L. Aires, pp. 291–306. London: Springer. ISBN 978-3-319-701998.

Dimitri, C., and L. Oberholzer. 2009. *Marketing US Organic Foods. Recent Trends from Farms to Consumers.* Washington, DC: United States Department of Agriculture.

Drewnowski, A. 2002. Taste, genetics and food choices. In *From Genes to Culture*, eds. H. Anderson, J. Blundell, and M. Chiva. Levallois-Perret: Danone Institute, http://publications.danoneinstitute.org/boutique/images_produits/fFOODSELEC_1.pdf).

Drewnowski, A. 2003. Fat and sugar: An economic analysis. *Journal of Nutrition* 133: 838S–840S.

Drewnowski, A. 2004. Obesity and the food environment: Dietary energy density and diet costs. *American Journal of Preventive Medicine* 27: 154–162.

Drewnowski, A., and B.M. Popkin. 1997. The nutrition transition: New trends in the global diet. *Nutrition Reviews* 55(2): 31–43.

Drewnowski, A., and S.E. Specter. 2004. Poverty and obesity: The role of energy density and energy costs. *American Journal of Clinical Nutrition* 79: 6–16.

Easterly, W. 2009. How the millennium development goals are unfair to Africa. *World Development* 37(1): 26–35.

Eijl, P.J. van, A. Pilot, and de P. Voogd. 2005. Effects of collaborative and individual learning in a blended learning environment. *Education and Information Technologies* 10(1–2): 49–63.

European Commission. 2014. Communication from the Commission to the European Parliament, the Council, the European Economic and Social Committee and the Committee of the regions—towards a circular economy: A zero waste programme for Europe. Brussels.

FAO. 2009. *How to Feed the World in 2050*. Rome: FAO.

FAO. 2012. Sustainable diets and biodiversity. Directions and solutions for policy, research and action. *Report of the International Scientific Symposium*: Biodiversity and sustainable diets against hunger. FAO, Rome, 3–5 November.

Ficapal-Cusí, P., and J. Boada-Grau. 2015. e-learning and team-based learning. Practical experience in virtual teams. *Procedia - Social and Behavioral Sciences* 196: 69–74.

Fischer, C.G., and T. Garnett. 2016. *Plates, Pyramids and Planets Developments in National Healthy and Sustainable Dietary Guidelines: A State of Play Assessment*. Food Climate Research Network, FAO and the University of Oxford.

Friel, S., A.D. Dangour, T. Garnett, K. Lock, Z. Chalabi, I. Roberts, A. Butler, C.D. Butler, J. Waage, A.J. McMichael, and A. Haines. 2009. Public health benefits of strategies to reduce greenhouse-gas emissions: Food and agriculture. *Lancet* 374(9706): 2016–2025. doi: 10.1016/S0140-6736(09)61753-0.

Garrison, D.R. 2006. Online collaboration principles. *Journal of Asynchronous Learning Networks* 10(1): 25–34.

Garnett, T. 2013. Food sustainability: Problems, perspectives and solutions. *Proceedings of the Nutrition Society* 72(1): 29–39.

Garnett, T. 2014a. What is a sustainable healthy diet? A discussion paper. Food Climate Research Network – Oxford Martin School – ECI – CCAFS.

Garnett, T. 2014b. Changing what we eat: A call for research and action on widespread adoption of sustainable healthy eating. Food Climate Research Network.

Goetzke, B., S. Nitzko, and A. Spiller. 2014. Consumption of organic and functional food. A matter of well-being and health? *Appetite* 77: 94–103. doi: 10.1016/j.appet.2014.02.012.

Grunert, K.G. 2011. Sustainability in the food sector: A consumer behavior perspective. *International Journal on Food System Dynamics* 2(3): 207–218.

Hay, A., J. Peltier, and W. Drago. 2004. Reflective learning and on-line education: A comparison of traditional and on line MBA students. *Strategic Change* 13(4): 169–182. doi: 10.1002/jsc.680.

Hansen, D.E. 2008. Knowledge transfer in online learning environments. *Journal of Marketing Education* 30(2): 93–105. doi: 10.1177/0273475308317702.

Holden, E., and K. Linnerud. 2007. The sustainable development area: Satisfying basic needs and safeguarding ecological sustainability. *Sustainable Development* 15: 174–187.

Jackson, T. 2009. *Prosperity without Growth? The Transition to a Sustainable Economy.* Surrey: Sustainable Development Commission. Available at http://epubs.surrey.ac.uk/id/eprint/745916.

Johnson, C., L. Hill, J. Lock, N. Altowairiki, C. Ostrowski, L. da Rosa dos Santos, and Y. Liu. 2017. Using design-based research to develop meaningful online discussions in undergraduate field experience courses. *International Review of Research in Open and Distributed Learning* 18(6). doi: 10.19173/irrodl.v18i6.2901.

Krebs-Smith, S.M., and L.S. Kantor. 2001. Choose a variety of fruits and vegetables daily: Understanding the complexities. *The Journal of Nutrition* 131(2s-1): 487s–501s. doi: 10.1093/jn/131.2.487S.

Kittler, P.G., and K. Sucher. 2004. *Food and Culture.* 4th edition. Toronto: Thomson Wadsworth.

Lairon, D. 2012. Biodiversity and sustainable nutrition with a food-based approach. In *Sustainable Diets and Biodiversity,* eds. B. Burlingame, and S. Dernini. Directions and Solutions for Policy an Action. Rome: FAO.

Le Blanc, D. 2015. Towards integration at last? The sustainable development goals as a network of targets. *Sustainable Development* 23(3): 176–187. doi: 10.1002/sd.1582.

Lee, H-J., and Z-S. Yun. 2015. Consumers' perceptions of organic food attributes and cognitive and affective attitudes as determinants of their purchase intentions toward organic food. *Food Quality and Preference* 39: 259–267. doi: 10.1016/j.foodqual.2014.06.002.

Magnusson, M.K., A. Arvola, U.K. Hursti, L. Aberg, and P.O. Sjoden. 2003. Choice of organic foods is related to perceived consequences for human health and to environmentally friendly behaviour. *Appetite* 40(2): 109–117.

Moldan, B., S. Janoušková, and T. Hák. 2012. How to understand and measure environmental sustainability: Indicators and targets. *Ecological Indicators* 17: 4–13. doi: 10.1016/j.ecolind.2011.04.033.

Morinaka, T., M. Wozniewicz, J. Jeszka, J. Bajerska, P. Nowaczyk, and Y. Sone. 2013. Westernization of dietary patterns among young Japanese and Polish females: A comparison study. *Annals of Agricultural and Environmental Medicine* 20(1): 122–130.

Moura, A.P., and L. Aires. 2012. How to learn about individual food choice criteria subject? A case study on e-learning environment module. In *Investigação e variantes curriculares do ensino online: desafios da interculturalidade na Era Tecnológica,* eds. J. Ribeiro, and L. Aires, Porto: CEMRI, Universidade Aberta. ISBN: 978-972-674-726-0. https://repositorioaberto.uab.pt/bitstream/10400.2/2196/1/ebook_interativo_investiga%c3%a7%c3%a3o%20e%20variantes.pdf.

Moura, A.P., and L. Aires. 2018. Food and sustainability: An emerging subject in sustainable environmental sciences education applying to the e-learning environment. In *Climate Literacy and Innovations in Climate Change Education,* eds. U.M. Azeiteiro, W. Leal- Filho, and L. Aires, 109–130, Switzerland: Springer.

Moura, A.P., and L.M. Cunha. 2005. Why consumers eat what they do: An approach to improve promoting new responses. In *Taking Responsibility. CCN Conference Proceedings,* eds. D. Tangen, and V.W. Thoresen, pp. 204–213, Norway: Forfatterne. ISBN: 82-7671-495-1.

Oliveira, B., A.P. Moura, and L.M. Cunha. 2016. Reducing food waste in the food service sector as a way to promote public health and environmental sustainability. In *Climate Change and Health Improving Resilience and Reducing Risks*, eds. W.L. Filho, U.M. Azeiteiro, and F.A. Alves, pp. 117–132. Switzerland: Springer. doi: 10.1007/978-3-319-24660-4_8.

Osborn, D., A. Cutter, and F. Ullah. 2015. Universal sustainable development goals. Understanding the Transformational Challenge for Developed Countries, Stakeholder Forum.

Pallof, R.M., and K. Pratt. 2007. *Building Online Learning Communities: Effective Strategies for the Virtual Classroom*. San Francisco, CA: Jossey-Bass.

Peltier, J.W., J.A. Schibrowsky, and W. Drago. 2007. The interdependence of the factors influencing the perceived quality of the online learning experience: A causal model. *Journal of Marketing Education* 29(2): 140–153. doi: 10.1177/0273475307302016.

Pereira, A., A.Q. Mendes, L. Morgado, L. Amante, and J. Bidarra. 2007. *Virtual Pedagogical Model: Model of Universidade Aberta for a Future University*. Lisbon: Universidade Aberta (in Portuguese). ISBN: 978-972-674-493-1.

Pinto de Moura, A., L. Miguel Cunha, U. Miranda Azeiteiro, L. Aires, P. Graça, and M.D. Vaz de Almeida. 2010. Food consumer science post-graduate courses: Comparison of face-to-face versus online delivery systems. *British Food Journal* 112(5): 544–556. doi: 10.1108/00070701011043781.

Popkin, B.M., and S. Du. 2003. Dynamics of the nutrition transition toward the animal foods sector in China and its implications: A worried perspective. *Journal of Nutrition* 133(11 Suppl 2): 3898s–3906s. doi: 10.1093/jn/133.11.3898S.

Quested, T., and H. Johnson. 2009. Household food and drink waste in the UK: A report containing quantification of the amount and types of household food and drink waste in the UK. Banbury: WRAP.

Sachs, J.D. 2012. From millennium development goals to sustainable development goals. *Lancet* 379(9832): 2206–2211. doi: 10.1016/s0140-6736(12)60685-0.

Shea, P., S. Hayes, S. Uzuner Smith, J. Vickers, T. Bidjerano, M. Gozza-Cohen, S-B. Jian, A. Pickett, J. Wilde, and C-H. Tseng. 2013. Online learner self-regulation: Learning presence viewed through quantitative content- and social network analysis. *International Review of Research in Open and Distributed Learning* 14(3): 35. doi: 10.19173/irrodl.v14i3.1466.

Shanley, E.L., C.A. Thompson, L.A. Leuchner, and Y. Zhao. 2004. Distance education is as effective as traditional education when teaching food safety. *Food Service Technology* 4(1): 1–8. doi: 10.1111/j.1471-5740.2003.00071.x.

Schaffner, D.J., W.R. Schroder, and M.D. Earle. 2003. *Food Marketing—An International Perspective*. 2nd edition. San Francisco, CA: McGraw-Hill.

Shepherd, R. 1999. Social determinants of food choice. *Proceedings of the Nutrition Society* 58(4): 807–12.

Smith, P., and M. Bustamante, eds. 2014. Agriculture, Forestry and Other Land Use (AFOLU). In *Climate Change 2014: Mitigation of Climate Change*. Contribution of Working Group III to the Fifth Assessment Report of the Intergovernmental Panel on Climate Change.

Sobocinski, M., J. Malmberg, and S. Järvelä. 2017. Exploring temporal sequences of regulatory phases and associated interactions in low- and high-challenge collaborative learning sessions. *Metacognition and Learning* 12(2): 275–294. doi: 10.1007/s11409-016-9167-5.

UN. 2000. Resolution adopted by the General Assembly, United Nations Millennium Declaration, 55/2, Sept 18, A/RES/55/2.

UN. 2012. *World Population Prospects, The 2012 Revision*. Rome: United Nations Department of Economic and Social Affairs, Population Estimate and Projections Section.

UN. 2014a. Prototype global sustainable development report. New York: United Nations Department of Economic and Social Affairs, Division for Sustainable Development, July 2014.

UN. 2014b. The road to dignity by 2030—Ending poverty, Transforming All Lives and Protecting the Planet: Synthesis Report of the Secretary-General on the Post-2015 Sustainable Development Agenda. New York.

UN. 2015a. The millennium development goals report 2015. New York.

UN. 2015b. Millennium development goal 8 taking stock of the global partnership for development, MDG gap task force report 2015, New York.

UN. 2015c. General assembly resolution A/RES/70/1. Transforming Our World, the 2030 Agenda for Sustainable Development.

UNEP. 2012. *Measuring Progress: Environmental Goals and Gaps*. Nairobi: UNEP.

UNSD. 2014. *Envstats: News and Notes*. New York: UN Statistical Division.

Urala, N., A. Arvola, and L. Lähteenmäki. 2003. Reasons behind consumers' functional food choices. *Nutrition & Food Science* 33(815): 826.

Varela-Moreiras, G., J.M. Ávila, C. Cuadrado, S. del Pozo, E. Ruiz, and O. Moreiras. 2010. Evaluation of food consumption and dietary patterns in Spain by the food consumption survey: updated information. *European Journal of Clinical Nutrition* 64: S37–43. doi: 10.1038/ejcn.2010.208.

Vega-Zamora, M., M. Parras-Rosa, M. Murgado-Armenteros Eva, and F.J. Torres-Ruiz. 2013. The influence of the term 'organic' on organic food purchasing behavior. *Procedia - Social and Behavioral Sciences* 81: 660–671. doi: 10.1016/j.sbspro.2013.06.493.

Vermeulen, S.J., B.M. Campbell, and J.S.I. Ingram. 2012. Climate change and food systems. *Annual Review of Environment and Resources* 37(1): 195–222. doi: 10.1146/annurev-environ-020411-130608.

Vygotsky, L.S. 1978. *Mind in Society: The Development of Higher Psychological Processes*, eds. M. Cole, V. John-Steiner, S. Scribner, and E. Souberman. Cambridge, MA: Harvard University Press.

Webb, N.M. 1995. Group collaboration in assessment: Multiple objectives, processes, and outcomes. *Educational Evaluation and Policy Analysis* 17(2): 239–261.

Westhoek, H., J.P. Lesschen, T. Rood, S. Wagner, A. De Marco, D. Murphy-Bokern, A. Leip, H. van Grinsven, M.A. Sutton, and O. Oenema. 2014. Food choices, health and environment: Effects of cutting Europe's meat and dairy intake. *Global Environmental Change* 26: 196–205. doi: 10.1016/j.gloenvcha.2014.02.004.

Willer, H., and J. Lernoud. 2017. The world of organic agriculture, statistics and emerging trends. *FiBL-IFOAM Report*. Bonn: Research Institute of Organic Agriculture (FiBL), Frick, and IFOAM Organics International.

chapter fifteen

Enhancing the university outreach through collaborations and sharing of resources

Experiences from India

Umesh Chandra Pandey
Indira Gandhi National Open University

Chhabi Kumar
Rani Durgavati University

Contents

15.1 Background

Historically, universities had a culture of functioning in an isolated work environment with little connection with the outside world. However, modern demands have gradually changed the universities from *Cathedrals of Learning & Research* to *Managerial & Corporate-like Structures* (Oertel and Soll, 2017). Rise of mega universities with their profit motives and international marketing operations have given an altogether different working culture to the university education (Mukerji and Tripathi, 2004). These changes have been driven primarily by paradigm shifts on socio-economic fronts across the world (Kalam, 2005).

The role of a university as the agent of mass education is a new phenomenon. It has emerged primarily due to enhanced role that education plays in development. Sustainable Development Goals (SDGs) have given a renewed thrust to university system. The systems of higher education are increasingly being viewed as a force that can support the implementation of each of the SDGs (SDG Knowledge Hub, 2017). Furthermore, engaging with SDGs can also benefit universities in different ways (Bhowmik, Samia and Huq, 2018).

The Open and Distance Learning (ODL) system in India has undergone remarkable growth during the past few decades (UGC Web Site). In 2010, it had more than 22%–23% of total the enrolment of higher education in the country.

Indira Gandhi National Open University (IGNOU) is the largest player in ODL system in India and also bears the distinction of being the largest open university of the world. Starting with just 4,528 students and two academic programmes in 1987, it has now grown in to a gigantic system of more than three million students and over 232 academic programmes. IGNOU presents a unique model based on openness and sharing of resources and works in close partnerships with more than 3,500 educational institutions in India and also through its overseas partner institutions in 13 countries currently.

This unique approach of collaboration and sharing as practiced by the university for more than three decades carries special relevance within the perspective of SDG 17 of Post 2015 Development Agenda (World Economic

Forum, 2017). SDG 17 strongly advocates for revitalizing partnerships and sharing of knowledge, expertise, technologies and financial resources among various stakeholders in all countries, particularly developing countries (UN-DESA, 2016).

It is quite interesting to understand how the educational model of IGNOU based on partnerships has worked for more than three decades. The lessons learnt will be crucially important for the global pursuit of sustainable development. This chapter showcases the strategic framework of IGNOU, its experiences and lessons learnt.

The chapter is structured as follows:

a. In Part I, the authors have given a background of how the role of universities is changing in the emerging knowledge economy. Further, various international developments have been described.
b. In Part II, the need for university outreach programmes has been described.
c. In Part III, a case study of IGNOU is presented that includes the strategic framework of the university, its experiences and lessons learnt.
d. The concluding remarks have been given in Part IV of the chapter.

15.1.1 Changing roles of universities

Growth of knowledge economy depends primarily on how well its knowledge-generating institutions can create and preserve knowledge and put in place mechanisms for universal accessibility. Easy availability of knowledge will ease the vertical mobility, empower those who are unfavourably placed in social ladder and thereby smoothen out disparities created by birth and other circumstances.

Twentieth century has witnessed paradigmatically different role that knowledge plays in our lives. It sells like a commodity, sustains livelihoods and can take the people out of poverty traps. We have observed the mind-boggling pace of knowledge generation and its equally fast obsolescence. The education at the work places is not a matter of choice but a compulsion of our age. Knowledge economy will cease to function if skills and competencies of work force are not continuously updated. Prosperity of knowledge society will be decided by its ability to create and maintain the knowledge infrastructure, develop knowledge workers and enhance their productivity through creation, growth and exploitation of new knowledge (Kalam, 2005). Such a scenario has raised serious challenges for universities.

Another major development of our times is the rise of democratic value system – a truly 20th century phenomenon. Knowledge is a promising democratizing force that can cut across socio-economic barriers and smoothen out inequalities imposed by birth and other circumstances.

Universities survive with the public money, and the knowledge generated by them should be treated as public good. The knowledge transfer to community is therefore a democratic obligation. Universities can justify their presence only if they become effective partners in the process of development. According to Cortese (2003), universities are in fact bound by their moral responsibilities of spreading awareness, disseminating knowledge and providing skill-based training for people in order to prepare them to develop, lead and manage the development process. Thus, universities have to be seen as agents of change and partners in enhancing the overall well-being of the society.

The functioning of the universities needs to be fine-tuned with these changing socio-economic realities. Universities are under obligation to come out of the four walls of their campuses, develop functional linkages with outside world and reposition themselves for social transformation.

15.1.2 *Embedding sustainability in university – a historical perspective*

The role of universities for sustainable development is not new. In fact, International Association of Universities (IAU) continued to maintain sustainable development as one of its key action areas since 1990. It adopted a policy statement in 1993 known as the Kyoto Declaration with its opening clause urging the universities to seek, establish and disseminate a clearer understanding of sustainable development. The IAU also developed an online portal on "Higher Education for Sustainable Development" to encourage and facilitate higher education institutions (HEIs) across the world to network and showcase their activities.

The concerns of the United Nations continued with the Declaration of 2005–2014 as the Decade of Sustainable Development. At the end of the decade, United Nations Educational, Scientific and Cultural Organization (UNESCO) prepared a summary report about assessment of progress achieved during the decade and the challenges encountered. With regard to higher education, the report takes a positive note of the efforts made by HEIs. The major highlights being efforts made by the HEIs to address sustainability in campus operations (commonly referred to as campus greening), introduction of new programmes and courses related to education for sustainable development and extension of teaching and research to respective communities. Major outcome during the decade is the creation of networks of institutions in different regions around the world – MESA in Africa, ProsPER.Net in Asia-Pacific, COPERNICUS Alliance in Europe and ARIUSA in Latin America and the Caribbean. These networks have created conducive atmosphere in respective regions to build capacity, share experiences and expand the influence of education for sustainable development. The Global Mega University Network (GMUNET) – a major network of open universities across the world has created a worldwide platform for mutual cooperation. In a recent initiative, United Nations Environment Programme (UNEP) under Global Universities Partnership on Environment for Sustainability (GUPES) started a network of 370 universities across the globe to implement environment and sustainability practices into the curricula. Higher education played an important role to promote sustainable development during the decade 2005–2014. However, report outlines several challenges also, e.g., the lack of a coordinated approach at all levels of HEIs to implement necessary changes, insufficient staff development activities to empower staff to transform curricula and pedagogy towards a sustainable development perspective and the persistence of disciplinary boundaries that inhibit the potential to address complex sustainable development issues (Mohamedbhai, 2015). The encouraging initiatives and achievements of HEIs during the decade gave rise to encouraging possibilities for a strong role of higher education to meaningfully address Post 2015 Development Agenda.

The United Nations, in 2015, declared the development pathway in terms of SDGs. The 17 "Global Goals" along with the 169 targets have been adopted to transform the world as we know it today and take it forward towards the path of all-inclusive and equitable development so that the benefits of the development process are reaped even by the poorest of the poor (UNESCO, 2012). The nations are expected to assess the direction and measure the level of development by looking at the extent of realization of these goals. Thus, development has to be seen in terms of a continuous and participatory exercise by the community and other stakeholders keeping in view the context and the needs of the community while maintaining a fine balance with the environment (Pandey and Kumar, 2017). Institutions now have the responsibility, more than ever before, to integrate sustainable development into all their teaching, research, community engagement and campus operations. The Post 2015 Development Agenda advocates for an entirely new perspective of working in partnerships. It has given rise to an urgent need to inculcate skills, mindsets and attitudes to truly transform organizations, societies, embody sustainable campus practices and promote innovative research (HESI, 2012). HEIs are hubs of innovations and critical thinking which favorably positions them to explore paradigmatically new ways of addressing SDGs. Thus, these institutions can nurture a new generation of leaders, policymakers, entrepreneurs, scientists, researchers and educators who will be better equipped to deal with the challenges of SDGs. Therefore, there has been an increasing realization to integrate the SDGs into research, teaching, pedagogy and campus practices of HEIs and thereby position them as the key drivers for achieving the SDGs. Further, there is a felt need for the developmental agencies to meaningfully engage with academia to utilize their knowledge and expertise on sustainability issues. Higher Education Sustainability Initiative (HSEI) is a major initiative in this direction.

15.2 University outreach – emerging perspectives

15.2.1 Why university outreach?

University outreach programmes carry a special relevance in the context of sustainable development. The developing countries present an alarming situation as most of socio-economically vulnerable communities live in geographically isolated regions and remain nearly cut-off from urban-based centres of excellence. In Indian context, most of the rural areas are still faced with infrastructural constraints, access to basic healthcare, clean drinking water and sanitation facilities among others that makes them highly vulnerable.

In order to bring about integrated development of such remote and underdeveloped areas, HEIs can meaningfully involve with various agencies like government organizations, non-governmental organizations and local bodies to address the problems. The idea is to utilize the available expertise in the universities to address grass-root issues/problems. It leads to an indirect benefit of practical-based learning experience for students/faculty. In India, the Kothari Commission on educational reforms first articulated the concept of "extension" as early as 1960 which was subsequently incorporated into its policy statement by the University Grants Commission. The Commission stated that "extension" was important for making education more relevant and contextual. At the same time, extension programmes were essential for preventing the alienation of the learners from society and for developing their sense of responsibility towards society. The incorporation of "Extension" as the 'Third Dimension' apart from teaching and research fulfils the university's obligation to work for the overall well-being of the community. It was conceived as a two-way process between the experts and the people. The idea was to have an intellectual intervention in the community's perceived problems which could be overcome or solved through an educational process.

15.2.2 Issues in outreach

The university outreach programmes in Indian context have to face obvious socio-economic barriers. The target groups to be covered through such outreach programmes are too vulnerable, poorly capacitated to adopt new practices and nearly cut-off from the urban-centred educational facilities. It is a challenge for the universities to build rapport with such communities, sensitize them and then inculcate the required skills and scientific knowhow to strengthen their livelihoods practices.

15.2.3 *Why ODL institutions are favourably placed?*

Successive Five-Year Plans in India underline the need of outreach activities for rural communities for skill upgradation to give them more diversified options of livelihoods and to capacitate them without any clash with their existing livelihood pursuits. However, despite sincere efforts of the government, the situation on the ground leaves much to be desired. The areas requiring immediate attention are poorly connected by the communication infrastructure, and the intended beneficiaries find it difficult to travel long distances to get educated. It is well realized and understood that capacity building systems for such target groups need to be radically different from "business as a usual approach".

Hence, there is an immediate need to employ "out of box approaches" primarily aimed at helping the people to get know-how right at their doorsteps at an affordable cost. "Innovative and Flexible Capacity Building Systems" have a major role to play in the livelihoods promotion. Though the universities have taken initiatives to reach out to such communities through their Outreach/Extension Education Programmes, such programmes have limited impact. The emerging systems of ODL having a focus on "Innovation and Flexibility" are ideally suited to meet such requirements.

15.2.4 *Way Forward*

University outreach programmes have significant implications for implementation of SDGs. Most of the issues and concepts related to development as envisaged by the UN-SDGs require a scientific outlook and understanding which is often not easy to grasp by the masses. Universities can contribute by spreading awareness and knowledge about the same. They can, at the same time, act as a linking pin between the government and the community at large. They may also provide research-based information to identify various training needs and capacity-building measures for the community. University personnel who represent a plethora of specialized fields of knowledge may also act as possible resource persons for providing training and capacity building measures for the community.

They may at the same time contribute in evolving comprehensive work programme, developing suitable tools, strengthening synergistic partnerships with various stakeholders, developing new approaches towards adaptation and mitigation activities for realizing the development goals. Universities may contribute through working across generations for promoting more sustainable lifestyles which is based on need-based consumption patterns in order to promote optimum utilization of resources (Kumar, 2017). Facilitating a multi-stakeholder engagement for development issues and advocating on behalf of the masses and representing them at the organizational and policy-making levels are other important contributions of universities to initiate community outreach activities.

However, universities may achieve the aforesaid only if they bring about certain philosophical, structural and procedural reforms in order to facilitate such activities. Universities, especially in developing countries such as India, would have to move from their traditional roles to repositioning themselves in developmental scenario. They would have to put a new organizational structure in place that would provide support to various development-related problems by way of interdisciplinary approaches to problem solving. The community outreach activities would have to be considered as the core academic mission and accordingly reformulate their planning and budgeting processes. Institutional policies and procedures would have to be made supportive of outreach involvement. Internal selection and evaluation procedures of personnel need to be broadened so as to include an assessment of capacity and performance related to outreach leadership. Most importantly, the methodology for teaching needs to be changed from being classroom centric to a more practical oriented one and the curriculum content is made more relevant to the emerging needs and realities (Pandey and Kumar, 2018).

15.3 *Case study of IGNOU*

Two major factors have led to the interest in distance learning across the world particularly in developing countries. Firstly, the growing need for continual skills upgrading and retraining, and another, the technological advances that have made it possible to teach more and more subjects at a distance (UNESCO, 2002).The task for open universities in developing countries such as India is all the more challenging as their target groups belong to most deprived segments of society (Pandey, 2017). Despite initial resistance and skepticism from its conventional peers, the ODL in India is here to stay. In terms of enrolment and its geographical spread, the system has made significant impact on Indian Higher Education (UGC Web Site). ODL system in India has demonstrated that expansion through collaboration, networking and judicious sharing of resources works quite well. The ODL system has been able to design and implement new areas of capacity building initiatives and evolve strategies for enhancing community participation and sensitization at every level of its outreach programmes. IGNOU has the distinction of being the largest university and a centre of excellence in ODL. The university has made perceptible impact on the national development in India (Panda, 2006) besides making significant outreach at international level.

15.3.1 Organization of case study

This case study gives an account of outreach activities being carried out by IGNOU, its unique approaches, outcomes and the challenges (IGNOU, 2018). The data has been taken from various secondary sources, several convocation reports and annual reports of the university since its inception and other publications of the university. The experiences of the university can be of significant relevance for the upcoming open universities of the developing world where the ODL systems have an important role to play. The lessons learnt have been listed, and the way forward has been explored.

15.3.2 The institutional culture

IGNOU has a unique institutional culture. The university has a network of more than 3,500 collaborative Learner Support Centers (LSCs) and more than 50,000 Part-Time Experts drawn from different HEIs in the country. These experts are inter alia involved in various activities such as providing academic support services to students, management of LSCs, development of content, preparation of question papers, and evaluation of students. Further, IGNOU has acted as an apex body and regulator for ODL system in the country for almost three decades. The fast-growing student enrolment, national jurisdiction, diverse clientele in terms of age, educational background, socio-economic backgrounds and prior qualifications gives IGNOU a distinct character. It is not easy for an organization having global spread to serve its clientele through diverse groups of partners ranging from governmental organizations, NGOs, corporate organizations, charitable organizations each having different mandates, organizational structures, goals, incentives and work culture. Therefore, management, monitoring, funding and training of part-time functionaries of such a diverse partners' network require paradigmatically different work culture, attitudes and professional ethics. Simultaneously, such a collaborative work environment leads to several constraints also. The schedules of university have to be carefully aligned with the priorities of collaborating institutions. Further, it requires the practitioners of university to acquire professional skills to tactfully involve diverse stakeholders, give leadership to keep them motivated and monitor the activities carefully. It gives IGNOU a very unconventional institutional culture and raises interesting professional challenges to the planners of the university.

15.3.3 *A new model*

According to Narasimharao (2009), university outreach is an all-encompassing strategy which incorporates various concepts like ODL, online learning, virtual universities and corporate universities among others. In a country like India, where the resources are limited, target groups are too vulnerable and change is immediately needed, the conventional strategies to impart education cannot work.

Further, it is not a viable idea to generate permanent infrastructure for providing academic support services to students. Enhancing the reach of the educational support at affordable cost and still maintaining quality through sustainable systems and processes has always been a challenge for the university. The planners of the university utilized a unique approach of philosophy of openness and sharing of resources to serve the educational requirements of diverse social groups at an affordable cost (Panda and Santosh, 2017).

Different models were adopted by the university on collaborative front. For example, Memorandum of Collaboration (MOC) was a generic term used for cooperation and joint action for specific purposes. MOCs were initiated by the university in several areas of mutual interest (IGNOU Web Site). The major areas of cooperation involve sharing of academic expertise, are design and implementation of academic programmes, are joint research and development projects, are information and communications technology (ICT)-enabled education, are development of interactive learner support, etc. The terms and conditions for the MOC are agreed upon on a case-to-case basis by signing separate agreement. The second collaborative approach is popularly known as "Memorandum of Understanding". Under this scheme, educational institutions are identified to act as LSC of IGNOU. The overseas institutions can also enter into such agreements to act as Partner Institutions. The parties share the administrative and financial responsibilities as per the standard agreement formats.

15.3.4 *Openness: a developmental imperative*

Knowledge can act as an effective force to smoothen out disparities imposed by birth and other circumstances. If everybody does not get benefitted by the knowledge then such disparities will continue. Therefore, democratization of knowledge has become a developmental imperative. The philosophy of openness in education has become more pressing in view of SDGs. The openness has three major dimensions, namely access, content and technology (Kanwar, 2017). There is an urgent need to provide free access to education, make the quality content easily available as per their contextual requirements and the technology used for delivery should be usable by all the categories of target groups alike.

ODL systems have taken pioneering initiatives to enhance accessibility to knowledge. However, there is a need for more and more openness in our freedom to change, remix, repurpose and redistribute it (UNESCO, 2014a,b; Baraniuk, 2008; Peter and Deimann, 2013). It has evolved a paradigmatically new perspective for the teaching and learning practices in the universities (Peters, 2008). Knowledge Commission of India took note of these developments and recommended for effective policies to enhance quality open access (OA) materials and "Open Educational Resources (OERs)" (National Knowledge Commission, 2005).

15.3.5 *Converging with governmental concerns: major priority*

As IGNOU is funded by the Ministry of Human Resource Development (MHRD) Government of India, it has to follow the mandate of the ministry. Moreover, the university has to work within governmental rules and regulations. The planners of the university have an obvious inclination towards creating convergence with government's programmes and policies. Further, due to its organizational culture, collaborative ventures are much more easily implementable if carried out with governmental organizations. Hence, collaborative ventures of IGNOU have been more inclined towards governmental organizations.

Right from its inception, the university has been instrumental in the implementation of ambitious developmental projects of Government of India. These projects have been in the area of Teacher Education under Sarva Shiksha Abhiyaan, Satellite-Based Networking of Rural Schools, Audio-Visual Content Generation for School Education, Women Empowerment Project of Ministry of Social Welfare and Justice, Unnat Bharat Abhiyaan (UBA), Educational Development of Educationally Backward Blocks, Anganwadi Workers' Capacity Building Programme, Educational Radio Network, Pan-African E-Learning Project, etc. Most of these projects involved the key expertise of the IGNOU's faculty in self-learning materials, core expertise of IGNOU for electronic communication, the state-of-the-art infrastructure in IGNOU for audio/video content generation and IGNOU's nationwide Students Support Services. In 2016, IGNOU entered into ambitious outreach programme to enroll Scheduled Caste/Scheduled Tribe (SC/ST) people in selected academic programme. As per the agreement, Government of India has agreed to pay the entire fee of these disadvantaged categories of students' infrastructure.

IGNOU's role is to identify such students and get them enroled. Similarly, in 2017, the Government of India agreed to pay the fees of all such prospective students from the community of weavers. Further, IGNOU has also been making its own efforts to support developmental issues of Government of India. In a significant move, the university has signed an MOU with Ministry of Home Affairs to provide free education to jail inmates. All the consequential expenses are being met by the university out of its own resources. The North East Region of India, which is relatively less developed, has got special attention under collaborative arrangements with IGNOU. The large-scale sensitization operations required for cashless economy drive of India was implemented through the network of IGNOU. The IGNOU's students were trained to be volunteers to sensitize the general public for cashless transactions and digital economy. There have been a range of collaborative arrangements between IGNOU and different ministries of Government of India which have been mutually enriching for university and government (IGNOU Web Site).

15.3.6 Exploring the convergence with corporate world

IGNOU started some of the most innovative collaborations between the corporate organizations. Some of the important ventures are Hero Honda Project, Aditya Birla Group, etc. (IGNOU Web Site). This convergence with the corporate organizations is quite significant keeping in view the Corporate Social Responsibility (CSR) Act of the Government of India. ODL system can prove to be instrumental in implementing large-scale CSR project primarily due to their extensive reach up to rural and remote areas. The corporate organizations are particularly interested to reach out to such areas largely inhabited by socio-economically disadvantaged population (Parveez and Pandey, 2012). The CSR is now getting prominently involved in the planning processes of IGNOU. However, corporate organizations' involvement with IGNOU is not new. It had a scheme of recognized study centres under which National Thermal Power Corporation (NTPC) collaborated with IGNOU way back in the 90s. With the advent of CSR activities, many corporate organizations showed interest to collaborate with IGNOU. Some of the prominent collaborative ventures with corporate organizations include Aditya Birla Group and Hero Honda Group.

IGNOU launched collaborative venture with Aditya Birla Nuevo Limited's owned subsidiary, Madura Fashion and Lifestyle Limited to impart necessary skills and training in garment stitching to the community members who are unemployed/underemployed and are residing mostly in the rural, suburban and disadvantaged locations. In order to implement this project, an MOU was signed between IGNOU and Aditya Birla Group to launch Short-Term Non-Credit (STNC) programmes of

6 months duration. The programmes were Basic Certificate in Garment Stitching for 1 month and Advance Certificate in Garment Stitching for 2 months. Advance Certificate in Garment Stitching and Quality Control for 2.5 months was also launched.

The corporate organization's interest to enter into a tie-up with IGNOU was precisely due to IGNOU's core competency to develop quality content, recognition of its programmes and long experience of working with disadvantaged social groups. As the training was approved by IGNOU, the subjects who passed out had a better chance of getting a job and to establish their own business ventures and compete in market. As per the MOU, Madura Fashion and Lifestyle was to enrol at least 200 students per year as per IGNOU-approved eligibility conditions and conduct programmes at approved learning centres only. The entire delivery strategy was planned by the university and the CSR department of corporate group. The programme implementation included a Joint Coordination Committee. It was the core body which was given the responsibility to oversee the teaching and training methodology, for industry exposure through internship and evaluation processes. Similarly, another innovative initiative taken by IGNOU was a collaborative venture with Hero Honda Motors, Ltd. It was much hyped programme of the university which won the best innovation award in the convocation function of the university. IGNOU's School of Engineering Technology has explored a collaborative venture with Hero Honda Motors, Ltd. to initiate a competency-based skill development programme for the motorcycle technicians of the country. Another important initiative of the university is about empowering teachers through collaborative ventures with Intel. The university and Intel joined hands for a national initiative for empowering teachers to develop 21st century skills, to create new models of teaching and learning and to help upgrade the capabilities of current and future teachers. In order to implement the project, an MOU was signed between Registrar (Administration), IGNOU, and Director, Corporate Affairs Group, Asia-Pacific, Intel. The university is also planning to work towards training manpower for CSR. The need for such a programme is being felt very strongly.

15.3.7 Quest for international presence

The Distance Education was little known in India at the time of IGNOU's inception. The conventional system did not accept it as a viable system of education delivery. Recognition and acceptance from conventional peers was the major issue at that point of time. There were several other misconceptions which adversely affected smooth sailing for this system. The planners of the university realized that a strong professional network of

Distance Educators was essential for further development of this mode of education – Distance Education. The university initially launched a Diploma Programme in Distance Education that was offered across the world. In the courses of time, several other collaborative projects namely Rajiv Gandhi Fellowship Scheme (RGFS); International Institute for Capacity Building in Africa (IICBA) – UNESCO and STRIDE-IGNOU Collaboration and Distance Education Modernization Project (DEMP) under collaborative arrangement between Sri Lanka and STRIDE-IGNOU were launched. These collaborative arrangements developed a firm foundation for professional expertise in Distance Education. Further, it made a beginning towards international outreach of IGNOU's programmes. The university started offering its academic programmes on international level from 1996 onwards through a fourfold approach of collaboration, coordination, cooperation and competition. Within less than two decades, IGNOU reach extended up to 35 countries across the globe through well-networked 62 Partner Institutions in U.A.E (Abu Dhabi, Dubai, Sharjah, Ras Al Khaimah) Qatar, Kuwait, kingdom of Saudi Arabia, kingdom of Bahrain and Sultanate of Oman among the Gulf countries; Mauritius, Singapore, Papua New Guinea and Seychelles among the Island nations; Ethiopia, Kenya and Ivory Coast in Africa, Nepal, Afghanistan and Sri Lanka among the South Asian Association for Regional Cooperation (SAARC) countries, Mongolia and Kyrgyzstan in the Central Asian countries. Though, the university has closed many of them, and currently, 18 Partner Institutions are collaborating with IGNOU in 13 countries. The open universities in several countries have shown interest in IGNOU's self-learning materials for adoption and adaptation. IGNOU also licenced Self-Instructional Materials as per certain terms and conditions to open universities at Hong Kong, Nigeria, Sri Lanka and Tanzania.

15.3.8 *Choice of technology: the game changer*

The educational development in India has been highly uneven. There are centres of excellence located in few urban pockets with world-class infrastructure and expertise. However, most of the educational institutions in deep rural interiors still strive for basic infrastructural facilities and expertise. Distance Education with the deployment of user-friendly technologies can be a real changer. Choice of technology is a crucial issue for the delivery of content. Satellite communication can prove to be a promising medium to link the centres of excellence to the rural interiors. During 90s, IGNOU embarked upon the satellite communication as a major media to enhance its reach up to deep rural interiors. Indian Space Research Organization (ISRO) has been one of the major collaborators of IGNOU since then. IGNOU created a nationwide network of ground-level infrastructure and practiced video teleconferencing (two-way audio and one-way video) with students across the country. Besides this, Satellite Interactive Terminals (SITs) with the facility of two-way audio and two-way video communication were also established in selected institutions. The video teleconferencing facility over the direct-to-home (DTH) platform was initiated around 2007 which significantly enhanced the reach of IGNOU's video teleconferencing programmes.

With the increasing policy thrust of government, IGNOU initiated the process of developing an Online Repository named eGyanKosh in October 2005 (National Knowledge Commission, 2005). The objective was to store, index, preserve, distribute and share the quality-learning resources for free use by students and faculty. It has emerged as one of the most popular and largest Educational Resource Repositories in the country. eGyanKosh gives free and open access to quality educational resources to self-learners and empowers educators through 2,200 courses and 2,000 video lectures. The self-learning materials are available on this repository as pdf files, whereas the video programmes are provided through a special channel of IGNOU on tube. The access to eGyanKosh was initially restricted only to its faculty, staff and students. However, it was made open access in 2008, and consequently, its popularity increased several folds. It had received more than 2 million hits by 2012 with an average of 1,000 visits per day across the world. Besides, eGyanKosh has the provision of wiki for collaborative content generation. The repository was closed down by the university for some time but was reactivated in 2017 (IGNOU, 2017) keeping in view two major initiatives of Government of India namely National Digital Library (NDL) and the SWAYAM (India Massive Open Online Courses (MOOCs)) launched to enhance the educational outreach up to remotest corners of the country. As per the available statistics, there were more than 300,000 active registered users of the repository having free access to its content (IGNOU Web Site).

IGNOU initiated a major ICT initiative in 2009 which is popularly known as Flexi Learn platform. The Flexi Learn provided the facility to register and explore courses for free regardless of their experience and prior qualifications. However, for certification, the learner had to fulfil the required eligibility criteria, pay the requisite fee through online payment gateway and undergo programme as per prescribed procedure. Flexi Learn provided complete learning experience with built-in Web 2.0 Tools that stimulated tremendous public response. Further, it had links to live educational channels with web cast. Presently, Gyan Darshan-1, Gyan Darshan-2 and Gyan Vani FM Radio, Delhi are linked to eGyanKosh. The Flexi Learn initiative of the university was later discontinued for want of a government policy on online programmes.

Major collaborative "ICT-Enabled Initiatives" of IGNOU to enhance reach

Sl no	The initiative	Collaborating agencies	Target groups and purpose
1	The broadcast of IGNOU's audio programmes through AIR Shillong	All India Radio Shillong and IGNOU	To provide content-based educational support to students of IGNOU
2	Telecast of IGNOU's programmes through Doordarshan	Doordarshan and IGNOU	To provide content-based educational support to students of IGNOU
3	The pilot experiment for video teleconference	ISRO and IGNOU	To provide content-based educational support to students of IGNOU
4	The Interactive Radio Counseling Programme of IGNOU	Prasar Bharati and IGNOU	To provide interactive content-based educational support to students of IGNOU
5	The Gyan Vani Educational Project	Prasar Bharati, Ministry of Information and Technology and IGNOU	To provide content-based educational support to students of IGNOU and specific programmes for general public
6	Rajiv Gandhi Project for EduSat Supported Elementary Education	MHRD Government of India, state governments of Hindi-speaking states and ISRO	To train rural teachers, create content generation for rural students and telecast educational programmes
7	One-way and two-way video teleconferencing programmes	ISRO and IGNOU	To provide interactive content-based educational support to students of IGNOU
8	Gyan Darshan educational TV channels	Doordarshan and IGNOU	To provide interactive content-based educational support to students of IGNOU and specific educational programmes for general public
9	eGyan Kosh and web cast of IGNOU's TV channels	MHRD, Government of India and IGNOU	To provide open access for educational content, content generation on wiki platform and interactive programme for IGNOU students
10	Multimedia eLearning Terminal for mobile Educational Support	MHRD, Government of India and IGNOU	To provide interactive content-based educational support to students of IGNOU

Despite apparently popular appeal of the technology-enabled methods, its utilization has not been encouraging, specifically from the disadvantaged segments. Though such an educational delivery has evoked relatively better response from urban areas, the rural interiors have not been able to reap the intended benefits (IGNOU Web Site). However, it is not the failure of technology per se but more due to the management of technology. Major reasons for such differential response could be due to low level of awareness about such facilities, a culture of face-to-face education and insignificant penetration of technology in disadvantaged communities. It is well realized and understood that the choice of technology is quite significant for the Distance Education. The recent policy developments have given rise to more viable opportunities for web-based educational delivery in Indian Distance Education system. The Government of India's major policy thrusts like "National eGovernance Programme" and "Digital India" have led to immense possibilities for web-based educational delivery. IGNOU has been trying to align its delivery operations in rural areas with the government's eGovernance Plans.

15.3.9 *Role for regional cooperation*

The increasing reach of ODL programmes has created conducive situation for Regional Cooperation in South Asian region (SAARC web site). There are emerging opportunities for a greater partnership between the educational institutions, exchange of academic expertise, joint offering of academic programmes and greater degree of people to people contact. There centres of excellences in all these countries can be networked and expertise can be shared for offering academic programmes of mutual interest in this region. It will further enhance peace initiatives and will help in strengthening sustainable development in this region.

Efforts have been made under the auspices of SAARC to promote cooperation, encourage collaborations and to promote the use of Open and Distance Education as a viable and cost-effective method of imparting quality education at all levels. A platform called SAARC Consortium on Open and Distance Learning (SACODiL) was created for this purpose (SAARC Web Site). SACODiL is operated through rotational secretariat across the member states. IGNOU plays a major role for such collaborative efforts through ODL mode in South Asia.

15.3.10 *Village outreach programmes*

As per the Gandhian concept of rural development, villages have to be developed by making them self-sufficient in every manner possible and bringing down their dependency on the urban areas. The idea is to create lasting and sustainable changes to the lives of the people by making optimum use of the local resources and involving the local community in every step of the development process (Pyarelal, 1977). At present, the prevailing developmental gaps and disconnects between the rural and urban areas have created a number of problems including economic inequity, discontent among local population, large-scale migration to urban centres and lack of employment opportunities among others. Recognizing the importance and role of universities in the outreach activities, the Government of India initiated an ambitious and nationwide programme to link universities with the rural areas directly and to use their expertise and knowledge to bring about holistic development in the areas. The programme aims to link the institutions of higher education with the rural community and initiate a dialogue of identifying development challenges and evolving strategies to bring about sustainable development of the region. It aims to provide an opportunity to the academicians, the civil society at large and the policymakers to understand various aspects of the village life and in the process get sensitized on the needs and aspirations of the people, their context and the social dynamics of the community. They are expected to act as change agents and facilitators for building cohesive and sustainable communities. The broad areas covered under the programme include infrastructure development, rural housing, water conservation and watershed management, microfinance, livelihood, health and sanitation and education among others. The nodal agencies are also responsible for coordinating with the governmental and non-governmental agencies, local bodies and corporate sector for facilitating the developmental activities.

Village Adoption Programme (VAP) has been conceived as part of the UBA initiated by the Government of India in November 2014 (Unnat Bharat Abhiyaan, 2014). UBA is based on the seminal work of Mahatma Gandhi *Hind Swaraj*, which talks about the concept of bringing self-sufficiency to the villages by making use of available local resources and eco-friendly technologies in order to meet the basic requirements of the community (UBA Web Site). The larger goals of the programme include developing a synergistic partnership between the rural areas and the universities. In this way, it creates conditions of mutual cooperation and benefits for all the stakeholders involved in the process.

Village adoption involves inspiring the local communities to initiate a process of change and work for the betterment of their communities. It also involves bringing about development along the lines of the SDGs. The universities as the nodal agency for carrying out the Village Development Programme have been entrusted with a number of roles and responsibilities which include creating awareness on various development activities, preparing village development plans in order to ensure socio-economic advancement of the community, performing need analysis for capacity building programmes, assisting in infrastructure development of the village, promoting environmental conservation and ecological balance and finally, monitoring the implementation of the programme.

15.4 Lessons learnt

15.4.1 Issue of sustainability of impact under VAP

The initiatives taken by the university have resulted in enhanced participation of people from rural and remote areas in IGNOU's academic programmes. However, sustainability of such initiatives is a major issue of concern. The university is perceived as an outside agency by the villagers, and therefore, confidence building initiatives need to be taken as a first step. In the areas where IGNOU has tried to take the community-based organizations such as Village Panchayats, local NGOs and elderly people into confidence, the impact of university's initiatives has been sustainable (Pandey, 2017). What follows is that the university needs to take representative organizations of the community into confidence and information should pass through them. The IGNOU's involvement in the meetings of *village panchayats* (local self-government in Indian villages) is worth replicating. Being the platforms of grassroots democracy in India, *village panchayats* are ideal platforms to assess educational requirements and to bring about people's involvement to fulfil those demands (Pandey, 2017).

15.4.2 Leadership issues

The LSCs of IGNOU which constitutes the interface with students are managed entirely by part-time functionaries of IGNOU. Hence, administrative and leadership patterns are vital for the growth and effectiveness of IGNOU's system (Khan, 2001, p. 154). The large number of collaborating organizations with different kinds of objectives and work cultures come together to serve the mandate of IGNOU. It gives rise to an unconventional work environment, expectations and priorities. Most of the functionaries are drawn from conventional institutions and have only just part-time commitments for IGNOU's system. To effectively manage such a diverse organizational system, the practitioners of this system need to acquire leadership and administrative skills to handle human resource. It is a huge administrative challenge for the system to keep such part-time functionaries stay motivated. The leaders need to acquire skills to communicate with diverse clientele, implement their vision and lead effectively in multicultural workplaces.

The issues of multicultural work environment, though, well researched in several other contexts (as cited by Chuang, 2013) have not been sufficiently examined in the context of mega universities.

These issues need to be sufficiently investigated, its leadership patterns need to be examined and explored to develop this sector further.

15.4.3 Ethical issues of collaboration

The ethical issues of collaboration and networking play a crucial role in organizations like IGNOU where the major chunk of functionaries is drawn from diverse organizations quite often competing with each other. The sanctity of the system rests on the conduct, behaviour and level of motivation of such part-time functionaries who join IGNOU's system. Further, it requires a very high level of ethical basis of collaborations. The issues are sensitive as such functionaries are involved in core work of IGNOU such as schedule planning, organizations of the academic services, conduct of examinations and evaluation work. The administrative leadership responds to their availability, monitors their functioning and keeps them motivated.

The part-time functionaries coordinate on behalf of their organizations and don't fit into a rigid hierarchical framework within IGNOU. Moreover, the staff drawn from collaborating organizations has mostly part-time commitments, and their terms are renewed on yearly basis. As the primary commitments of these functionaries rest with their host institutions, the IGNOU's functioning has to adjust with the preferences and priorities of these part-time functionaries. It is a challenge for the administrative leadership to ensure that functionaries should not develop a feeling of compromising their own organizational interest. The administrative leadership has to accommodate diverse preferences and communicate with them effectively with a clear vision. Identification of goals with the effective involvement of its collaborators is a major challenge for administrators in IGNOU. Despite the constraints, IGNOU's management practices have been doing quite well and the experiences are worth replicating elsewhere. Across the world, ODL system confronts with crucial issues related to its policy development, quality assurance, accreditation and assessment systems (Bozkurt, Kocdar and Buyuk, 2018), and experiences of IGNOU can be crucial to address such issues.

15.4.4 Responding to fast change

ODL institutions in India have been expanding with mind-boggling pace of change. IGNOU as the biggest role player in India and in recent times has grown to more than 3 million in 2016.With more than 3,500 collaborating institutions across the country and presence in 43 different countries, it has evolved into a gigantic network with peculiarly different organizational system. The system has to respond to this fast pace of change, explore collaborations with competing organizations for service delivery to its clientele and monitor the network in a cordial atmosphere. The administrative leaders of IGNOU have to monitor its fast-growing network, changing aspirations of clientele and accordingly change the service delivery strategies. It offers enormous challenges as such an objective has to be realized with in a multi-organizational environment.

15.4.5 *Incentivizing the collaborators and sustenance of the partnerships*

IGNOU's organizational set-up does not provide for major financial incentives to collaborating institutions for setting up LSCs. Under the collaborative scheme of LSCs, the host institutions have to provide infrastructural facilities without any cost and spare their academic expertise on weekends for providing academic support to IGNOU students. This model based on "Sharing of Resources" has largely worked satisfactorily with public sector organizations wherever they have sufficient infrastructure and expertise. Such organizations are generally interested to serve their mandate towards community rather than any commercial benefits.

However, educational institutions from public sector are generally not well equipped with infrastructural facilities and remain hard pressed for funds in rural interiors which are IGNOU's major target area. The competitions for the same resources and revenue sources create competing situations which tend to jeopardize the concepts of partnership and collaboration. Such situations are demotivating factors for the collaborators. Better incentives for the collaborating organizations can resolve such sources of conflicts in a befitting manner.

The collaboration of IGNOU with private players has not given very encouraging results. Such collaborations have not been able to develop a sustainable business model for its operations. The private players quite often lack the effective involvement for lack of any significant financial interest in its operations. Several major collaborations ended up in legal complications due to commercialized activities of collaborators. Sustainability emanates out of meaningful convergence in the mandates of the collaborating organizations. ODL institutions need to plan their outreach operations based on genuine interests of collaborating organizations. It will keep them motivated to continue the collaboration.

15.5 *Recommendations and concluding remarks*

Universities have the most prominent role to play in bringing about socio-economic development, institutions of higher education can substantially contribute to the realization of SDGs. Firstly, universities act as a linking pin between knowledge generation and knowledge transfer/ dissemination (Sandmann, 1996). Secondly, they can directly contribute to the development process through outreach programmes. However, there is a need to transport knowledge to the doorsteps of the people rather than transporting people to knowledge-generating institutions. Moreover, life-long education is a developmental imperative. As knowledge is literally exploding, it is an economic compulsion to update people right at their workplaces. Knowledge has to transcend the four walls of the classrooms and reach the aspirants at the time and place of their convenience. The mechanisms should be innovative and flexible enough to suit the convenience of the target groups.

The case study of IGNOU presented here is specifically relevant in the perspective of SDGs of Post 2015 Development Agenda. The university's experiences are quite significant particularly keeping in view the fact that Post 2015 Development Agenda advocates for institutional partnerships. The three decades of experience can be immensely useful for the replication of the IGNOU's model in developing world. The case study presented here brings out following few recommendations to enhance outreach of IGNOU's programmes.

a. **Local actors to enhance outreach**

   Existing mechanisms to involve the community-based organizations are just fragmented in approach. They are in the form of local-level initiatives to enhance enrolment in IGNOU's academic programmes. Such mechanisms should be comprehensive, aimed at making them sustainable, and efforts should be done to make them scalable. The locally customized mechanisms should be developed to assess educational needs, to involve local actors and financially incentivize such actors to enhance reach (Pandey, 2017). The multi-stakeholder participation will be absolutely necessary in this regard (Pandey and Kumar, 2017). It will help IGNOU to take benefit of their rapport with the community which will help to prepare inroads in rural interiors. Though some of those initiatives are successful, such initiatives are not sustainable and not well documented. What is essentially required is that good practices should be documented, critically examined and mainstreamed in the organizational pattern of IGNOU.

b. **Paradigmatically new information dissemination strategies**

   There is a need for paradigmatically different outreach strategies for rural areas. Despite the fact that there are schemes such as special study centers and distance learning facilitators for rural areas, it is observed that these schemes have not been able to make any perceptible difference in enrolment patterns. The information dissemination strategies are primarily based on newspaper advertisements which have differential reach in rural, tribal and urban areas. The penetration of the newspapers in deep rural interiors is very low which deprives timely information to the native people. In some of the pockets isolated, initiatives have been taken to involve local community-based organizations for information dissemination and have yielded favourable outcomes in terms of people's participation. However, majority of such initiatives have not been mainstreamed in university's main organizational set-up. Moreover, the sustenance of such initiatives is an issue. The innovative initiatives to enhance local level participation can be strengthened if university can financially incentivize such local actors.

c. **OERs on multiple formats for physically challenged**

One of the major areas which have not received adequate attention of ODL professionals is the area of disability. Though, IGNOU had signed MOU with Rehabilitation Council of India, but it was subsequently cancelled. There has been significant increase in the number of people living with disability across the world. It had increased from 10% in 1970 to 15% in 2010.This number turns out to be around 1 billion across the world. There are apprehensions that this number will grow even further due to increasing incidences of conflicts, poverty, ageing population, etc. (Kanwar, 2017). ODL systems are favourably positioned to reach out to such a large segment of physically challenged people spread across the world. However, IGNOU has not done enough to utilize the technological advances to reach out to such people. Lot of possibilities are awaiting Indian Distance Education system to make meaningful interventions in this regard. The technological developments have made it possible to overcome physical disability to a great extent. For example, OERs are a major development which gives opportunities to contextualize the learning materials as per the requirements of physically challenged persons. Further, social media has given rise to interesting new possibilities of development and sharing of content. IGNOU being the largest open university of the world should take a lead in this direction. The OER can be made available on multiple formats to make them easily accessible by physically challenged persons (Kanwar, 2017).

d. **Developing contextual solutions**

Despite its tremendous success in urban areas, the IGNOU's model ceases to work in pockets where conventional education set-up is poor, population is sparsely distributed and transport network is poor. India's sizeable population lives in rural interiors largely devoid of basic communication infrastructure, transport facilities, educational infrastructure and academic expertise. Moreover, the heterogeneity in Indian context is a major issue of concern. The customized solutions for rural areas can address these issues.

IGNOU experimented with several schemes such as Admission-cum-Information Centres, Sub-Study Centres, Distance Learning

Facilitators and Mobile Study Centres to provide academic support services in rural interiors. In 2015, the university ventured into very small aperture terminal (VSAT)-enabled Mobile eLearning Terminals in selected regions. Local community involvement through their representative institutions has also been tried at several places. Despite several inconsistencies, such isolated initiatives have led to encouraging results in most of the places. However, such islands of successes are the results of local leaderships at several places and need to be carefully nurtured for their sustenance. Successful community-based interventions to enhance outreach have been documented in several studies (Pandey, 2017). Mobile Study Centre experiment initiated in Dindori tribal district in India is worth mentioning which prompted IGNOU to notify the Mobile Study Centre Scheme for the entire country. Similarly, interesting local-level initiatives have been taken in Maoist insurgency-affected areas which have resulted in significant participation of the rural population in job-oriented programmes. IGNOU's initiatives to involve jail inmates in Tihar Jail were later on replicated across the country once it was found a successful experiment. IGNOU's educational interventions in tsunami-hit islands of Andaman and Nicobar have been documented which have led to a noticeable impact on education of island dwellers. These educational interventions were made possible through Internet-compatible kiosks installed in such islands as no other way of communicating with the communities was possible. Such contextualized solutions should be taken on large scale keeping in view the heterogeneous nature of Indian society.

Thus, IGNOU has grown into a gigantic institution with international recognition and presence. The university presents a successful example of "Collaboration and Networking" to enhance reach among disadvantaged communities and to promote the Sustainable Development in South Asian region. The networking of the educational institutions through ODL mechanisms will positively impact people-to-people contacts and peace initiatives in the region. The experiences of the university can have significant implications for replication of this model in other developing countries.

References

Baraniuk, R. G. (2008): Challenges and opportunities for the open education movement: A case study, in Iiyoshi, T. and Vijay Kumar, M.S. (Eds.), *Opening up Education: The Collective Advancement of Education through Open Technology, Open Content and Open Knowledge* (pp. 229–246), Cambridge, MA: MIT Press. https://oerknowledgecloud.org/sites/oerknowledgecloud.org/files/0262033712pref1.pdf (accessed on 30th March 2017).

Bhowmik, J., Samia, A., and Huq, S. (2018): The Role of Universities in achieving SDGs: A Policy Brief, published by University of Liberal Arts Bangladesh and International Centre for Climate Change and Development. www.icccad.net/wp-content/uploads/2015/12/Policy-Brief-on-role-of-Universities-in-achieving-SDGs.pdf (accessed on 31st March 2018).

Bozkurt, A., Kocdar, S., and Buyuk, K. (2018): Administrative Leadership in Open and Distance Learning Programs (PDF Download Available). www.researchgate.net/publication/317668693_Administrative_Leadership_in_Open_and_Distance_Learning_Programs (accessed 10th October 2017).

Chuang, S. (2013): Essential Skills for Workplace Effectiveness in Diverse Workplace Development. http://opensiuc.lib.siu.edu/cgi/viewcontent.cgi?article=1133&context=ojwed (accessed on 10th October 2017).

Cortese, D. A. (2003): The critical role of higher education in creating a sustainable future, *Planning for Higher Education*, 31(3), 15–22.

HESI (2012): Higher Education Institutions –Key Drivers of Sustainable Development. https://sustainabledevelopment.un.org/content/documents/17043HESI_Summary_2017.pdf (accessed on 22nd March 2018).

IGNOU (2017): Reactivation of IGNOU eGyan Kosh. http://ignou.ac.in/userfiles/eGyankosh-reactivation.pdf (accessed on 29th March 2018).

IGNOU (2018): IGNOU Web site at http://www.ignou.ac.in/ignou/aboutignou/profile/4 (accessed on 04th April 2018).

Kalam, A. (2005): How to add value to Distance Education, Address at the 16th Convocation of Indira Gandhi National Open University on 5th March 2005. http://abdulkalam.nic.in/sp050305.html (accessed on 25th March 2018.)

Kanwar, A. (2017): Making Open and Distance Learning Inclusive the Role of Technology. http://oasis.col.org/bitstream/handle/11599/2827/2017_Kanwar-Cheng_Making-ODL-Inclusive_Transcript.pdf?sequence=6&isAllowed=y (accessed on 31st March 2018).

Khan, A. W. (2001): Indira Gandhi National Open University and Distance Education Council-Institution and System Building in Hanna, D. E. and Latchem, C. (Eds), India in *Leadership for 21st Century Learning: Global Perspectives*, https://books.google.co.in/books?id=nTr_AQAAQBAJ&pg=PA147&lpg=PA147&dq=leadership+issues+in+IGNOU&source=bl&ots=bhtVyysLyA&sig=cZPfU1nBRyF4Dd-HDw-JHpkHI84&hl=en&sa=X&ved=0ahUKEwiNxo-rju7WAhUQTY8KHYrhCO0Q6AEISTAF#v=onepage&q=leadership%20issues%20in%20IGNOU&f=false (accessed on 13th October 2017).

Kumar, C. (2017): Strategizing social work response to sustainable development goals through open and distance learning, in Pandey, U.C. et al., (Eds.), *Open and Distance Learning Initiatives for Sustainable Development*, Hershey, PA: IGI Global Publishing Company.

Mohamedbhai, G. (2015): What role for higher education in sustainable development?, University World News, The Global Window of Higher Education, Issue No:349. www.universityworldnews.com/article.php?story=20150108194231213 (accessed on 22nd March 2018).

Mukerji, S. and Tripathi, P. (2004): Academic program life cycle: A redefined approach to understanding market demands, *The Journal of Distance Education/Revue de l'ducation Distance*, 19(2), 14–27. https://www.learntechlib.org/p/102767/. Athabasca University Press (Retrieved 2nd April 2018).

Narasimharao, B. P. (2009): Knowledge economy and knowledge society—Role of university outreach programmes in India, *Science, Technology and Society Journal*, 14(1), 119–151. Sage Publications.

National Knowledge Commission (2005): https://nationalknowledgecommission.wordpress.com/category/open-educational-resources/ (accessed on 29th March 2018).

Oertel, S. and Soll, M. (2017): Universities between traditional forces and modern demands: The role of imprinting on the missions of German universities, *Higher Education*, 73(1), 1–18.

Panda, S. and Santosh, S. (2017): Faculty perceptions of openness and attitude to open sharing at Indian National Open University, *International Review of Research in Open and Distributed Learning*, 18(7), 89–111. www.irrodl.org/index.php/irrodl/article/view/2942/4463 (accessed on 30th March 2018).

Panda, S. (2006): Higher education at a distance and national development: Reflections on the Indian experience, *Distance Education*, 26(2), 205–225. doi: 10.1080/01587910500168868.

Pandey, U. C. (2017): Strategizing the role of local actors to enhance outreach, in Narasimharao, B. P. et al., (Eds.), *Handbook of Research on Science Education and University Outreach as a Tool for Regional Development*, Pennsylvania: IGI Global Publishing House.

Pandey, U. C. and Kumar, C. (2018): A SDG compliant curriculum framework for social work education: Issues and challenges, in Leal Filho, W. et al., (Eds.), *Implementing Sustainability in the Curriculum of Universities*. World Sustainability Series, Cham: Springer International Publishers.

Parveez, M. and Pandey, U. C. (2012): Corporate social responsibility in education: Promises of open and distance learning systems, in *International Conference on Corporate Social Responsibility (CSR) organized by Xavier Institute of Management Jabalpur (XIMJ) an educational wing of Xavier Institute of Development Action and Studies (XIDAS)*, Jabalpur (Madhya Pradesh), 4th–5th October 2012.

Peter, S. and Deimann, M. (2013): On the role of openness in education: A historical reconstruction, *Open Praxis*, 5(1), 7–14.www.openpraxis.org/~openprax/ojs-2.3.7/files/journals/1/articles/23/public/23-147-3-PB.pdf (accessed on 30th March 2018).

Peters, M. A. (2008): The history and emergent paradigm of open education in Peters, M. A. and Britez, R. (Eds.), *Open Education and Education for Openness* (pp. 3–16). Rotterdam & Taipei: Sense Publishers. www.sensepublishers.com/media/729-open-education-and-education-for-openness.pdf (accessed on 30th March 2018).

Pyarelal (1977): *Mahatma Gandhi on Human Settlements*, Ahemadabad: Navajivan Publishing House.

SAARC Web Site. www.saarc-sec.org (accessed on 4th April 2018).

Sandmann, L. R. (Ed.) (1996): Fulfilling higher education's covenant with society: The emerging outreach Agenda, in *Summary of the Capstone Symposium of the W.K. Kellogg Foundation MSU Lifelong Education Grant*, Michigan State University.https://engage.msu.edu/upload/documents-reports/capstone.pdf (accessed on 15th February 2018).

SDG Knowledge Hub Reports (2017): Focus on Role of Universities in Achieving SDG. http://sdg.iisd.org/news/reports-focus-on-role-of-universities-in-achieving-sdgs/ (accessed on 31st March 2017).

UGC (2013): University Grant Commission.www.ugc.ac.in/deb/pdf/growthDEB.pdf (accessed on 4th April 2018).

UN-DESA (2016): Partnerships for SDGs-2016, *Special Report*. https://sustainabledevelopment.un.org/sdinaction/2016report (accessed on 31st March 2018).

UNESCO (2002): Open and Distance Learning: Trend Policy and Strategy Consideration. http://unesdoc.unesco.org/images/0012/001284/128463e.pdf (accessed on 30th March 2018).

UNESCO (2012): Education for Sustainable Development: Source Book. http://unesdoc.unesco.org/images/0021/002163/216383e.pdf (accessed on 24th December 2017).

UNESCO (2014a): Shaping the Future We Want: UN Decade of Education for Sustainable Development (2005-2014). UNESCO, France (accessed on 24th December 2017).

UNESCO (2014b): How Openness Impacts on Higher Education: A Policy Brief. http://iite.unesco.org/pics/publications/en/files/3214734.pdf (accessed on 30th March 2018).

Unnat Bharat Abhiyaan (2014): Official website for Indian Institute of Technology, Delhi. unnat.iitd.ac.in/index.php/Pages/display/introduction (accessed on 15th March 2018).

World Economic Forum (2017): Why Collaborations will be key to achieving Sustainable Development Goals. www.weforum.org/agenda/2017/01/realising-the-potential-of-cross-sector-partnerships/ (accessed on 31st March 2018).

Index

For Product Safety Concerns and Information please contact our EU representative GPSR@taylorandfrancis.com Taylor & Francis Verlag GmbH, Kaufingerstraße 24, 80331 München, Germany

Printed and bound by CPI Group (UK) Ltd, Croydon, CR0 4YY
08/05/2025
01864325-0002